Bioaugmentation, Biobarriers, and Biogeochemistry

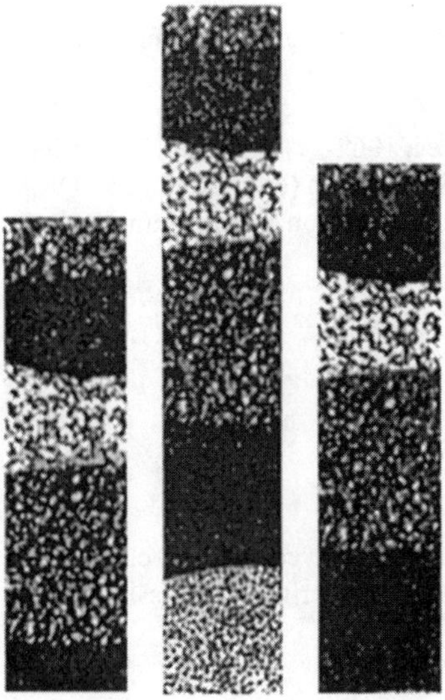

Editors

Andrea Leeson, Bruce C. Alleman, Pedro J. Alvarez, and Victor S. Magar

The Sixth International In Situ and On-Site Bioremediation Symposium

San Diego, California, June 4–7, 2001

BATTELLE PRESS
Columbus • Richland

Library of Congress Cataloging-in-Publication Data

International In Situ and On-Site Bioremediation Symposium (6th : 2001 : San Diego, Calif.)
Bioaugmentation, biobarriers, and biogeochemistry : the Sixth International In Situ and On-Site Bioremediation Symposium : San Diego, California, June 4-7, 2001 / editors, A. Leeson ... [et al.].
p. cm. -- (The Sixth International In Situ and On-Site Bioremediation Symposium ; 8)
Includes bibliographical references and index.
ISBN 1-57477-118-3 (hc. : alk. paper)
1. Bioremediation--Congresses. I. Leeson, Andrea, 1962- . II. Title. III. Series: International In Situ and On-Site Bioremediation Symposium (6th : 2001 : San Diego, Calif.). Sixth International In Situ and On-Site Bioremediation Symposium ; 8.
TD192.5.I56 2001 vol. 8
628.5 s--dc21
[628.5]

2001037941

Printed in the United States of America

Battelle Press
505 King Avenue
Columbus, Ohio 43201, USA
614-424-6393 or 1-800-451-3543
Fax: 1-614-424-3819
Internet: press@battelle.org
Website: www.battelle.org/bookstore

For information on future environmental conferences, write to:
Battelle
Environmental Restoration Department, Room 10-123B
505 King Avenue
Columbus, Ohio 43201-2693
Phone: 614-424-7604
Fax: 614-424-3667
Website: www.battelle.org/conferences

CONTENTS

Foreword vii

Bioaugmentation

Explanation of Bacterial Transport Enhancement by Starvation. *R. Gerlach, A. Cunningham, and F. Caccavo* 1

Engineered Full Scale Bioremediation of Chlorinated Ethylenes. *M.J.C. Henssen, A.W. van der Werf, S. Keuning, C. Hubach, R. Blokzijl, E. van Kulen, B. Alblas, C. Haasnoot, H. Boender, and E. Meijerink* 11

Preventing Contaminant Discharge to Surface Waters: Plume Control with Bioaugmentation. *J. Lendvay, P. Adriaens, M. Barcelona, C.L. Major, J. Tiedje, M. Dollhopf, F. Loeffler, B. Fathepure, E. Petrovskis, M. Gebhard, G. Daniels, R. Hickey, R. Heine, and J. Shi* 19

Successful Field Demonstration of Bioaugmentation to Degrade PCE and TCE to Ethene. *D.W. Major, M.L. McMaster, E.E. Cox, B.J. Lee, E.E. Gentry, E. Hendrickson, E. Edwards, and S. Dworatzek* 27

In Situ Biotreatment of Chlorinated Hydrocarbons in Groundwater Using Gel Beads. *R. Govind and C. Tian* 35

Using a Molecular Approach to Monitor a Bioaugmentation Pilot. *E.R. Hendrickson, M.G. Starr, M.A. Elberson, H.B. Huang, E.E. Mack, M.L. McMaster, and D.E. Ellis* 43

Bioaugmentation with *Burkholderia cepacia* $PR1_{301}$: Immobilization for Activity Retention Enhancement. *D.J. Adams and K.F. Reardon* 53

Biobarrier Design Concepts and Case Studies

In Situ Bioremediation of MTBE Using Biobarriers of Single or Mixed Cultures. *J.P. Salanitro, G.E. Spinnler, P.M. Maner, D.L. Tharpe, D.W. Pickle, H.L. Wisniewski, P.C. Johnson, and C. Bruce* 61

Organic Mulch Biowall Treatment of Chlorinated Solvent-Impacted Groundwater. *C.E. Aziz, M.M. Hampton, M. Schipper, and P. Haas* 73

Biofilm Barriers for Groundwater Containment. *G. James and R. Hiebert* 79

Biowall In Situ Groundwater Treatment. *P.W. Becker, B.B. Archibald, J.H. Higinbotham, and P.C. Madden* 87

Biodegradation of a Naphthalene Plume in a Funnel-and-Gate™ System. *P. Lamarche, F. Lauzon, M. Tetreault, and J.F. Barker* 95

Activated Carbon and Other Support Media Used for Biobarriers

Biological Activated Carbon Barriers for the Removal of Chloroorganics/BTEX Mixtures. *A. Tiehm, M. Gozan, A. Muller, K. Bockle, H. Schell, H. Lorbeer, and P. Werner* 105

Biobarrier Comprised of Soil and BAC: Suppression of Greenhouse Gases. *Y. Sakakibara and D. Kamimura* 113

Comparative Cost and Performance of Two Novel Biological Permeable Barriers (BPBs). *F.R. Shirazi* 121

Sorption and Microbial Degradation of Toluene on a Surfactant-Modified Zeolite Support. *A.M. Fuierer, R.S. Bowman, and T.L. Kieft* 131

Biologically Enhanced Iron Barriers, and Iron-Reducing Processes

Abiotic and Biotic Cr(VI) Reduction in a Laboratory-Scale Permeable Reactive Barrier. *C. Henny, L.J. Weathers, L.E. Katz, and J.D. MacRae* 139

Degradation of TCE, Cr(VI) NO^{-}_{3}, and SO^{2-}_{4} Mixtures in Columns Mimicking Bioaugmented Fe^{0} Barriers. *S. Gandhi, B.-T. Oh, J.L. Schnoor, and P.J.J. Alvarez* 147

Dissolved Plume PCE Remediation Using a Combination of Zero-Valent Iron and a Hydrogen-Release Compound. *N.M. Fischer, T. Reed, C. Madsen, and T. Mascarenas* 157

Combination of Iron and Mixed Anaerobic Culture for Perchloroethene Degradation. *X. Luo and G.W. Sewell* 167

RDX Degradation with Bioaugmented Fe(0) Filings: Implications for Enhanced PRB Performance. *B.-T. Oh and P.J.J. Alvarez* 175

Probabalistic Design of a Combined Permeable Barrier and Natural Biodegradation Remedy. *J.E. Vidumsky and R.C. Landis* 185

Biotic and Abiotic Dechlorination in Iron-Reducing and Sulfidogenic Environments. *P. Adriaens, M.J. Barcelona, K.F. Hayes, M.L. McCormick, and K.L. Skubal* 193

Enhancement of Dissimilatory Iron(III) Reduction by Natural Organic Matter. *W. Burgos, R. Royer, A. Fisher, and R. Unz* 201

A Bioavailable Ferric Iron Assay and Relevance to Reductive Dechlorination. *P.J. Evans and S.S. Koenigsberg* 209

Author Index 217

Keyword Index 245

FOREWORD

The papers in this volume correspond to presentations made at the Sixth International In Situ and On-Site Bioremediation Symposium (San Diego, California, June 4-7 2001). The program included approximately 600 presentations in 50 sessions on a variety of bioremediation and supporting technologies used for a wide range of contaminants.

This volume, *Bioaugmentation, Biobarriers, and Biogeochemistry*, presents exciting progress in the development of new types of biobarriers and advances in the use of bioauagmentation for the treatment of contaminated groundwater. Activated carbon and other support media such as zeolites, mulch, and even soil are demonstrated for use in biobarriers. A number of papers document the continued use of iron barriers for groundwater impacted by chlorinated solvents. Bioaugmentation is discussed not only as it related to biobarriers, which are primarily used for plume containment, but also as an approach for aggressive plume treatment.

The author of each presentation accepted for the symposium program was invited to prepare an eight-page paper. According to its topic, each paper received was tentatively assigned to one of ten volumes and subsequently was reviewed by the editors of that volume and by the Symposium chairs. We appreciate the significant commitment of time by the volume editors, each of whom reviewed as many as 40 papers. The result of the review was that 352 papers were accepted for publication and assembled into the following ten volumes:

Bioremediation of MTBE, Alcohols, and Ethers — 6(1). Eds: Victor S. Magar, James T. Gibbs, Kirk T. O'Reilly, Michael R. Hyman, and Andrea Leeson.

Natural Attenuation of Environmental Contaminants — 6(2). Eds: Andrea Leeson, Mark E. Kelley, Hanadi S. Rifai, and Victor S. Magar.

Bioremediation of Energetics, Phenolics, and Polycyclic Aromatic Hydrocarbons — 6(3). Eds: Victor S. Magar, Glenn Johnson, Say Kee Ong, and Andrea Leeson.

Innovative Methods in Support of Bioremediation — 6(4). Eds: Victor S. Magar, Timothy M. Vogel, C. Marjorie Aelion, and Andrea Leeson.

Phytoremediation, Wetlands, and Sediments — 6(5). Eds: Andrea Leeson, Eric A. Foote, M. Katherine Banks, and Victor S. Magar.

Ex Situ Biological Treatment Technologies — 6(6). Eds: Victor S. Magar, F. Michael von Fahnestock, and Andrea Leeson.

Anaerobic Degradation of Chlorinated Solvents— 6(7). Eds: Victor S. Magar, Donna E. Fennell, Jeffrey J. Morse, Bruce C. Alleman, and Andrea Leeson.

Bioaugmentation, Biobarriers, and Biogeochemistry — 6(8). Eds: Andrea Leeson, Bruce C. Alleman, Pedro J. Alvarez, and Victor S. Magar.

Bioremediation of Inorganic Compounds — 6(9). Eds: Andrea Leeson, Brent M. Peyton, Jeffrey L. Means, and Victor S. Magar.

In Situ Aeration and Aerobic Remediation — 6(10). Eds: Andrea Leeson, Paul C. Johnson, Robert E. Hinchee, Lewis Semprini, and Victor S. Magar.

In addition to the volume editors, we would like to thank the Battelle staff who assembled the ten volumes and prepared them for printing: Lori Helsel, Carol Young, Loretta Bahn, Regina Lynch, and Gina Melaragno. Joseph Sheldrick, manager of Battelle Press, provided valuable production-planning advice and coordinated with the printer; he and Gar Dingess designed the covers.

The Bioremediation Symposium is sponsored and organized by Battelle Memorial Institute, with the assistance of a number of environmental remediation organizations. In 2001, the following co-sponsors made financial contributions toward the Symposium:

Geomatrix Consultants, Inc.
The IT Group, Inc.
Parsons
Regenesis
U.S. Air Force Center for Environmental Excellence (AFCEE)
U.S. Naval Facilities Engineering Command (NAVFAC)

Additional participating organizations assisted with distribution of information about the Symposium:

Ajou University, College of Engineering
American Petroleum Institute
Asian Institute of Technology
National Center for Integrated Bioremediation Research & Development (University of Michigan)
U.S. Air Force Research Laboratory, Air Expeditionary Forces Technologies Division
U.S. Environmental Protection Agency
Western Region Hazardous Substance Research Center (Stanford University and Oregon State University)

Although the technical review provided guidance to the authors to help clarify their presentations, the materials in these volumes ultimately represent the authors' results and interpretations. The support provided to the Symposium by Battelle, the co-sponsors, and the participating organizations should not be construed as their endorsement of the content of these volumes.

Andrea Leeson & Victor Magar, Battelle
2001 Bioremediation Symposium Co-Chairs

EXPLANATION OF BACTERIAL TRANSPORT ENHANCEMENT BY STARVATION

Robin Gerlach and Al Cunningham (Center for Biofilm Engineering, Bozeman, Montana)
Frank Caccavo Jr. (Whitworth College, Spokane, WA)

ABSTRACT: The economic feasibility of bioaugmenting large volumes of soil *in situ* depends on the effective delivery of bacteria with the desired metabolic traits into the subsurface. Bacterial starvation can significantly improve the delivery of bacterial strains into the subsurface; however, the reasons for the enhanced transport of starved bacteria remain unclear. We investigated the change of a number of transport-related bacterial cell properties during starvation. A decrease in cell numbers (colony forming units, direct cell counts, and optical density at 600 nm), apparent buoyant density, and adhesivity to quartz sand over a starvation period of 7 weeks was accompanied by an increase in diffusivity and increased transportability through 1 ft columns (measured as fractional recovery). The zeta potential and the hydrophobicity did not significantly change during starvation. It does not appear that the change of the investigated cell properties can explain the improved transport behavior of starved cells but rather a combination of the investigated and potentially other parameters.

INTRODUCTION

The development of strategies for the enhancement of bacterial transport through porous media is of high interest since subsurface bioaugmentation strategies rely upon the delivery of bacteria into the contaminated subsurface. Unfortunately, widely applicable, effective strategies for the delivery of bacteria into the subsurface are lacking.

The examination of parameters influencing bacterial adhesion to surfaces and transport through porous media (Bouwer et al., 2000) makes evident that only a very limited number of parameters can be effectively manipulated in field scale applications. Strategies to change the solution chemistry or the physicochemical properties of the porous medium are likely to become an economical and technological challenge. Thus, manipulating parameters falling into these two categories might be extremely difficult or not advisable. The manipulation and control of the bacterial inoculum appear to have greater potential of economical and technological success in field scale subsurface bioaugmentation strategies.

We and others have demonstrated that bacterial transport through porous media can be enhanced by starvation (Bouwer et al., 2000; Cusack et al., 1992; Gerlach et al., 1998; Lappin-Scott and Costerton, 1992; Lappin-Scott et al., 1988a; Lappin-Scott et al., 1988b; MacLeod et al., 1988; Sharp et al., 1999). Starvation is believed to be a survival mechanism for bacteria which cannot form spores or cysts (Hood and MacDonell, 1987; Tabor et al., 1981). Starvation of bacteria can result in radical size reduction and a rapid decrease in metabolic activity until the bacteria approach complete dormancy (Kjelleberg, 1993). Improved transport of metabolically dormant cells is believed to contribute towards the presence of bacteria in the deep subsurface (Lappin-Scott and Costerton, 1990). Starved bacteria can survive for years in the absence of nutrients (Amy and Haldeman, 1997; Kjelleberg, 1993), but can be resuscitated relatively rapidly by the addition of suitable nutrients (Amy and Morita, 1983; Cunningham et al., 1997; Kjelleberg, 1993; Novitsky and Morita, 1978).

The injection, resuscitation, and subsequent plugging of high permeability zones in oil bearing formations using starved bacteria was demonstrated a number of times (Cusack et al., 1992; Lappin-Scott and Costerton, 1990; Lappin-Scott et al., 1988a; Shaw et al., 1985). The use of starved bacteria resulted in significant improvements in secondary oil recovery since the bacteria traveled deeper into the high permeability zones and formed bacterial plugs upon resuscitation, which forced the injected water through the low permeability zones to recover residual oil more efficiently.

Starved bacteria were also used in a series of laboratory- and field-scale experiments demonstrating the possibility to form hydraulic barriers for groundwater containment. Starved cells derived from bacteria known to produce large amounts of extracellular polysaccharides (EPS) were injected into porous media and resuscitated. The bacteria formed thick biofilms upon resuscitation and decreased the porous media permeability dramatically (Cunningham, 2000; Cunningham et al., 1997).

However, a complete explanation for the increased bacterial transport and decreased tendency of starved cells to adhere to surfaces is lacking. Changes in cell size and cell shape have been correlated with transport through porous media and smaller spherical cells are transported better than larger and elongated cells (Bitton et al., 1974; Fontes et al., 1991; Gannon et al., 1991; Weiss et al., 1995). However, the change in cell size of *Shewanella algae* BrY from approximately 2.2 µm x 0.6 µm (length x width) to more spherical cells with a diameter of 1.04 x 0.6 µm after 7 weeks of starvation (Caccavo et al., 1996) does not explain the significant improvement in bacterial transport. Based on the one-dimensional filtration theory (Logan et al., 1995), the change in cell size during starvation would result in approximately equal effluent concentrations of starved and vegetative cells during a passage through 30 cm of quartz sand if all other parameters are assumed to be the same for starved and vegetative cells.

A number of cell properties have been shown to influence bacterial adhesion and transport through porous media and are suspected to change depending on the growth state of the bacteria and thus during bacterial long-term starvation (Caccavo et al., 1996; Dawson et al., 1981; Grasso et al., 1996). We are reporting the change of a number of transport-related cell properties during starvation of *S. algae* BrY in an attempt to identify parameters that may contribute towards the transport enhancement of starved bacteria.

MATERIALS AND METHODS

Cultures of *Shewanella algae* BrY (formerly *Shewanella alga* BrY) were maintained and grown as described in (Caccavo et al., 1996 and Caccavo et al.,

1992). After washing, *S. algae* BrY cells were starved by aseptically stirring on a magnetic stir plate at room temperature. Cells were harvested by centrifugation (5860 x g, 20 minutes, 4 °C) and resuspended in Phosphate Buffered Saline (PBS) for all assays, except microscopic direct counts, viable cell counts, and absorption measurements. Changes in cell hydrophobicity were estimated using the bacterial adhesion to hydrocarbons (BATH) assay (Rosenberg, 1984; Rosenberg et al., 1980), which is based on the reduction in cell numbers in an aqueous suspension in the presence of a hydrocarbon. The hydrophobicity index (HI) was calculated according to equation 1:

$$HI = \frac{A_{600nm,\,initial} - A_{600nm,\,final}}{A_{600nm,\,initial}} \qquad (1)$$

Zeta potentials and the apparent diffusion coefficients were determined from cell suspensions in PBS at pH 7.0 and 25 °C using a Zetasizer 2c (Malvern Instruments). The apparent buoyant density was calculated using the density gradient centrifugation procedure described by Harvey et al. (1997). Adhesion studies were performed as described previously (Caccavo et al., 1997). In brief, cells were allowed to adhere to quartz sand and the fraction of cells adhered to the sand was calculated by comparison to controls lacking quartz sand. The transportability of cells was determined in 30 cm long porous media columns and is expressed in terms of fractional recovery of *S. algae* BrY in the effluent. The fractional recovery is defined as the number of cells recovered in the effluent of a column divided by the total number of cells injected.

RESULTS AND DISCUSSION

Viable cells from starvation cultures were estimated as colony forming units on tryptic soy agar. The concentration of viable cells slightly increased during the first week of starvation from initially $3.83 \times 10^9 \pm 1.33 \times 10^8$ CFU/mL to $4.93 \times 10^9 \pm 4.84 \times 10^8$ CFU/mL after one week (Figure 1). In the following weeks, the number of cells decreased to $2.10 \times 10^8 \pm 1.04 \times 10^7$ CFU/mL after seven weeks of starvation. The concentration of total cells was estimated using microscopic direct counts. The microscopically detected cell concentration initially was approximately ten times higher

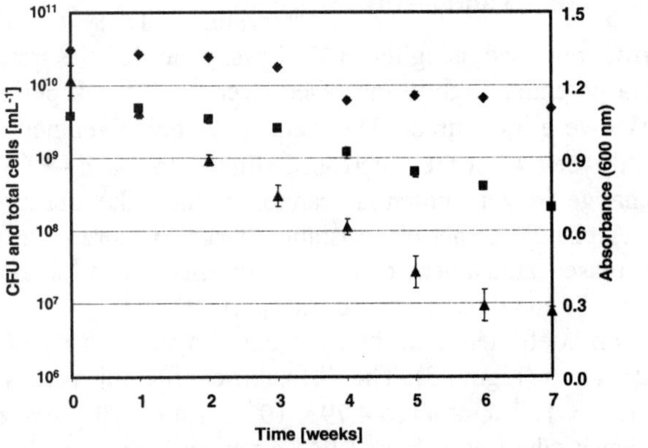
10¹¹
10¹⁰
10⁹
10⁸
10⁷
10⁶
CFU and total cells [mL⁻¹]
1.5
1.2
0.9
0.6
0.3
0.0
Absorbance (600 nm)
0 1 2 3 4 5 6 7
Time [weeks]

FIGURE 1. Concentration of total cells (◆), culturable cells (■), and absorbance at 600 nm (▲) during starvation of *Shewanella algae* BrY. Error bars represent the standard error of the means (n = 3).

($3.15 \times 10^{10} \pm 2.05 \times 10^{9}$ cells/mL) than the concentration of colony forming units. Unlike the colony forming units, the total concentration of cells decreased slightly during the first week to $2.77 \times 10^{10} \pm 2.02 \times 10^{9}$ cells/mL and was thus only approximately 2.5 times higher than the number of viable cells. The numbers further decreased over time and reached $4.98 \times 10^{9} \pm 1.05 \times 10^{8}$ cells/mL after 7 weeks, which was approximately 25 times higher than the concentration of colony forming units at that time. The absorbance (600 nm) of the cell suspensions decreased almost linearly from 1.22 ± 0.06 to 0.172 ± 0.01 over 7 weeks of starvation. These results concur with the observations by Caccavo et al. (1996) who reported a decrease in viable cell numbers during starvation.

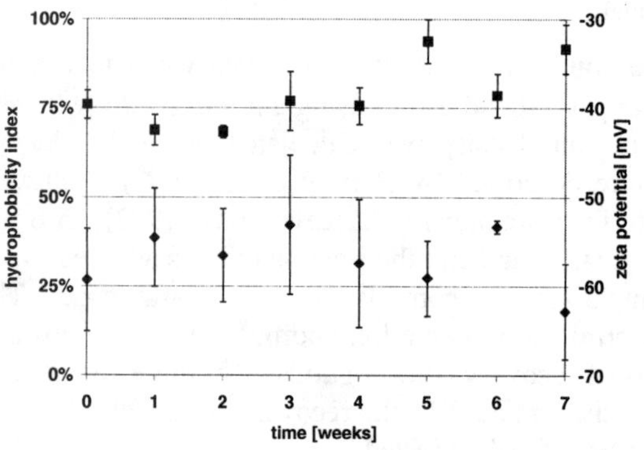
100%
75%
50%
25%
0%
hydrophobicity index
-30
-40
-50
-60
-70
zeta potential [mV]
0
1
2
3
4
5
6
7
time [weeks]

FIGURE 2. Change in hydrophobicity index (◆), and zeta potential (■) during starvation of *Shewanella algae* BrY. Error bars represent the standard error of the means (n = 3 and n=10).

Cell hydrophobicity and cell net electrostatic charge have been shown to influence the transport behavior of bacteria (Fontes et al., 1991; Gilbert et al., 1991). The change in hydrophobicity and zeta potential is shown in Figure 2. The hydrophobicity index appeared to slightly increase during the first 3 weeks of starvation. The HI remained between 27 % and 42 % until week 6 and dropped slightly to 18 ± 13 % after 7 weeks of starvation. Due to the relatively high standard errors obtained using the BATH assay, no definite trends could be detected. The zeta potential slightly increased over the 7 week period. Initially –39.5 ± 1.6 mV were measured. The zeta potential remained at approximately –40 mV until week 4 but then increased slightly to reach –33.3 ± 2.8 mV. However, the change in zeta potential cannot explain the observed improvement in transport. A change of net electrostatic charge to more positive values should result in increased attachment of cells to the negatively charged surface of the quartz sand (pH 7) and thus decreased transport.

The apparent diffusion coefficient and the apparent buoyant density both changed slightly during starvation (Figure 3). The diffusion coefficient increased from initially $2.7 \times 10^{-9} \pm 1.11 \times 10^{-10}$ $cm^2\ s^{-1}$ to $4.79 \times 10^{-9} \pm 1.43 \times 10^{-10}$ $cm^2\ s^{-1}$ and the apparent buoyant density decreased from 1.044 g cm^3 to 1.020 g cm^3. An increase in apparent diffusivity can potentially lead to an increase in the number of collisions of bacteria with surfaces and thus to higher adhesion. However the observed increase of the apparent diffusion coefficient from 2.7×10^{9} cm^2/s to approximately 4.5×10^{9} cm^2/s appears to be negligible in light of the work by R. Ford and others who stated that bacterial mobility seems to have a negligible influence on bacterial transport through porous media at groundwater relevant flow velocities (Barton and Ford, 1995; Barton and Ford, 1997). The decrease in buoyant density from 1.044 g/cm^3 to 1.020 g/cm^3 after 7 weeks of starvation was also not sufficient to explain the significant improvement in bacterial transport using the filtration model by Logan et al (1995, calculations not shown).

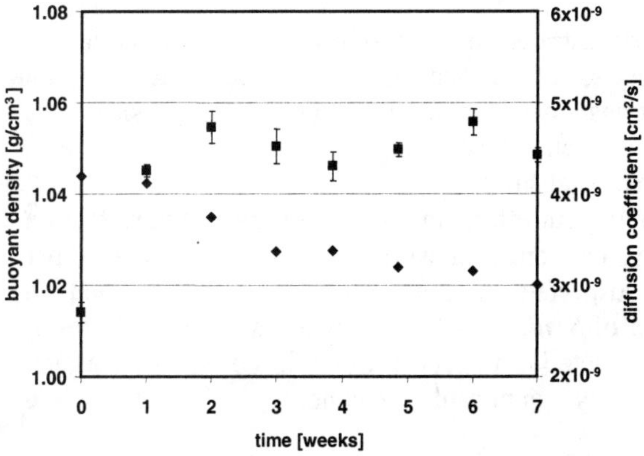

1.08
1.06
1.04
1.02
1.00
buoyant density [g/cm³]
6x10⁻⁹
5x10⁻⁹
4x10⁻⁹
3x10⁻⁹
2x10⁻⁹
diffusion coefficient [cm²/s]
0
1
2
3
4
5
6
7
time [weeks]

FIGURE 3. Change of apparent buoyant density (◆), and apparent diffusion coefficient (■) during starvation of *Shewanella algae* BrY. Error bars represent the standard error of the means (n = 3 and n=10).

Not one single parameter of the ones investigated so far appears to explain the increased transport of bacteria through porous media. Thus, we compared the tendency of *S. algae* BrY cells to adhere to quartz sand in batch systems to the fractional recovery of cells after passage through 1 ft quartz sand columns (Figure 4). The adhesion assay showed a rapid decrease in the fraction of cells adhered to quartz sand after one week of starvation. The number of adhered cells appeared to decrease further after 2 weeks and remained below 10 % in average until 7 weeks starvation time. It should be noted that the currently employed adhesion assay yielded fairly high standard errors and the assay yielded negative adhesion, i.e. an increase in cell numbers at the end of the assay, after 2 and 5 weeks. The number of cells adhered to quartz sand seemed to inversely correlate with the fractional recovery of cells. A decrease in the fraction of cells adhered to quartz sand in batch experiments resulted in an increase in fractional recovery after transport

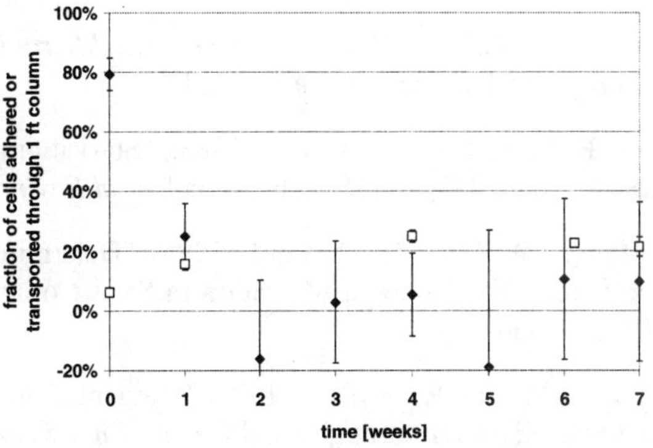

100%
80%
60%
40%
20%
0%
-20%
fraction of cells adhered or transported through 1 ft column
0
1
2
3
4
5
6
7
time [weeks]

FIGURE 4. Change in adhesivity (fraction of cells adhered to quartz sand, ◆), and transportability (fractional recovery of viable cells after passage through 30 cm of quartz sand, □) during starvation of *Shewanella algae* BrY. Error bars represent the standard error of the means.

through 30 cm of quartz sand. These results imply that it might be possible to use simple and less time- and cost-intensive batch experiments to predict the transport behavior of cells through porous media, as soon as better reproducibility is achieved in the adhesion assay.

CONCLUSIONS

We and others have demonstrated that long-term nutrient starvation can significantly improve our ability to deliver and distribute bacterial activity into porous media. The improved transport of starved bacteria and the possibility to resuscitate starved bacteria to metabolically highly active cells has been demonstrated for a number of bacterial strains.

We are currently continuing our efforts in further improving the starvation transport enhancement strategy. Our ongoing work attempts to elucidate other bacterial cell characteristics suspected to change during starvation and to influence the transport behavior of *S. algae* BrY. This work is expected to result in a better understanding of bacterial transport through porous media and potentially result in a widely applicable technology for subsurface bioaugmentation.

ACKNOWLEDGMENTS

Thanks to Kara Boettcher for help in the laboratory. This research was supported by the National Science Foundation (NSF Cooperative Agreement EEC-8907039) and the Inland Northwest Research Alliance (INRA). Partial support was obtained from the Undergraduate Research Opportunities Program at Montana State University-Bozeman.

REFERENCES

Amy P. S. and D. L. Haldeman. 1997. *The Microbiology of the Terrestrial Deep Subsurface*. CRC Press, Boca Raton, FL.

Amy P. S. and R. Y. Morita. 1983. "Starvation-Survival Patterns of Sixteen Freshly Isolated Open-Ocean Bacteria." *Appl.Environ.Microbiol. 45*:1109-1115.

Barton J. W. and R. M. Ford. 1995. "Determination of Effective Transport Coefficients for Bacterial Migration in Sand Columns." *Appl.Environ.Microbiol. 61*:3329-3335.

Barton J. W. and R. M. Ford. 1997. "Mathematical model for Characterization of Bacterial Migration through Sand Cores." *Biotechnol.Bioeng. 53*:487-496.

Bitton G., N. Lahav, and Y. Henis. 1974. "Movement and retention of *Klebsiella aerogenes* in soil columns." *Plant and Soil 40*:373-380.

Bouwer E. J., H. H. M. Rijnaarts, A. B. Cunningham, and R. Gerlach. 2000. "Biofilms in Porous Media." In: J. D. Bryers (Ed.), *Biofilms II: Process Analysis and Applications*. pp. 123-158. Wiley-Liss, Inc., New York.

Caccavo F., Jr., N. Birger Ramsing, and J. W. Costerton. 1996. "Morphological and metabolic responses to starvation by the dissimilatory metal-reducing bacterium *Shewanella alga* BrY." *Appl.Environ.Microbiol. 62*:4678-4682.

Caccavo F., Jr., R. P. Blakemore, and D. R. Lovley. 1992. "A hydrogen-oxidizing, Fe(III)-reducing microorganism from the Great Bay Estuary, New Hampshire." *Appl.Environ.Microbiol. 58*:3211-3216.

Caccavo F., Jr., P. C. Schamberger, K. Keiding, and P. H. Nielsen. 1997. "Role of hydrophobicity in adhesion of the dissimilatory Fe(III)-reducing bacterium *Shewanella alga* to amorphous Fe(III) oxide." *Appl.Environ.Microbiol. 63*:3837-3843.

Cunningham, A. B. 2000. “Subsurface Biofilm Barriers: An Emerging Technology for Containment and Remediation of Contaminated Groundwater.” Keynote Presentation at the Hazardous Waste Research 2000 Conference in Denver, CO, 23-25 May 2000.

Cunningham A. B., B. Warwood, P. Sturman, K. Horrigan, G. James, J. W. Costerton, and R. Hiebert. 1997. "Biofilm processes in porous media - practical applications." In: P. S. Amy, and D. L. Haldeman (Eds.), *The Microbiology of the Terrestrial Deep Subsurface*. pp. 325-344. CRC Press, Boca Raton, FL.

Cusack F., S. Singh, C. McCarthy, J. Grecco, M. De Rocco, D. Nguyen, H. M. Lappin-Scott, and J. W. Costerton. 1992. "Enhanced oil recovery - three dimensional sandpack simulation of ultramicrobacteria resuscitation in reservoir formations." *J.Gen.Microbiol. 138*:647-655.

Dawson M. P., B. Humphrey, and K. C. Marshall. 1981. "Adhesion: A Tactic in the Survival Strategy of a Marine Vibrio During Starvation." *Curr.Microbiol. 6*:195-198.

Fontes D. E., A. L. Mills, G. M. Hornberger, and J. S. Herman. 1991. "Physical and Chemical Factors Influencing Transport of Microorganisms Through Porous Media." *Appl.Environ.Microbiol. 57*:2473-2481.

Gannon J. T., V. B. Manilal, and M. Alexander. 1991. "Relationship Between Cell Surface Properties and Transport of Bacteria Through Soil." *Appl.Environ.Microbiol. 57*:190-193.

Gerlach, R., Cunningham, A. B., and Caccavo, F., Jr. 1998. Formation of Redox-Reactive Subsurface Barriers Using Dissimilatory Metal-Reducing Bacteria. In Ericksen, L. E. and Rankin, M. M.: *Proceedings of the 1998 Conference on Hazardous Waste Research - Bridging Gaps in Technology and Culture. May 18-21, 1998. Snowbird, UT*. pp. 209-223. Kansas State University. Manhattan, KS.

Gilbert P., D. J. Evans, I. G. Duguid, and M. R. W. Brown. 1991. "Surface characteristics and adhesion of *Escherichia coli* and *Staphylococcus epidermidis*." *J.Appl.Bacteriol. 71*:72-77.

Grasso D., B. F. Smets, K. A. Strevett, B. D. Machinist, C. J. van Oss, R. F. Giese, and W. Wu. 1996. "Impact of physiological state on surface thermodynamics and adhesion of *Pseudomonas aeruginosa.*" *Environ.Sci.Technol. 30*:3604-3608.

Harvey R. W., D. W. Metge, N. Kinner, and N. Mayberry. 1997. "Physiological considerations in applying laboratory-determined buoyant densities to predictions of bacterial and protozoan transport in groundwater: results of in-situ and laboratory tests." *Environ.Sci.Technol. 31*:289-295.

Hood M. A. and M. T. MacDonell. 1987. "Distribution of ultramicrobacteria in a gulf coast estuary and induction of ultramicrobacteria." *Microbiol.Ecol. 14*:113-127.

Kjelleberg S. (Ed.) 1993. *Starvation in Bacteria.* Plenum Press, New York.

Lappin-Scott H. M., J. W. Costerton. 1990. "Starvation and penetration of bacteria in soils and rocks." *Experientia 46*:807-812.

Lappin-Scott H. M., J. W. Costerton. 1992. *Current Opinion in Biotechnology 3*:283-285.

Lappin-Scott H. M., F. Cusack, F. A. MacLeod, and J. W. Costerton. 1988a. "Nutrient Resuscitation and Growth of Starved Cells in Sandstone Cores: A Novel Approach to Enhanced Oil Recovery." *Appl.Environ.Microbiol. 54*:1373-1382.

Lappin-Scott H. M., F. Cusack, F. A. MacLeod, and J. W. Costerton. 1988b. "Starvation and Nutrient Resuscitation of *Klebsiella pneumoniae* isolated from oil well waters." *J.Appl.Bacteriol. 64*:541-549.

Logan B. E., D. G. Jewett, R. G. Arnold, E. J. Bouwer, and C. R. O'Melia. 1995. "Clarification of clean-bed filtration models." *Journal of Environmental Engineering 121*:869-873.

MacLeod F. A., H. M. Lappin-Scott, and J. W. Costerton. 1988. "Plugging of a Model Rock System by Using Starved Bacteria." *Appl.Environ.Microbiol. 54*:1365-1372.

Novitsky J. A. and R. Y. Morita. 1978. "Starvation induced barotolerance as a survival mechanism of a psychrophilic marine vibrio in the waters of the Arctic Convergence." *Mar.Biol. 49*:7-10.

Rosenberg M. 1984. "Bacterial adherence to hydrocarbons: a useful technique for studying cell surface hydrophobicity." *FEMS Microbiology Letters 22*:289-295.

Rosenberg M., D. Gutnick, and E. Rosenberg. 1980. "Adherence of bacteria to hydrocarbons: A simple method for measuring cell-surface hydrophobicity." *FEMS Microbiology Letters 9*:29-33.

Sharp, R. R., Gerlach, R., and Cunningham, A. B. 1999. Bacterial Transport Issues Related to Subsurface Biobarriers. In Leeson, A. L. and Alleman, B. C.: *Engineered approaches for in situ bioremediation of chlorinated solvent contamination*. pp. 211-216. Battelle Press, Columbus, OH.

Shaw J. C., B. Bramhill, N. C. Wardlaw, and J. W. Costerton. 1985. "Bacterial Fouling of a Model Core System." *Appl.Environ.Microbiol. 49*:693-701.

Tabor P. S., K. Ohwada, and R. R. Colwell. 1981. "Filterable marine bacteria found in the deep sea: Distribution, taxonomy, and response to starvation." *Microbiol.Ecol. 7*:67-83.

Weiss T. H., A. L. Mills, G. M. Hornberger, and J. S. Herman. 1995. "Effect of Bacterial Cell Shape on Transport of Bacteria in Porous Media." *Environ.Sci.Technol. 29*:1737-1740.

ENGINEERED FULL SCALE BIOREMEDIATION OF CHLORINATED ETHYLENES

M.J.C. Henssen, A.W. van der Werf, S. Keuning (Bioclear Environmental Technology, Groningen, The Netherlands)
C. Hubach, R. Blokzijl, E. van Keulen (DHV Environment and Infrastructure, Groningen, The Netherlands)
B. Alblas, C. Haasnoot, H. Boender (Logisticon Water Treatment, Groot-Ammers, The Netherlands)
E. Meijerink (Province of Drenthe, Assen, The Netherlands)

ABSTRACT: At the Evenblij site in Hoogeveen, The Netherlands, soil and groundwater are contaminated with the chlorinated ethylenes perchloroethylene (PCE) and trichloroethylene (TCE) due to industrial activities. Complete removal of the contamination in the soil and groundwater using excavation and pump-and-treat would cost approximately US $ 40,000,000, while controlling the plume area only is estimated at US $ 9,000,000. Bioremediation techniques are used to reduce cost of remediation. However, complete biodegradation of PCE into ethylene does not seem possible with the native bacterial population at the site. Therefore a new engineered bioremediation concept was developed using dechlorinating bioreactors for bioaugmentation to improve the soil degradation capacity. The effluent of the bioreactor is infiltrated into the site in order to increase the soil degradation capacity. Full scale experiments in 1999 and 2000 showed the feasibility of the concept. It was possible to infiltrate the effluent of the bioreactors continuously for more than 5 months without any problems. Besides this continuous biomass infiltration, carbon source was injected discontinuously. Due to infiltration of biomass and a suitable carbon source, complete dechlorination of PCE into ethylene and even ethane occurred at the site. Non-stimulated areas showed no changes compared to the situation in 1998 and before. The capacity of the infiltrated biomass seems to last for a longer period, since dechlorination is ongoing in areas in which infiltration had taken place months ago.

INTRODUCTION

Due to industrial processes at the site and spills of perchloroethylene (PCE), both soil and groundwater up to 40 m below ground level have been contaminated. Since conventional remediation techniques for this site would mean huge investments the possibilities for biological treatment were examined. Field data and degradation tests indicated that complete biodegradation of PCE into ethylene does not occur. PCE and trichloroethylene (TCE) was degraded into *cis*-1,2-dichloro-ethylene (c-DCE), no other degradation products (e.g. vinylchloride (VC) or ethylene) are found. Adding only carbon sources over more than one year did not result in further degradation of c-DCE.

This limited capacity is presumably related to the sandy soil type and lack of carbon source, resulting in only moderate reduced conditions unfavourable for (development of) dechlorination reactions.

Based on groundwater characterisations at more than 100 contaminated sites in The Netherlands, this limited dechlorination capacity seems not to be unique. From these results it was concluded that bioreactors would be usable for bioaugmentation at sites where natural attenuation of PCE into ethylene is not occurring due to lack of suitable biomass or where degradation is rather slow. Dechlorinating biomass in the effluent from bioreactors can serve as inoculating material to improve the degradation capacity. Since several bioremediation techniques are combined this bioremediation is called Total Concept Evenbij (Figure 1).

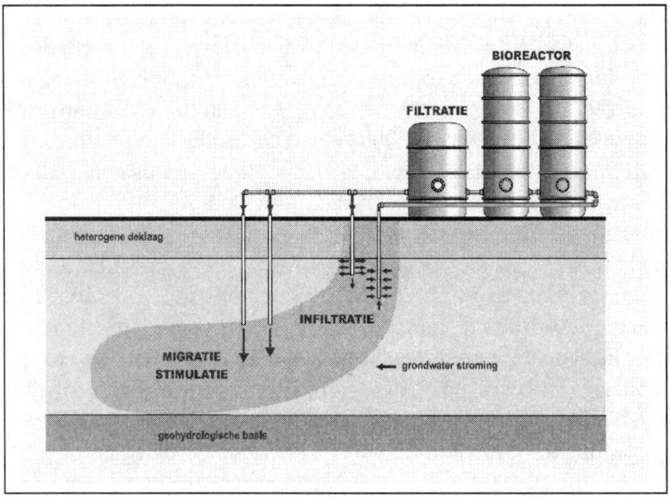
BIOREACTOR
FILTRATIE
heterogene deklaag
INFILTRATIE
MIGRATIE
STIMULATIE
grondwater stroming
geohydrologische basis

FIGURE 1. Schematic draft of Total Concept Evenblij

At the end of 1998, four 25 m^3 full-scale bioreactors were started up using sludge capable of dechlorination. Over the last 1.5 years the full scale bioremediation at Evenblij has been tested. In this concept dechlorinating bioreactors are used to improve the soil capacity for bioremediation. This improvement consists of transferring dechlorination capacity from the bioreactors to the soil.

Objective. Based on the results of this project the effectiveness of the bioaugmentation and possibility for enhancement of the biological processes were evaluated and design parameters for this kind of remediation were defined.

MATERIALS AND METHODS

Field scale experiments and monitoring at the site are used to evaluate the effects and performances of the in situ bioremediation concept. In phase 1 of the project, the natural degradation capacity of the soil, including the possibility of enhancement by addition of carbon sources, and bioaugmentation were already investigated by laboratory scale experiments.

In this phase the focus is mainly on the stability of the bioreactors, the infiltration of the effluent, and the stimulation of the soil by bioaugmentation at full scale using monitoring at the site.

Bioreactor. Samples of the water and sludge phase were taken periodically and analyzed for chlorinated ethylenes and dechlorination capacity of the sludge. Information was obtained about the stability of the bioreactor process and capacity of the water phase used for inoculation of the soil.

Pretreatment Tests. Before the effluent of the bioreactors can be infiltrated a pretreatment of this water phase is necessary to avoid well clogging. Different filtration techniques were tested at pilot scale at the Evenblij site.

Infiltration Tests. During several infiltration experiments the possibilities for infiltration of the effluent were tested using different kinds of preventative regeneration steps and carbon source addition techniques. A mixture of acetate and lactate was used as carbon source. The resulting concentration in the groundwater was approximately 200 mg/L DOC.

(Ground)water. Groundwater samples and samples from the pretreatment systems were taken and analyzed in order to determine the efficiency of the pretreatment systems and the dechlorination efficiency due to infiltration of biomass and carbon source. Besides, molecular analyses (DGGE) were performed to obtain information on the migration of bacteria in the soil.

Laboratory Tests. The biodegradation capacity of the groundwater was tested using degradation tests with solely groundwater. Performing tests before and after infiltration gave insight in the migration of bacteria and increase of degradation potential.

RESULTS AND DISCUSSION

BIOREACTOR

At the end of 1998, four anaerobic bioreactors (Figure 2) were started up and were fed with groundwater containing PCE and TCE from the site. The bioreactors are capable of transforming PCE into non-chlorinated ethylene using adapted biomass. Adaptation of the system took approximately 4 months after which the effluent contained 40-60% ethylene.

Continuous monitoring of the bioreactor systems showed very good stability, both technical and biological, over a period of several years.

Degradation tests showed that dechlorinating biomass is present in the effluent of the bioreactors; within 28 days complete transformation of PCE to ethylene occurs in an almost clear water sample from the bioreactor indicating large dechlorination potential in this water phase.

With the bioreactors, an effluent flow of 5 to 15 m^3/h was prepared which can be used for infiltration.

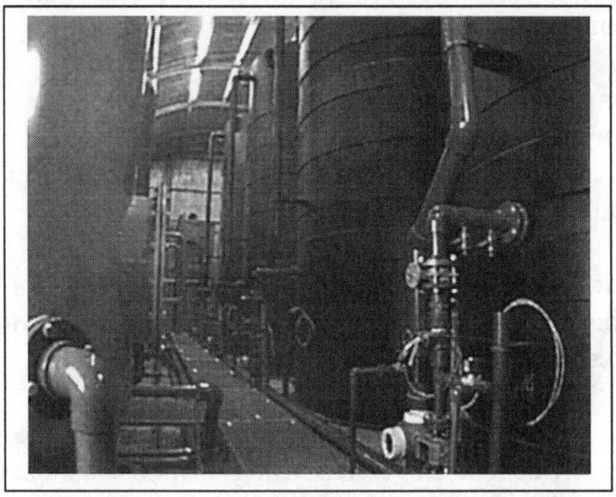

FIGURE 2. Full scale anaerobic bioreactors used at the Evenblij location.

INFILTRATION.

Infiltration tests performed in 1999 and 2000 revealed that direct infiltration of the effluent was not possible, mainly due to solids in the effluent consisting of biomass and inorganic precipitates like iron sulfide (FeS). Precipitates are formed within the anaerobic bioreactor due to the presence of sulfate and iron(II) in the groundwater.

Pilot tests showed that a two layer (fine sand and anthracite) filtration step was very effective to obtain infiltration water with less than 0.5 mg/L of solids. Since composition of the water phase was critical (e.g. phosphates) and keeping anaerobic conditions was urgent, changes were made on the bioreactor system.

After the changes on the system, mainly consisting of pre-treatment of the effluent, changes in the dosing of nutrients and carbon source, and changes in the process control of the bioreactors, it was possible to infiltrate the effluent of the bioreactors continuously for more than 5 months without any problem (figure 3, first 2.5 months). Infiltration flow was approximately 3-4 m^3/h using 3 wells. Besides this continuous biomass infiltration carbon source was injected discontinuously. Preventive regeneration with citric acid was performed every week to clean the wells. Besides, citric acid can be used as a carbon source.

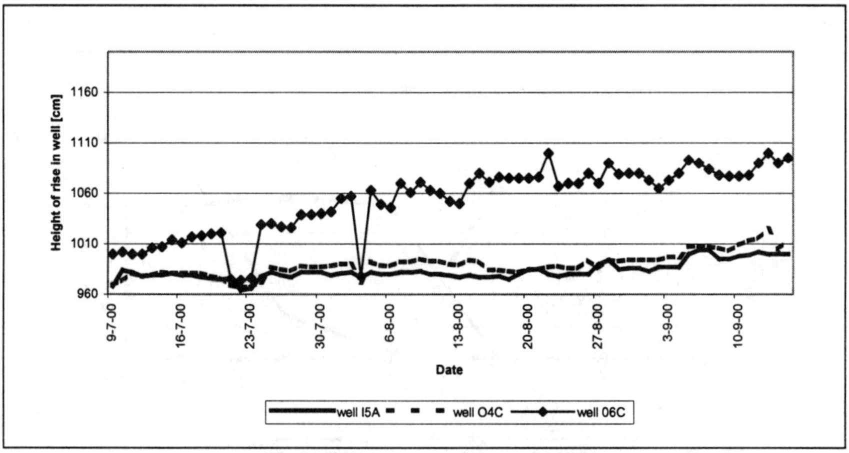
Height of rise in well [cm]
1160
1110
1060
1010
960
9-7-00
16-7-00
23-7-00
30-7-00
6-8-00
13-8-00
20-8-00
27-8-00
3-9-00
10-9-00
Date
well I5A
well O4C
well 06C

FIGURE 3. Infiltration experiments with three different wells. Reference well for old situation: LC03-O6C (upper line) and two new wells LC01-15A and LC02-O4C (lower two lines).

ENHANCED BIOREMEDIATION.

Within 4 weeks after starting the injection, positive effect became clear: up to 3.5 meters (well A) from the injection well, the dechlorination capacity had improved.

Former incomplete dechlorination to c-DCE (Figure 4) was transformed into complete dechlorination capacity to ethylene (Figure 5). This was measured using degradation tests with groundwater fed artificially with PCE and carbon source.

After 8 weeks infiltration monitoring wells at 7 meters from the infiltration well also showed an improvement to complete dechlorination, indicating a large region of influence for migration of bacteria (Table 1). The region of influence is suspected to be at least 7-10 m (20-30 ft).

Information about migration of dechlorinating bacteria was also gathered by using molecular monitoring tools (denaturant gradient gel electrophoresis, DGGE). Similarity of DGGE patterns increased comparing the effluent of the bioreactors and the groundwater at different distances from the infiltration well in course of time.

Both from the degradation tests and the molecular analyses it was concluded that migration of bacteria from the dechlorinating bioreactors had occurred, leading to an increased dechlorination capacity of the soil.

Most important of all, due to infiltration of biomass and suitable carbon source complete dechlorination of PCE into ethylene and even ethane occurred at the site in stimulated areas. Non-stimulated areas showed no changes compared to the situation in 1998 and before.

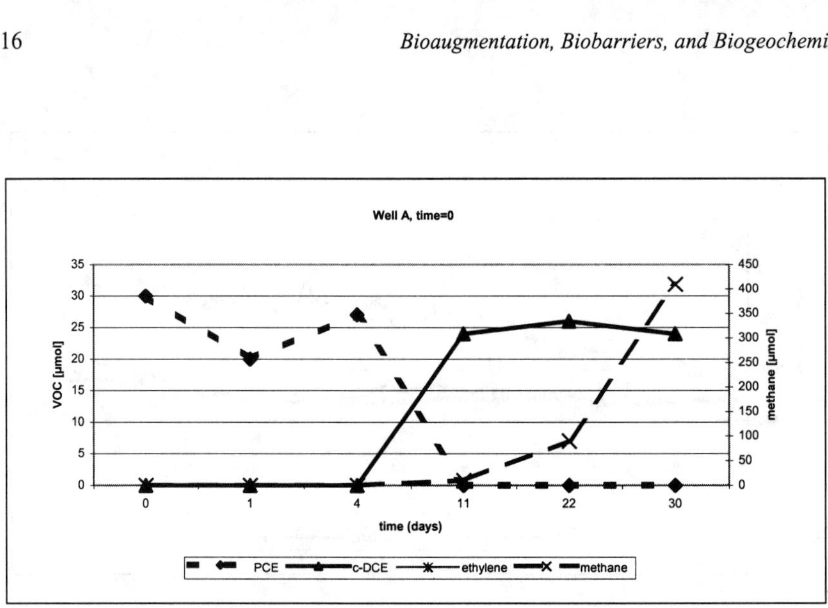
Well A, time=0
VOC [µmol]
methane [µmol]
time (days)
PCE
c-DCE
ethylene
methane

FIGURE 4. Incomplete degradation of PCE into c-DCE, situation in well A before infiltration of effluent of bioreactors (t=0).

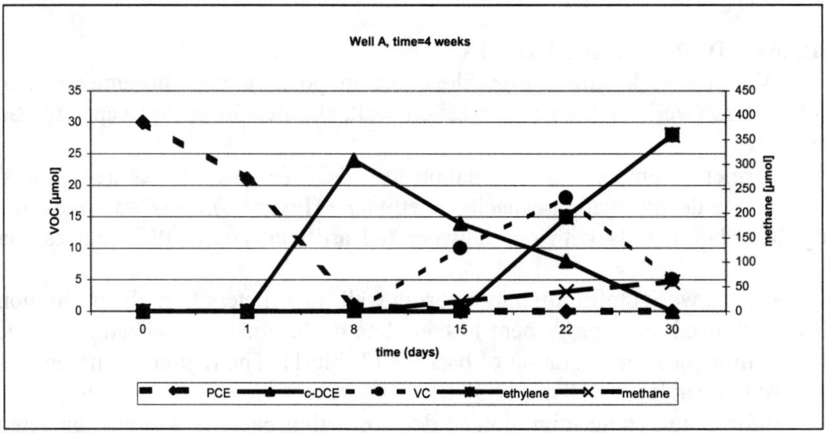
Well A, time=4 weeks
VOC [µmol]
methane [µmol]
time (days)
PCE
c-DCE
VC
ethylene
methane

FIGURE 5. Complete degradation of PCE into ethylene within 28 days due to infiltration, situation in well A after 4 weeks of infiltrating effluent of bioreactors (t=4 weeks).

TABLE 1. Summary results of stimulating effect due to infiltration

	distance from injection well	Dechlorination capacity at time : 0 weeks	4 weeks	8 weeks
Well A	3 m (10 ft)	PCE→c-DCE	PCE→ ethylene	PCE→ ethylene
Well B	7.5 m (22 ft)	PCE (no degr.)	PCE→ c-DCE	PCE→ ethylene

Besides, capacity of the infiltrated biomass seems to stay for a longer period since dechlorination is ongoing in areas in which infiltration has taken place months ago, indicating good survival of the bacteria.

CONCLUSIONS

The results show good stability of the anaerobic bioreactors. Approximately 60% of the effluent consisted of ethylene indicating the capacity for complete dechlorination within the bioreactors.

The effluent contains dechlorinating biomass which degrades PCE completely into ethylene. Infiltration of the effluent showed to be critical. However, after adaptations to the system and the way of infiltration continuous infiltration of the effluent combined with discontinuous carbon source addition was possible without any problem for more than 5 months.

Based on the monitoring it was concluded that complete dechlorination of PCE into ethylene and even ethane occurred at the site in stimulated areas. In addition, dechlorination is ongoing in areas in which infiltration has taken place months ago.

The TCE concept with bioaugmentation seems to be very effective in treatment of locations with limited biological capacity.

ACKNOWLEDGEMENTS

This project is partly funded by NOBIS/SKB, the Dutch research program on biotechnological in situ soil remediation.

PREVENTING CONTAMINANT DISCHARGE TO SURFACE WATERS: PLUME CONTROL WITH BIOAUGMENTATION

John Lendvay (University of San Francisco, San Francisco, CA)
Peter Adriaens, Michael Barcelona, and C. Lee Major, Jr. (University of Michigan, Ann Arbor, MI)
James Tiedje and Michael Dollhopf (Michigan State University, East Lansing, MI)
Frank Löffler (Georgia Institute of Technology, Atlanta, GA)
Babu Fathepure (Oklahoma State University, Stillwater, OK)
Erik Petrovskis, Michael Gebhard, and Gary Daniels (GeoTrans, Ann Arbor, MI)
Robert Hickey, Robert Heine, and Jing Shi (EFX Systems, Inc., Lansing, MI)

ABSTRACT: Widespread use of chlorinated solvents has resulted in contamination of groundwater requiring effective source and plume containment strategies to prevent contaminant migration and mitigate adverse environmental impacts. The Bachman Road Residential Wells Site is contaminated with predominantly tetrachloroethene (PCE) and the contaminant plume flows into Lake Huron. Due to natural dechlorination processes, some of the PCE has been transformed to cis-1,2-dichloroethene (cis-DCE), with lesser amounts of trichloroethene (TCE) and chloroethene (VC). Laboratory and field investigations demonstrated the presence of a PCE to cis-DCE dechlorinating *Desulfuromonas* species and the presence of a *Dehalococcoides* species, which completely dechlorinates chloroethenes to ethene. Both populations are known to use chloroethenes as metabolic electron acceptors. A bioaugmentation strategy using these microorganisms at the pilot scale has resulted in complete dechlorination of the groundwater in less than 50 days. Additionally, a 16S rDNA-based molecular approach was developed to track the PCE to cis-DCE and the chloroethene dechlorinating species. This methodology allows us to quantitatively track the populations of interest, monitor their migration and to develop cause and effect relationships with contaminant degradation. This field demonstration illustrates that bioaugmentation using halorespiring microorganisms is a viable alternative for preventing contaminant migration into associated surface water systems.

INTRODUCTION

The Great Lakes region of the United States is highly industrialized. One result of this industrialization has been extensive contamination of groundwater with priority pollutants, including chlorinated ethenes. The extent of this problem is significant in that priority pollutants have been identified by the U.S. EPA as the leading cause of impaired shorelines in the region (U.S. EPA, 2000). Specifically, the U.S. EPA recently reported to Congress that after assessing 90% of the Great Lakes' shoreline, 28% of the assessed shoreline is impaired due to priority pollutants (U.S. EPA, 2000). The State of Michigan is further impacted by this problem because it has the longest shoreline of any of the Great Lakes' states and has a geologic surface structure consisting of many unconfined aquifers. The consequence is that many unconfined aquifers in the state have been contaminated with priority pollutants and, due to the hydrogeology of these formations, discharge the contaminated groundwater either directly into the Great Lakes or into other surface waters of the Great Lakes watershed. Regardless of the flow path, the result is deterioration of the groundwater and surface water quality of the region (Ricci et al., 2001). Therefore, the Michigan Department of Environmental Quality (MDEQ) has supported developing technologies that may be used to remediate contaminant plumes and mitigate their environmental impact. The Bachman Road Residential Wells site is one such project focusing on the remediation of a PCE contaminant plume in a surface aquifer along Lake Huron's shoreline (Figure 1).

The Bachman Road Residential Wells site consists of a PCE contaminant plume undergoing in-situ dechlorination. The halorespiration plume shows evidence of naturally occurring dechlorination as indicated by the appearance of lesser-chlorinated daughter products, trichloroethene (TCE), cis-dichloroethene (cis-DCE), and chloroethene (VC). Current in situ degradation processes at this site are insufficient to stop migration of the chlorinated ethenes to Lake Huron. Therefore, we have selected this site to enhance natural degradation by augmenting with halorespirers, previously reported as present at this site (Fathepure et al., 2000; Löffler et al., 1999a). Halorespirers are those microorganisms that use chlorinated ethenes as a terminal electron acceptor for respiratory processes.

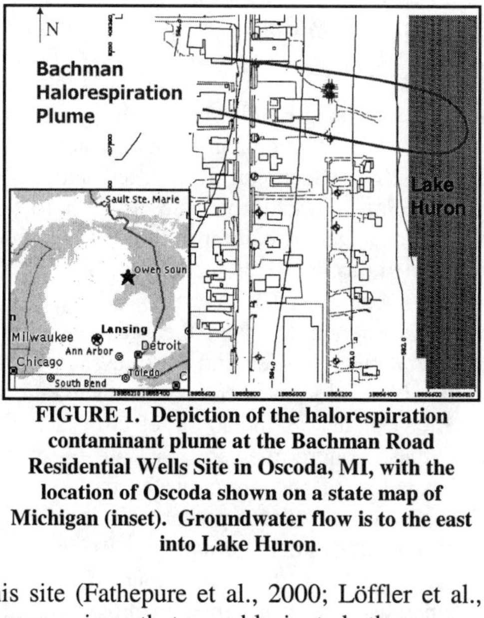
N
Bachman
Halorespiration
Plume
Lake
Huron
Sault Ste. Marie
Owen Soun
Milwaukee
Lansing
Ann Arbor
Detroit
Chicago
Toledo
South Bend
FIGURE 1. Depiction of the halorespiration contaminant plume at the Bachman Road Residential Wells Site in Oscoda, MI, with the location of Oscoda shown on a state map of Michigan (inset). Groundwater flow is to the east into Lake Huron.

The primary purpose of this pilot scale field project is to demonstrate the effectiveness of bioaugmenting the aquifer with halorespiring bacteria to efficiently dechlorinate the contaminant (PCE) and daughter products (TCE, cis-DCE, and VC) to ethene. Preliminary data from the pilot phase of the project is the focus of this paper.

BACKGROUND

Several research studies have reported the reductive dechlorination of chlorinated ethenes via metabolic processes (e.g. Holliger et al., 1993; Holliger and Schumacher, 1994), generating much excitement in the field of bioremediation towards the possible use of these microorganisms to efficiently remediate contaminated aquifers. Additionally, a recent publication has described one field demonstration of this technology at the Dover Air Force Base, Delaware (Ellis et al., 2000). This research reported complete degradation of TCE to ethene in an aerobic surface aquifer following reduction of the aquifer and injection of *Pinellas* culture into the contaminated zone. This dechlorination of the dissolved contaminants in the plume occurred in about 200 days, following injection of the halorespiring microorganisms.

Microbial halorespiration has been demonstrated to be among the fastest means of achieving complete dechlorination (and detoxification) of chlorinated solvents. In addition, metabolic reductive dechlorination is a highly efficient process. A recent study has shown that bacteria that respire with a chlorinated compound channel 60 to 70% of the electrons released in electron donor oxidation to reductive dechlorination (Löffler et al., 1999). Although one organism (*Dehalococcoides ethenogenes*) having the ability to dechlorinate PCE to ethene was described, complete dechlorination is most efficiently performed by more than one population (Maymo-Gatell et al., 1997). Thus, a consortium of halorespirers is generally required to fully degrade PCE to ethene. The first step is reduction of PCE to cis-DCE via TCE by one type of halorespiring microorganism. Following this step, *Dehalococcoides ethenogenes* and other unidentified bacteria reduce cis-DCE fully to ethene via VC. Ethene is the only degradation product in this process that is not a priority pollutant, and is therefore the desired product.

APPROACH

The goal of this project is to clearly show a cause and effect relationship between the inoculated halorespiring consortium and reductive dechlorination processes occurring in the PCE-contaminated aquifer. We duplicated hydraulic conditions in two plots, one a test plot and the other a control, at this site by pumping water from the extraction well back into the two injection wells starting on 5 September 2000. In the test plot, 0.1 mM lactate and nutrients were continuously added to the flow-stream starting on 25 September. The purpose of this electron donor addition was to reduce the aquifer in preparation for microbial inoculation on 24 October. No carbon source, nutrients, or microorganisms were added to the control plot. Thus, the control provided information on how groundwater circulation in the aquifer impacted actual and relative contaminant concentrations in the groundwater and the aquifer solids. It may also provide information on the significance of non-enhanced dechlorination rates at this site.

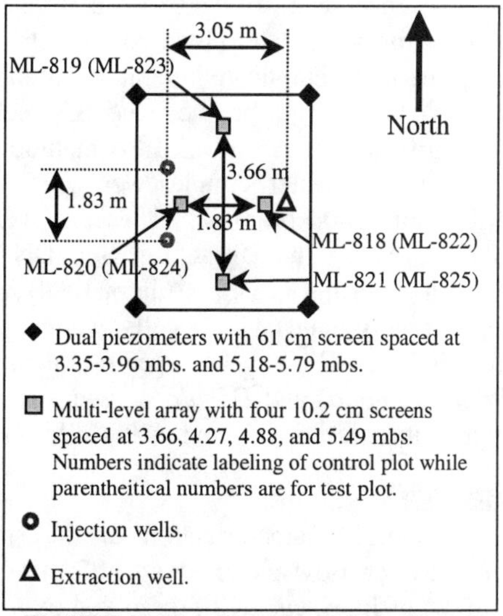
3.05 m
ML-819 (ML-823)
North
3.66 m
1.83 m
1.83 m
ML-818 (ML-822)
ML-820 (ML-824)
ML-821 (ML-825)
Dual piezometers with 61 cm screen spaced at 3.35-3.96 mbs. and 5.18-5.79 mbs.
Multi-level array with four 10.2 cm screens spaced at 3.66, 4.27, 4.88, and 5.49 mbs. Numbers indicate labeling of control plot while parentheitical numbers are for test plot.
Injection wells.
Extraction well.

FIGURE 2. Plan drawing of the wells and sampling points for a typical plot at the Bachman Road Residential Wells Site.

Construction of Plots. The test and control plots were constructed as shown in Figure 2. The two plots are spaced normal to groundwater flow with the control plot located to the north of the test plot. Groundwater flow is toward Lake Huron, located approximately 80 meters east of the plots. Each plot consists of dual piezometers with 61 cm screens at each corner. Screened depths are at 3.4 – 4.0 meters and 5.2 – 5.8 meters below surface (mbs) for the shallow and deep piezometers, respectively. The plot size is 5.5 meters north to south (normal to groundwater flow) and 4.6 meters east to west. Two 10 cm diameter injection wells, screened at 3.4 – 5.8 mbs, are located as shown in Figure 2 on the west side of the plots and a single extraction well (screened at 4.6 – 5.8 mbs) located toward the eastern border of the plot. As the injection wells are screened over most of the saturated thickness of the aquifer and the extraction well only in the deeper portion, we effectively moved microorganisms from shallow to deep portions of the aquifer. This design resulted from initial characterization of the aquifer microbial cultures, which indicated that the cis-DCE to ethene degraders were predominantly in the deep portion of the aquifer. It was necessary for them to be present in sufficient numbers in both the shallow and deep portions for this study to effectively dechlorinate PCE to ethene in the shallow zone as well as the deep zone. Finally, four separate multi-level arrays are positioned as shown with 10 cm screens at 3.7, 4.3, 4.9, and 5.5 mbs. These were positioned to provide discrete monitoring points with depth over the center of each plot.

Methods. Piezometers and multi-level arrays were sampled using peristaltic pumps. After purging the sample tubing, the pump effluent was attached to a QED flow cell (Ann Arbor, MI), which is equipped to measure temperature, pH, oxidation-reduction (redox) potential, specific conductance, and dissolved oxygen. All probes were calibrated prior to sampling. Flow cell readings were recorded when stable, usually 20-30 minutes after initiation of pumping regime.

Subsequent to the flow cell readings, dissolved oxygen and aqueous ferrous iron were determined colorometrically using Chemetrics (Calverton, VA) sampling kits. Duplicate laboratory samples were collected in 40 mL borosilicate VOA vials, preserved and capped. Collected samples were immediately stored at $4^{o}C$ until analysis using U.S. EPA method 8260.

Groundwater samples were filtered through 0.2 μm membrane filters to collect microbial biomass. The cells were suspended in buffer, and DNA was extracted using the UltraClean Soil DNA Kit from Mo Bio Laboratories, Inc. (Solana Beach, CA). Alternatively, the cell suspensions were diluted appropriately, and whole cells were used as template in Polymerase Chain Reaction (PCR). The specific detection of chloroethene dechlorinating *Desulfuromonas* and *Dehalococcoides* species was performed with a nested PCR approach as described previously (Löffler et al., 2000).

RESULTS

Initial characterization of the site consisted of a transect normal to groundwater flow sampled at 3.05, 4.57, and 6.10 mbs with the contaminant profiles of PCE, cis-DCE, dissolved oxygen, and soluble iron shown in Figure 3. These data showed that the plots were located in the contaminant plume and that

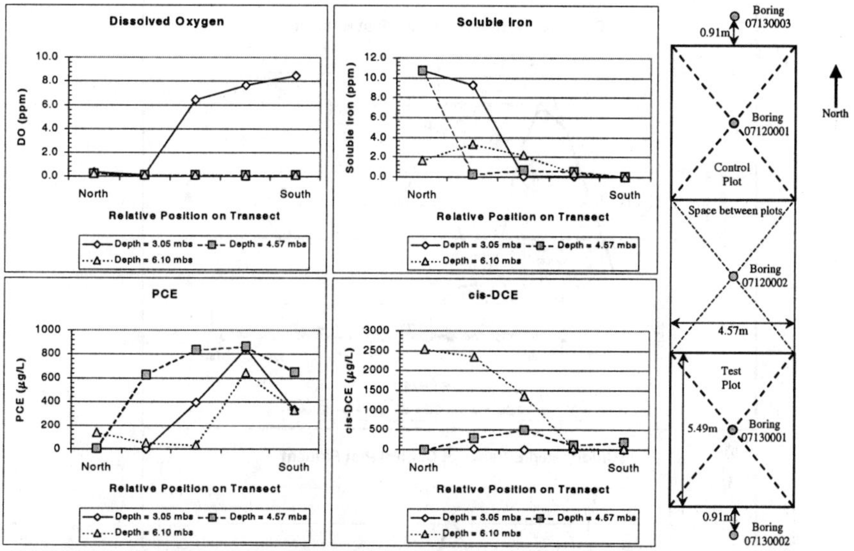
Dissolved Oxygen
DO (ppm)
Soluble Iron
Soluble Iron (ppm)
PCE
PCE (µg/L)
cis-DCE
cis-DCE (µg/L)
North
South
Relative Position on Transect
Depth = 3.05 mbs
Depth = 4.57 mbs
Depth = 6.10 mbs
Boring 07130003
0.91m
North
Boring 07120001
Control Plot
Space between plots
Boring 07120002
4.57m
Test Plot
5.49m
Boring 07130001
0.91m
Boring 07130002

FIGURE 3. Groundwater analyses for three separate depths along a transect bisecting the control and test plots. Location of each sample point is shown on plan view, right.

PCE predominated in the shallow portions of the aquifer while cis-DCE predominated in the deep zone.

Following installation of the equipment for the test and control plots; flow was initiated at 15 Lpm on 2 September 2000 to establish the flow field. Start of lactate addition to reduce the aquifer commenced on 25 September. Monitoring of oxidation-reduction potential and dissolved oxygen are shown for the four discrete depths of the multi-level array (ML-824) closest to the injection well of the test plot (Figure 4). These data provide evidence of aquifer reduction within 30 days of the start of lactate injection into the test plot.

Once the aquifer was sufficiently reduced, the test plot was inoculated with approximately 210 L of microbial suspension at a concentration of $1.08*10^8$ cells/mL on 24 October. This was followed by continuous injection of 0.1 mM lactate and nutrients (nitrogen and phosphorous) into the test plot. The results of contaminant transformation following the start of the test compared with those of

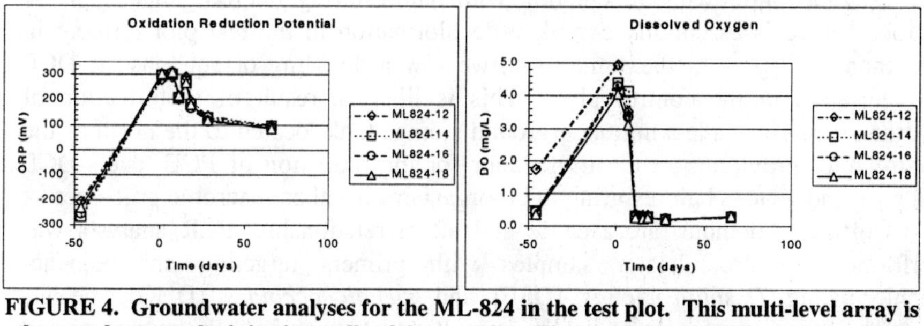
Oxidation Reduction Potential
ORP (mV)
Time (days)
ML824-12
ML824-14
ML824-16
ML824-18
Dissolved Oxygen
DO (mg/L)
Time (days)
ML824-12
ML824-14
ML824-16
ML824-18

FIGURE 4. Groundwater analyses for the ML-824 in the test plot. This multi-level array is the one closest to the injection wells of the test plot. Time zero days is 25 September 2000, the start of lactate injection.

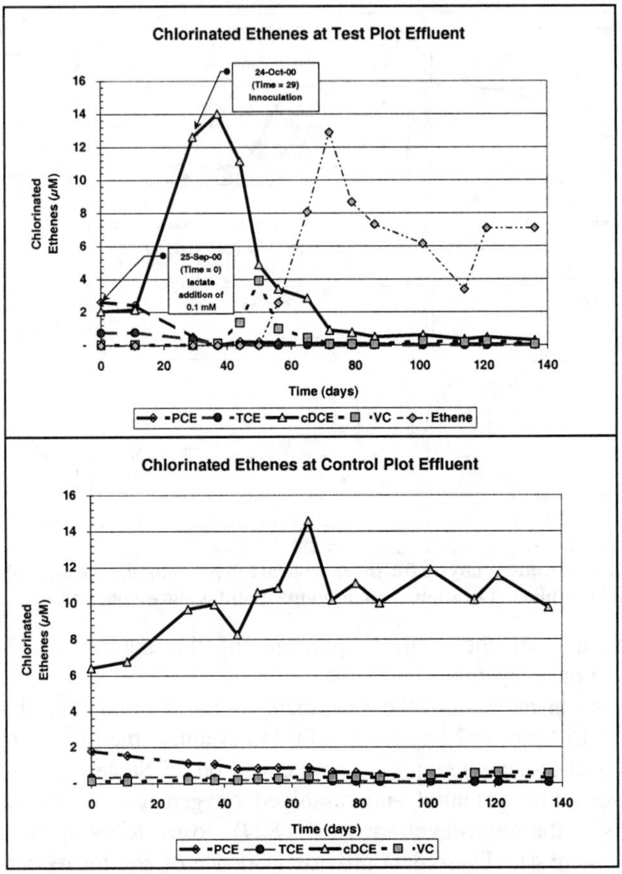
Chlorinated Ethenes at Test Plot Effluent
24-Oct-00
(Time = 29)
Innoculation
25-Sep-00
(Time = 0)
lactate
addition of
0.1 mM
Chlorinated Ethenes (µM)
Time (days)
PCE
TCE
cDCE
VC
Ethene
Chlorinated Ethenes at Control Plot Effluent
Chlorinated Ethenes (µM)
Time (days)
PCE
TCE
cDCE
VC

FIGURE 5. Contaminant profiles are shown for the test (top) and control (bottom) plots. Time zero is arbitrarily set to the start of lactate injection on 25 September 2000.

the control plot are shown in Figure 5 for water sampled from the extraction well of each plot. These data show a sharp increase in cis-DCE concentration between the start of groundwater circulation and inoculation in the test plot relative to the control. This is potentially a result of biostimulation of native microorganisms in the test plot. Following inoculation with halorespiring microorganisms on 24 October, there is essentially complete dechlorination in the test plot effluent in less than 50 days. At the same time, we saw a doubling of aqueous cis-DCE concentration in the control plot. This is likely a result of mobilization of hydrocarbons from a leaking underground storage tank located to the north of the control plot providing an electron donor for the reduction of PCE to cis-DCE either by indigenous halorespiring microorganisms or other anaerobic processes.

In an effort to demonstrate a cause and effect relationship, PCR analysis was performed on groundwater samples with primers targeting chloroethene-dechlorinating *Desulfuromonas* (DSF) and *Dehalococcoides* (DHC) species. Samples were taken from two depths of multi-level arrays in both the control and test plots. The PCR data, Table 1, suggest that both DSF and DHC type

TABLE 1. Positive detection for PCR analysis of *Dehalococcoides* (top) and *Desulfuromonas* (bottom) in the test and control plots on the dates specified.

Dehalococcoides			**Detection**	
Multi-Level Array	**Depth (m)**	**24 Oct 00**	**29 Nov 00**	**02 Jan 01**
ML-818	3.66	-	-	+
Control Plot Array	5.49	+	+	+
ML-820	3.66	+	+	+
Control Plot Array	5.49	-	-	+
ML-822	3.66	-	+	+
Test Plot Array	5.49	-	+	+
ML-824	3.66	-	+	+
Test Plot Array	5.49	-	+	+
Desulfuromonas			**Detection**	
Multi-Level Array	**Depth (m)**	**24 Oct 00**	**29 Nov 00**	**02 Jan 01**
ML-818	3.66	-	-	-
Control Plot Array	5.49	-	+	-
ML-820	3.66	-	-	-
Control Plot Array	5.49	-	+	+
ML-822	3.66	+	+	+
Test Plot Array	5.49	+	-	+
ML-824	3.66	-	-	+
Test Plot Array	5.49	-	+	+

microorganisms were introduced to the test plot with the inoculum. However, these data also show the presence of both populations at some wells in the control plot. The presence of DHC and DSF in the control plot is not surprising because both populations are indigenous to the Bachman aquifer. This may explain the partial reduction of PCE to cis-DCE in the control plot as previously described with the groundwater analysis (Figure 5).

CONCLUSIONS

The data clearly show reduction of groundwater in the test plot within 30 days following continuous injection of lactate. Contaminant data of the test plot effluent stream suggests rapid and complete dechlorination of all contaminants to ethene relative to the control plot in less than 50 days. These data represent average groundwater values as they are of the plot's effluent flow and not discrete profiles from the multi-level arrays. Finally, microbial screening using PCR with primers targeting DSF and DHC species confirm the presence of both populations in the test plot.

Future work on this project centers on two focal points. First, confirming the average groundwater data from the effluent wells with discrete samples from the multi-level arrays and aquifer solids analysis. Second, separate effects of biostimulation and bioaugmentation at this site will be evaluated. As other contaminated sites may not have indigenous populations of halorespiring microorganisms, it is necessary to show bioaugmentation is possible and effective to stimulate the reductive dechlorination process and thus detoxify chloroethene contaminated groundwater plumes.

ACKNOWLEGEMENTS

Funding for this project was provided by a grant from the Michigan Department of Environmental Quality, Peter Adriaens, Project Director. Additional funding was provided to John Lendvay from the Lily Drake Cancer Research Fund at the University of San Francisco.

REFERENCES

Ellis, D. E., E. J. Lutz, J. M. Odom, R. J. Buchanan, Jr., and C. L. Barlett. 2000. "Bioaugmentation for Accelerated In Situ Anaerobic Bioremediation." *Environ. Sci. Technol. 34*(11): 2254-2260.

Fathepure, B. Z., S. P. Beaver, M. E. Dollhopf, F. E Löffler, P. Adriaens, and J. M. Tiedje. 2000. "Optimization of Bioaugmentation Strategies for *In Situ* Treatment of Chloroethene-Contaminated Groundwater Using Halorespiring Bacteria at the Bachman Road Site, Oscoda, MI." In *Abstracts of the 100th Annual Meeting of the American Society for Microbiology 2000*. Los Angeles. Abstract Q-133, p. 570-571.

Holliger, C., and W. Schumacher. 1994. "Reductive Dehalogenation as a Respiratory Process." *Antonie van Leeuwenhoek 66*: 239-246.

Holliger, C., G. Schraa, A. J. M. Stams, and A. J. B. Zehnder. 1993. "A Highly Purified Enrichment Culture Couples the Reductive Dechlorination of Tetrachloroethene to Growth." *Appl. Environ. Microbiol. 59*(9): 2991-2997.

Löffler, F. E., M. E. Dollhopf, E. A. Petrovskis, J. M. Tiedje, and B. Z. Fathepure. 1999a. "Test Methods to Determine the Activity of Halorespiring Bacteria." In *Abstracts of the 5th International Symposium In situ and on-site Bioremediation*. San Diego, CA.

Löffler, F. E., J. M. Tiedje, and R. A. Sanford. 1999b. "Fraction of Electrons Consumed in Electron Acceptor Reduction and Hydrogen Thresholds as Indicators of Halorespiatory Physiology." *Appl. Environ. Microbiol. 65*(9): 4049-4056.

Löffler, F. E., Q. Sun, J. Li, and J. M. Tiedje. 2000. "16S rRNA Gene-Based Detection of Tetrachloroethene-Dechlorinating *Desulfuromonas* and *Dehalococcoides* Species." *Appl. Environ. Microbiol. 66*(4): 1369-1374.

Maymó-Gatell, X., Y.-T. Chien, J. M. Gossett, and S. H. Zinder. 1997. "Isolation of a Bacterium that Reductively Dechlorinates Tetrachloroethene to Ethene." *Science 276*: 1568-1571.

Ricci, P. F., L. E. Ricci, W. Smith, and R. Goldstein. 2001. "Water Demand, Supply and Quality in the US and an Overview on the Next 50 Years." Submitted to *Science*.

United States Environmental Protection Agency. 2000. "National Water Quality Inventory: 1998 report to Congress." *EPA841-R-00-011*. Washington, DC.

SUCCESSFUL FIELD DEMONSTRATION OF BIOAUGMENTATION TO DEGRADE PCE AND TCE TO ETHENE

D.W. Major, M.L. McMaster and E. E. Cox (GeoSyntec Consultants, Guelph, Ontario, Canada), B. J. Lee and E.E. Gentry (Science Applications International Corporation, San Antonio TX), E. Hendrickson (Dupont, Delaware) E. Edwards and S. Dworatzek (University of Toronto, Toronto, Ontario, Canada)

ABSTRACT: Field demonstrations were conducted at Kelly Air Force Base in Texas to evaluate the applicability of accelerated anaerobic bioremediation via bioaugmentation to treat chlorinated ethenes in groundwater. KB-1, a natural, non-pathogenic microbial consortium isolated by University of Toronto and GeoSyntec is being used for bioaugmentation. Results of laboratory studies and the field pilot tests show that complete dechlorination to ethene occurred only after bioaugmentation with KB-1.

INTRODUCTION

In most subsurface environments, the main biodegradation mechanism for chlorinated ethenes is reductive dechlorination, which involves the sequential replacement of chlorine atoms on the alkene molecule by hydrogen atoms. The chlorinated ethenes serve as electron acceptors and hydrogen produced from the anaerobic metabolism of sugars, alcohols, fatty acids or other complex carbohydrates or carbon sources, serves as the electron donor during dechlorination reactions.

Although a number of anaerobic microorganisms can reductively dechlorinate through co-metabolic reactions (likely due to the transfer of electrons from reduced co-factors involved in anaerobic metabolism), they do not necessarily derive energy from the reaction. In contrast, specific dehalorespiring microorganisms use chlorinated solvents as their terminal electron acceptors and gain energy from reductive dechlorination. Dehalorespiring bacteria that have been identified, include *Dehalospirillium multivorans* (Scholz-Muramatsu et al., 1995), *Dehalobacter restrictus* (Schumacher and Holliger, 1996) and *Dehalococcoides ethenogenes* (Maymo-Gatell et al., 1997). Of these microorganisms, *D. ethenogenes* is able to completely dechlorinate the chlorinated ethenes and possibly ethanes. *D. ethenogenes* does not appear to be ubiquitous at all sites, or alternatively, are present but are not active. As a result, dechlorination of teratchloroethene (PCE) and trichloroethene (TCE) stalls at *cis*-1,2-dichloroethene (cis-1,2-DCE) at many sites, resulting in a build up of this dechlorination product.

Several stable, natural microbial consortia containing *D. ethenogenes* strains have been isolated that are capable of mediating complete dechlorination of TCE to ethene. A field demonstration by the Remediation Technologies Development Forum (RTDF) at Dover Air Force Base in Delware showed that TCE dechlorination had stalled at cis-1,2-DCE despite continued electron donor addition. Following bioaugmentation, complete dechlorination of cis-1,2-DCE via vinyl chloride (VC) to ethene was observed (Ellis et al., 2000).

GeoSyntec, working with the University of Toronto (UT), has isolated a stable, natural microbial consortia (referred to as KB-1) capable of stimulating rapid dechlorination of PCE and/or TCE to ethene at sites where this activity is otherwise deficient. Experiments have shown that addition of KB-1 to microcosms in which PCE and/or TCE dechlorination had stalled at cis-1,2-DCE, despite continued electron donor addition, immediately stimulated dechlorination of cis-1,2-DCE via VC to ethene; cis-1,2-DCE was completely transformed to ethene in a matter of days (Cox et al., 1998, Wehr et al, 2001). Development of these and similar microbial cultures now provides the ability to accelerate bioremediation of TCE and related chlorinated solvents at sites where complete dechlorination reactions do not otherwise occur.

A laboratory and field test was conducted at Kelly AFB to assess if enhancement by electron donor addition or bioaugmentation with KB-1 was required to achieve complete reductive dechlorination of PCE to ethene. A key component of this demonstration was the ability to assess before the field demonstration the absence of *D. ethenogenes* at the site and track the spread of KB-1 strains of *D. ethenogenes.*

Site Description: Kelly AFB is located in south central Texas, approximately seven miles southwest of downtown San Antonio, Texas. The main base covers an area of 3,929 acres and is located in an area generally consisting of residential, commercial, and light industrial land usage. The pilot test area (PTA) is located in the courtyard of Building 360 of Kelly AFB. The geology in the vicinity of Building 360 consists of unconsolidated alluvial deposits overlying an undulatory erosional surface of Navarro Clay. The alluvial deposits consist of gravel, sand, silt and clay, ranging in thickness from 20 to 40 feet. From the surface downward, the geology typically consists of: 1 to 4 feet of black organic clay (denoted as fill/clay); 6 to 16 feet of tan silty, calcareous clay; and 4 to 20 feet of clayey limestone and chert gravel (denoted as clayey/gravel). The groundwater flow direction is typically to the southeast and the flow velocity is about 3.0 ft/day (SAIC, 1999). Volatile organic compounds (VOCs) in site groundwater consisted primarily of PCE, with lesser amounts of TCE, and cis-1,2-DCE. The site groundwater contained nitrate and sulfate at about 24 and 16 mg/L, respectively. Dissolved oxygen and redox measurements indicated that the groundwater was aerobic and oxidizing on a macro-scale; but elevated levels of dissolved manganese suggest anaerobic microenvironments. Methane, ethene and ethane were not detected in the sampled groundwater.

RESULTS FROM MICROCOSM STUDIES

A laboratory microcosm study was used to evaluate whether the intrinsic biodegradation activity could be accelerated through electron donor addition, or if bioaugmentation with KB-1 was required to completely degrade PCE and TCE to ethene. A series of treatment and control microcosms were constructed in triplicate using site soil and groundwater, and incubated at room temperature in an anaerobic glovebox. Microcosms were constructed using 250 mL nominal volume glass bottles filled with 60 g soil and 150 mL groundwater. Headspace and groundwater were sampled through Miniert™ valves for analysis of electron donors, anions, VOCs, and biogenic gases (ethene, methane). Microcosms amended with TCE and mercuric chloride or just TCE served as sterile and intrinsic (biotic) controls. Electron Donor treatment microcosms were amended with only electron donors (methanol or lactate) and TCE at approximately 1 mg/L. Bioaugmentation treatment microcosms were amended with methanol, KB-1 and TCE at three different concentrations (approximately 1, 8 and 80 mg/L). Table 1 shows distribution of TCE and its dechlorination products under each control and treatment condition at the end of the experiment. The results indicate that the indigenous microorganisms in the site groundwater can reductively dechlorinate TCE to cis-1,2-DCE but cannot mediate further dechlorination of cis-1,2-DCE to VC and ethene. Bioaugmentation of the microcosms with KB-1 promoted complete dechlorination to ethene at all initial TCE concentrations tested.

TABLE 1. Initial Mass of TCE and Final Distribution of VOC Mass in Microcosm Treatments and Controls

Treatment	Mass Distribution at Final Sampling Point (umoles/bottle)					Conversion to Ethene (%)
	TCE Initial	TCE	DCE	VC	Ethene	
Sterile Control	10.3	11.4	0.03	<0.05	<0.01	0%
Intrinsic Control	11.2	10.7	0.06	<0.05	<0.01	0%
Methanol	1.04	1.02	0.87	<0.05	<0.01	0%
Lactate	1.16	<0.01	1.51	<0.05	<0.01	0%
Low TCE & KB-1	1.16	<0.01	<0.01	<0.05	1.91	164%
Mid TCE & KB-1	10.4	<0.01	<0.01	<0.05	10.8	105%
High TCE & KB-1	117	0.02	0.24	44.7	104	89%

Notes: Values represent average of triplicate microcosms.
Final data are after 70 or 100 days of incubation
KB-1 ammended treatment receieved methanol as electron donor

FIELD DEMONSTRATION

The field pilot test consisted of a closed loop recirculation system, including 3 extraction wells, one injection well, and 5 biomonitoring wells. Figure 1 presents a schematic of the system in plan view and cross section. Groundwater was extracted, combined through a common header, amended with electron donors (methanol/acetate) and tracer (bromide) as required, and injected via the injection well. The electron donor/tracers were metered into injected groundwater to achieve the desired concentration based on the extracted groundwater measured flow rate.

FIGURE 1A . Plan View of Pilot Test Area

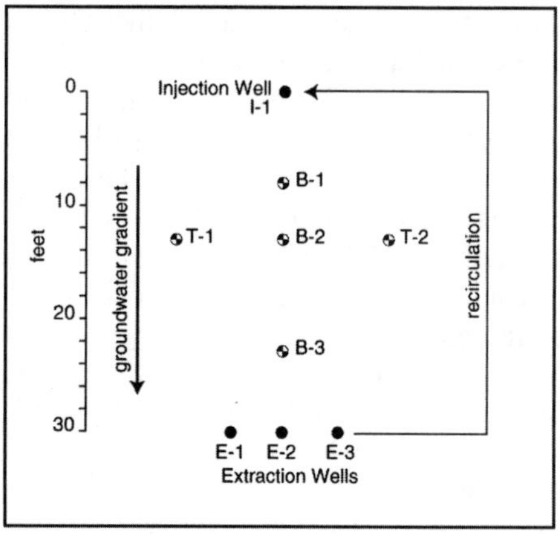
Injection Well
I-1
B-1
T-1
B-2
T-2
B-3
E-1 E-2 E-3
Extraction Wells
feet
0
10
20
30
groundwater gradient
recirculation

FIGURE 1B. Cross-Section Schematic of Pilot Test Area

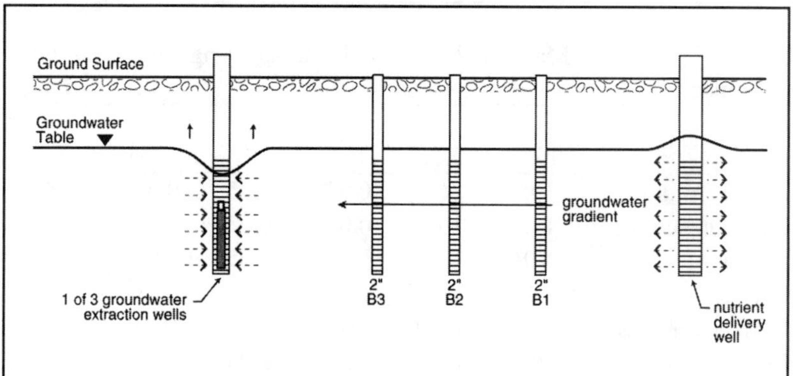
Ground Surface
Groundwater Table
groundwater gradient
1 of 3 groundwater extraction wells
2" B3
2" B2
2" B1
nutrient delivery well

A bromide tracer test was used to verify residence times of the entire system and between monitoring wells, and estimate mass capture efficiency and the groundwater pore volume (PV). The time to capture and recirculate one pore volume was approximately 6 days, and the time between the injection well and the first monitoring well (B1) was between 4 to 8 hours. The mass capture efficiency and number of groundwater PVs recirculated were estimated to be approximately 90% and 64 m^3, respectively. The extraction wells efficiency indicates that either background water or recirculated water that was taking longer flow paths, was being re-injected into the pilot test area. This resulted in the continuous addition of PCE and cis-1,2-DCE into the pilot test area over the test period. From Day 89 (electron donor started) to Day 318 (last sample event) a total of 39 PV were re-circulated through the pilot test area (PTA). Since bioaugmentation (Day 176), approximately 24 PVs were re-circulated.

Figure 2 presents the VOC and ethene concentrations over time at B1 (similar results were observed at all monitoring wells at the end of the test). When both acetate and methanol additions were made PCE concentrations in the PTA declined by more than 90%, with the dominant degradation products being cis-1,2-DCE. However, VC and ethene were not produced prior to bioaugmentation.

FIGURE 2. Micromoles of CVOC and Ethene Over Time at Monitoring Well B1

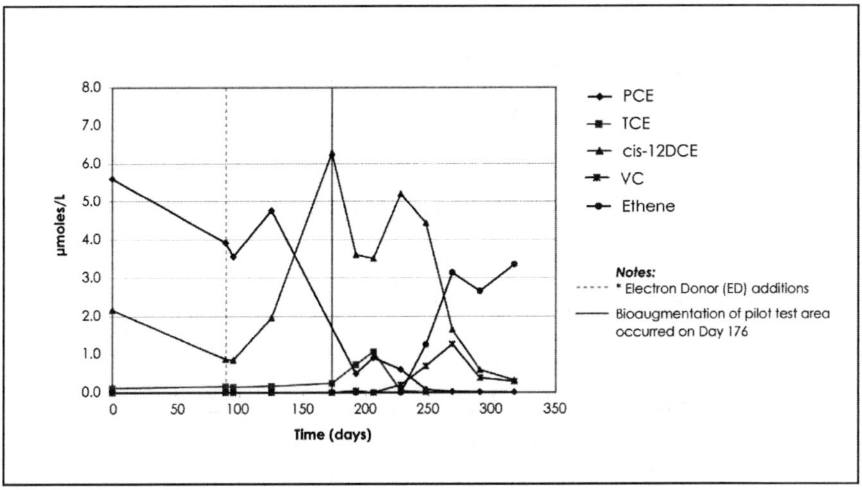

µmoles/L
8.0
7.0
6.0
5.0
4.0
3.0
2.0
1.0
0.0
0
50
100
150
200
250
300
350
Time (days)
PCE
TCE
cis-12DCE
VC
Ethene
Notes:
* Electron Donor (ED) additions
Bioaugmentation of pilot test area occurred on Day 176

Prior to the addition of KB-1, analysis of soil and groundwater samples through gene probes (16S rRNA analysis) showed that *D. ethenogenes* was not detected at the site (Hendrickson et al. 2001). On 6 May 2000 (Day 176) approximately 13 L of KB-1 was added to the PTA through a submerged delivery line in the injection well (IW). Sixteen days after bioaugmentation, trace amounts of VC were reported in well B1, and ethene was detected after 52 days. By day 318 (142 days after bioaugmentation), ethene was the dominant product in the PTA.

Table 2 compares the concentration of cVOCs and ethene between the injection and B1 wells over the last three sample periods, the associated integrated half-life (PCE to ethene), and the mass balance between cVOCs removed and ethene produced. Recall, the residence time between the injection well and B1 is 4 hours. Table 2 shows that we obtained very good agreement between ethene production and cVOC loss, and that the average integrated half-life of cVOCs to ethene over this time period is approximately 3.8 hours.

Groundwater samples collected for molecular analysis showed that detection of *D. ethenogenes* correlated with the production of VC and ethene in the PTA (Data not shown, see Hendrickson et al., 2001). At the completion of the study, *D. ethenogenes* was detected in all monitoring and extraction wells. *D. ethenogenes* was not detected in soil and groundwater samples taken from outside the PTA. In addition, 12/13C stable isotope analysis was also conducted on groundwater samples, and verified the loss of cVOCs was attributed to biodegradation due to 12/13 C stable isotope ratios in the cVOC parent and daughter products (data not shown, see Morrill et al., 2001).

TABLE 2. Mass Conversion of Chlorinated VOCs to Ethene in the Pilot Test Area

	μmoles							
	Total VOCs	Ethene	Total VOCs	Ethene	Total VOCs	Ethene	% of	Half-life
Well ID	IW		B1		Δ IW-B1		Expected*	(Hours)#
Days After KB-1 addition								
93	4.5	0.9	2.9	3.1	1.6	2.3	144%	6.5
115	2.9	1.1	1.0	2.7	1.9	1.5	79%	2.6
142	2.0	1.9	0.6	3.4	1.4	1.4	102%	2.3
Average	3.1	1.3	1.5	3.0	1.6	1.7	108%	3.8

Notes:
IW = Injection well
B1 = Monitoring well B1
* percentage of ethene expected given mass of total cVOCs loss
# Integrated half-life (i.e., conversion of PCE, TCE, cis-1,2-DCE and VC to ethene)

CONCLUSIONS

We demonstrated through a laboratory and a field pilot test that the indigenous microorganisms at the Kelly AFB were capable of reductively dechlorinating PCE to cis-1,2-DCE when electron donors were present. However, complete dechlorination to ethene was only observed when KB-1, a natural, non-pathogenic, dehalo-respiring, microbial consortium was added to microcosms or to the aquifer. KB-1 was shown to completely dechlorinate TCE up to 80 mg/L. Only a small volume of KB-1 was required to effectively inoculate the PTA and resulted in observed half-lives of cVOCs to ethene in the order of hours. The study also verified that 16S rRNA molecular probing techniques can be used to assist in the monitoring, tracking and verification of KB-1's establishment at field scale. Isotope analysis confirmed that the degradation was biologically mediated.

ACKNOWLEDGMENTS

The authors gratefully acknowledge funding provided by the Army Corp of Engineers (Tulsa District) and the Remediation Technologies Development Forum, and the Kelly AFB for providing access to the site. We also appreciate data review and input by RTDF members during the execution of this project.

REFERENCES

Cox, E.E., M.McMaster, T. McAlary, L. Lehmicke, E. Edwards and D.W. Major. 1998. "Accelerated bioremediation of trichloroethene: From field and laboratory studies to full scale." In: Remediation of Chlorinated and Recalcitrant Compounds: Volume 1. Battelle Press, Columbus, OH.

Ellis, D.E., Lutz, E., J.M. Odom, R. J. Buchanan, C.J. Bartlett, M. D. Lee, M. R. Harkness, K. A.. DeWeerd. 2000. "Bioaugmentation for Accelerated In Situ Anaerobic Bioremediation." *Environ. Sci. Technol.* **34(11)**:224-2260.

Hendrickson, E. R., M.G. Starr, M.A. Elbersson, J.A. Tabinowski, E.E. Mack, M.L. McMaster, D. E. Ellis. 2001. " Using a Molecular Approach to Monitor A Bioaugmentation Plot." In: In Situ and On-Site Bioremediation Sixth International Symposium Proceedings. June 4-7, 2001 San Diego. Battelle Press.

Maymo-Gatell, X., J.M. Gossett and S.H. Zinder. 1997. "*Dehalococcus Ethenogenes* Strain 195: Ethene production from halogenated aliphatics." In: In Situ and On-Site Bioremediation: Volume 3. Alleman, B.C. and Leeson, A. (Eds). Battelle Press, Columbus, OH.

Morrill, P. L., G. F. Slater, G. Lacrampe-Couloume, B. E. Sleep, E.A. Edwards, B. Sherwood Lollar M. McMaster and D. Major.2001. "Isotopic evidence of reductive dechlorination during a Field Demonstration of Bioaugmentation at Kelly AFB." In: In Situ and On-Site Bioremediation Sixth International Symposium Proceedings. June 4-7, 2001 San Diego. Battelle Press.

SAIC. 1999. Bioaugmentation Pilot Test Building 360 Area, Kelly AFB, Texas. Work Plan and 90% Design. September 1999.

Scholz-Muramatsu, H., A. Neumann, M., Mebmer, E. Moore and G. Diekert. 1995. Isolation and characterization of *Dehalospirillium multivorans* ge. nov., sp. nov., a tetrachloroethene-utilizing, strictly anaerobic bacterium. Arch. Microbiol. 163:48-56.

Schumacher, W. and C. Holliger. 1996. The proton/electron ratio of the menaquinone-dependent electron transport from dihydrogen to tetrachloroethene in "*Dehalobacter restrictus*". J. Bacteriol. 178:2328-2333.

Wehr, S., M. Duhamel, S. Dworatzek E. Edwards, E.Cox, M.McMaster, and D.W. Major.2001. "Microorganisms Required For Complete Dechlorination Of Chlorinated Ethenes". In: In Situ and On-Site Bioremediation Sixth International Symposium Proceedings. June 4-7, 2001 San Diego. Battelle Press.

IN-SITU BIOTREATMENT OF CHLORINATED HYDROCARBONS IN GROUNDWATER USING GEL BEADS

Rakesh Govind and F. Tian (Department of Chemical Engineering, University of Cincinnati, Cincinnati, OH)

ABSTRACT: Chlorinated solvents, consisting primarily of chlorinated aliphatic hydrocarbons (CAHs), have been used widely for degreasing of aircraft engines, automobile parts, electronic components, and clothing. Due to water solubilities exceeding drinking water standards and densities higher than water, CAHs migrate downward through soils contaminating ground water and penetrate deeply into aquifers forming dense non aqueous phase liquids (DNAPLs) on aquifer bottoms. Ground water toxicity problems associated with CAHs occur at over 358 major hazardous waste sites and many minor sites across the nation. Most of the CAHs are aerobically degradable. Some CAHs, such as trichloroethylene require cometabolites or specialized organisms for aerobic degradation. Full-scale field applications of cometabolic destruction of CAHs are greatly limited by the availability, cost, and potential adverse environmental impacts of the secondary substrates needed for induction of cometabolic activity. In this paper, the use of specially formulated silica gel beads with active biomass encapsulated within the bead, has been shown to biodegrade TCE without any organic substrates. Bench-scale experimental studies have shown high rates of TCE degradation without any release of intermediates, such as vinyl chloride. Kinetic models have been developed to obtain the kinetic coefficients from experimental data.

INTRODUCTION

Over the past several years, it has become evident that many groundwater supplies throughout the country have been affected by volatile organic chemicals (VOCs). Frequently found compounds include several chlorinated aliphatic hydrocarbons (CAHs) including trichloroethylene (TCE), tetrachloroethylene (PCE), carbon tetrachloride (CT), and 1,1,1-trichloroethane (TCA). These chlorinated solvents have, in many cases, been disposed into refuse sites, waste pits and lagoons, and storage tanks. They tend to migrate downward through soils, contaminating the groundwater with which they come into contact. CAHs have water solubilities in the range of 1 g/L, or several orders of magnitude higher than their associated drinking water standards. It is accepted that even very low levels could have long-term adverse health effects.

Treatment methods for contaminated groundwater can be generally classified into two main types: (1) *in-situ* techniques, which included injection of nutrients, oxidizing agents, such as peroxide, injection of co-substrates, such as methane, acetate, etc. or use of barriers; and (2) *ex-situ* methods, which include pump-and-treat processes.

Considerable work has been performed to-date toward evaluating treatment techniques for the removal and/or degradation of CAHs from water. At present, the most commonly used technologies include vapor extraction, air stripping in conjunction with catalytic oxidation, and adsorption. While they are effective in removing chemicals from water, each of these methods is either economically or environmentally unacceptable. A promising approach, termed biological activated carbon (BAC), integrates biological removal and granular activated carbon adsorption into a single unit process.

Bioremediation has been recognized as a very cost effective process for removing halogenated compounds in groundwater, Variations of biological technologies have the potential to completely destroy the contaminants by transforming them to less halogenated compounds (Wilson et al., 1986). TCE is biodegraded by pure and mixed microbial consortia (Fliermans *et al.*, 1988; Fogel *et al.*, 1986; Little *et al.*, 1988; Nelson *et al.*, 1987). Heterotrophic consortia obtained from contaminated sub-surface sediments degraded TCE at concentrations exceeding 100 mg liter^{-1} with methanol, propane, or yeast extract as the substrate (Fliermans *et al.*, 1988; Phelps *et al.*, 1989). Chlorinated alkenes were converted to known carcinogens, such as vinyl chloride, under anaerobic conditions (Bouwer and McCarty, 1983), and natural gas stimulated aerobic TCE mineralization (Wilson and Wilson, 1985); Strandberg *et al.*, 1989). TCE degradation in membrane biofilters in the absence of organic substrates has also been experimentally demonstrated (Parvatiyar *et al.*, 1996).

Objective. The objective of this study was to use an Alginate-Silica (AS) gel bead system as a bioreactor, to estimate the feasibility and potential of TCE bioremediation under groundwater conditions. The AS gel system was chosen for the immobilization of enzymes or microorganisms due to inherent advantages, which includes: (1) long-term stability of the gel; (2) ability to entrap viable microorganisms within the gel for extended periods of time; (3) reasonable diffusivities of contaminant (CAHs), nutrients, and degradation products within the gel matrix; and (4) ability to synthesize the gel under mild conditions, to maintain viability of the microorganims during the synthesis procedure.

MATERIALS AND METHODS

Master Culture Reactors. Several anaerobic master culture reactors (10 liters in volume each) were set-up in the laboratory. Biomass obtained from an anaerobic digester at a local municipal waste water treatment plant was used to seed the master culture reactors and nutrients were added for culture growth. Nutrient composition used was: KH_2PO_4 0.5 g/L; NH_4Cl 1.0 g/L; $MgSO_4$ $7H_2O$ 0.006 mg/L; $CaCl_2$ $6H_2O$ 0.006 g/L; $FeSO_4$ 0.004 g/L; Sodium acetate 7.0 g/L; Na_2SO_4 0.5 g/L; NaCl 2.0 g/L. The headspace of each master culture reactor was purged with oxygen-free nitrogen gas and gas production in the reactor was measured volumetrically to monitor reactor performance. Each master culture reactor was fed each day with 500 mL of nutrient solution and 500 mL of reactor liquid was withdrawn anaerobically to maintain reactor volume.

Synthesis of Gel beads. A mixture of anaerobic culture (15 g/L), 3% alginate solution and distilled water were added into the colloidal silica solution so that the final concentration of alginate, biomass and colloidal silica would become 1.5%, 5% and 5-20%, respectively. The pH of the colloidal silica solution was maintained between 6-7.The solution thus prepared was dropped into a 5% $CaCl_2$ solution to form beads of about 0.4 cm in diameter. Then the beads were cured for 5 hours in the solution containing equal concentration of biomass in gel in order to prevent bacteria from diffusing out into the liquid solution.

Batch Testing of Alginate-Silica (AS) Gel Beads. Batch testing of gel beads was conducted in 60 mL glass bottles sealed with a rubber stopper and aluminum seals. Each bottle contained 20 mL of beads and 20 mL of deionized, distilled water containing nutrient composition, given before. TCE was injected into each bottle resulting in an aqueous TCE concentration of about 25 mg/L. The initial TCE concentration in the nutrient solution was calculated using the equilibrium distribution of TCE between the gas and liquid phases. The bottles were then placed in a 25°C constant temperature environment and 5 μL gas samples were periodically withdrawn using a 1 mL Hamilton CR700-200 syringe from the headspace of each bottle for TCE analysis. These gas samples were analyzed on a Hewlett Packard 5730 A gas chromatograph (Hewlett Packard, Avondale, PA) equipped with a flame ionization detector and a 60/80 Carbosieve Column (5 ft long x 1/8 in diameter, Supelco Inc., Bellafonte, PA) was used for analysis. Samples were taken for 18 days. From the total amount of TCE degraded, the removal efficiency of TCE (g of TCE transformed per g of initial TCE) was determined.

Experiments were conducted with gel beads containing varying amounts of silica, to determine the "optimum" silica concentration corresponding to the maximum removal efficiency of contaminant.

Analysis of Aqueous Concentration of TCE. Liquid samples were injected into 60 mL glass vials sealed with stopper and aluminum seals and equipped with a Teflon^{T} coated stirrer. The samples were stirred for 2 hours at room temperature, and then allowed to equilibrate without mixing for 20 minutes. The concentration of TCE in gas phase was determined using gas chromatography. The Headspace Gas Chromatography Method (Parvatiyar *et al.*, 1996) which uses the equilibrium distribution of TCE between the gas and liquid phase was applied to determine the aqueous concentration of TCE. The operating conditions of the GC were: detector temperature: 300°C; injector and column temperature: 150°C; hydrogen flow rate: 30 mL/min, 15 psi; compressed air flow rate: 400 mL/min, 40 psi; nitrogen flow rate: 35 mL/min, 60 psi.

Operation of Bioreactor. A schematic of the experimental bioreactor system is shown in Figure 1. The packed-bed bioreactor was a 4 inch (100 mm) diameter glass column, 16 inches (406.40 mm) height, which was designed with three separate sections, with two glass sieve plates between the central and top and bottom sections of the bioreactor. The central section was packed with gel beads, with active bacterial cells entrapped within the gel matrix. The top and bottom sections were packed with 2 mm diameter glass beads. The glass beads were placed to mix the water as it flowed into and out of the bioreactor's central gel bead section. Table 1 lists the experimental conditions that were run using the bioreactor system.

Dissolved oxygen in the water reservoir was removed by bubbling nitrogen gas through the reservoir liquid. Liquid from the reservoir was pumped through the bioreactor's bypass line to mix the contaminant with the water in the reservoir. Measured amounts of TCE was injected into the water as it flowed through the by-pass line, using a syringe pump. After a specific concentration of TCE was achieved in the reservoir water, the by-pass line was closed and water contaminated with TCE was allowed to flow upwards through the bioreactor system at approximately 10 mL/minute flow rate.

In the control experiment, the bioreactor was operated without the gel beads, and the change in TCE concentration was experimentally measured by withdrawing aqueous samples from the reservoir liquid and using the Headspace Chromatograph method to determine the aqueous concentration of TCE.

TABLE 1. Summary of experimental runs with operating conditions.

Experimental Run Number	Contaminant	Operating Condition	Initial Concentration (mg/L)	Flow Rate of Groundwater (mL/min)
1	Toluene	Anaerobic	125.0	10.0
2	Toluene and	Anaerobic	Toluene: 125.0	10.0
	TCE		TCE: 25.0	10.0
3	TCE	Anaerobic	25.0	10.0
4	TCE	Anaerobic	35.0	10.0
5	TCE	Anaerobic	45.0	10.0
6	TCE	Anaerobic	27.0	10.0
7	TCE	Anaerobic	25.0	20.0

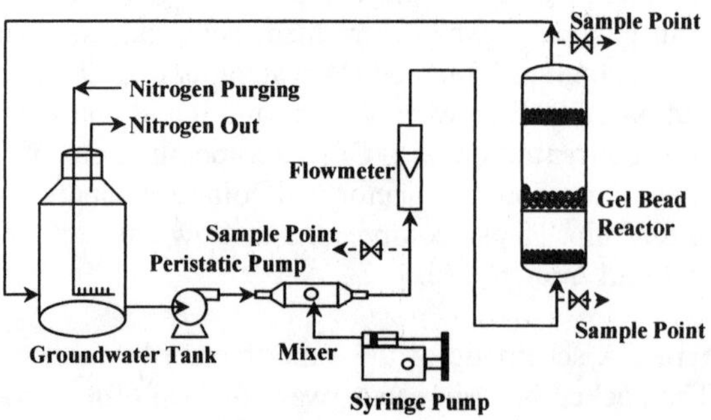
Sample Point
Nitrogen Purging
Nitrogen Out
Flowmeter
Gel Bead
Reactor
Sample Point
Peristatic Pump
Groundwater Tank
Mixer
Sample Point
Syringe Pump

FIGURE 1. Schematic of the experimental bioreactor system.

RESULTS AND DISCUSSION

Batch Testing of Gel Beads. Batch tests were conducted with gel beads using varying concentrations of colloidal silica. TCE concentration decreased as a function of time with TCE removal efficiency attaining a high value of 92% after 18 days using 20 wt % colloidal silica in the gel bead. This result demonstrated the ability of entrapped microorganisms to biodegrade TCE in water.

It was found that colloidal silica concentration of 20 wt % produced the highest TCE removal efficiency after 18 days of test duration. This "optimum" silica concentration of 20 wt % was then subsequently used in all bioreactor experiments.

Bioreactor Operation. In the control study, where no gel beads were used in the bioreactor system, less than 0.5% of initial TCE was lost after 10 days of reactor operation. This demonstrated that loss of TCE due to leakage or adsorption to tubing and glassware was less than 0.5% of TCE amount initially present in the reservoir water.

Experiments with toluene showed biodegradation of toluene during reactor operation. Figure 2 shows decreasing toluene concentration, from an initial value of about 125 mg/L to less than 5 mg/L after 32 days of operation. About 92% of the initial toluene present was biodegraded in 32 days. This result showed that anaerobic biodegradation of toluene was attained by the active microorganisms entrapped in the alginate-silica gel beads.

The next series of experiments involved cometabolic biodegradation of TCE with toluene as the organic substrate, with 2 ppm oxygen concentration in the water. TCE was introduced in the water reservoir at an initial concentration of 25 mg/L with 125 mg/L of toluene. Figure 3 shows the experimental results, with both toluene and TCE biodegrading simultaneously. After 10 days of reactor operation, about 50% of the initial TCE was biodegraded by the active gel beads.

Finally, TCE was used as the sole contaminant at three different initial concentrations of 25, 35 and 45 mg/L. Figure 4 shows the results achieved for TCE only, with no organic substrate present in the water. After 260 hours of reactor operation, with an initial TCE concentration of 25 mg/L, about 51% removal of TCE was attained. Application of the mathematical model (not presented in this paper) to the reactor system showed that the model equations were able to fit the experimental data quite well. This result confirmed that use of first-order kinetic equation for degradation of TCE is acceptable when TCE concentration is about 35 mg/L or lower. In Figure 4, the model fits are better at lower TCE concentration; however when TCE concentration is higher than 35 mg/L, the experimentally observed reaction rate is higher than the model calculated values. As the concentration of TCE decreases with time, the model fit improves, again indicating that use of the first-order kinetic equation is valid at TCE concentrations of 35 mg/L and lower. At higher TCE concentrations, a nonlinear kinetic equation may be more applicable for degradation of TCE.

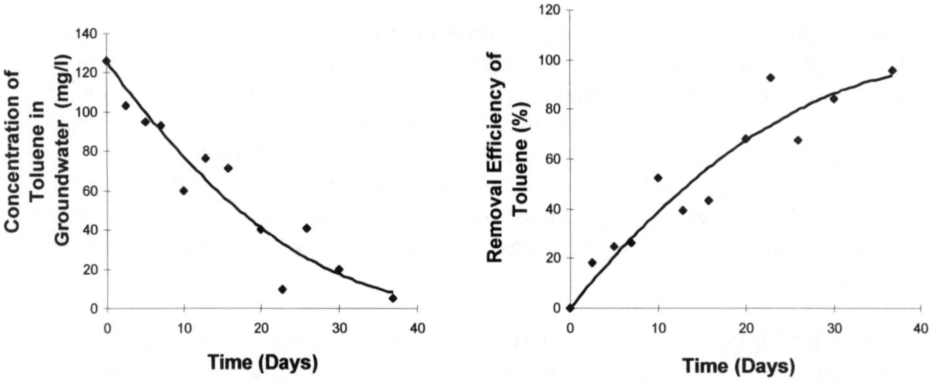
Concentration of Toluene in Groundwater (mg/l)
0
20
40
60
80
100
120
140
0
10
20
30
40
Time (Days)
Removal Efficiency of Toluene (%)
0
20
40
60
80
100
120
0
10
20
30
40
Time (Days)

FIGURE 2. Biodegradation of toluene by gel entrapped microorganisms.

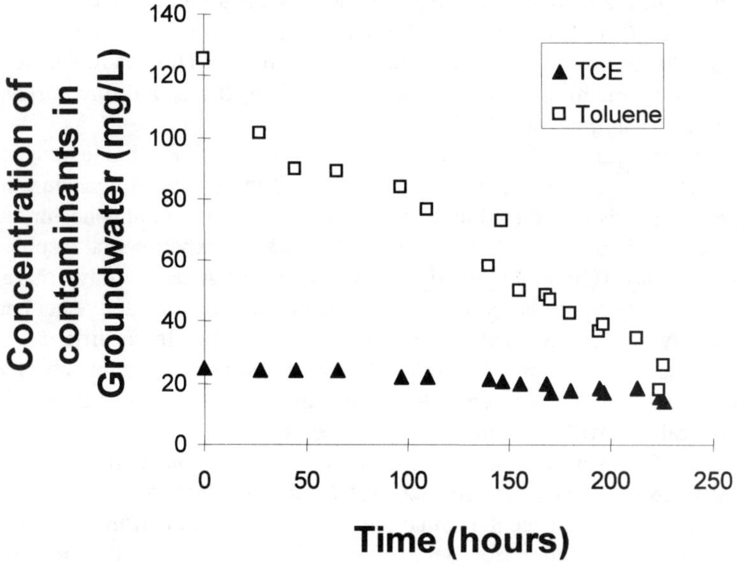
Concentration of contaminants in Groundwater (mg/L)
0
20
40
60
80
100
120
140
0
50
100
150
200
250
Time (hours)
TCE
Toluene

FIGURE 3. Cometabolic degradation of TCE by the gel beads using toluene as the organic substrate.

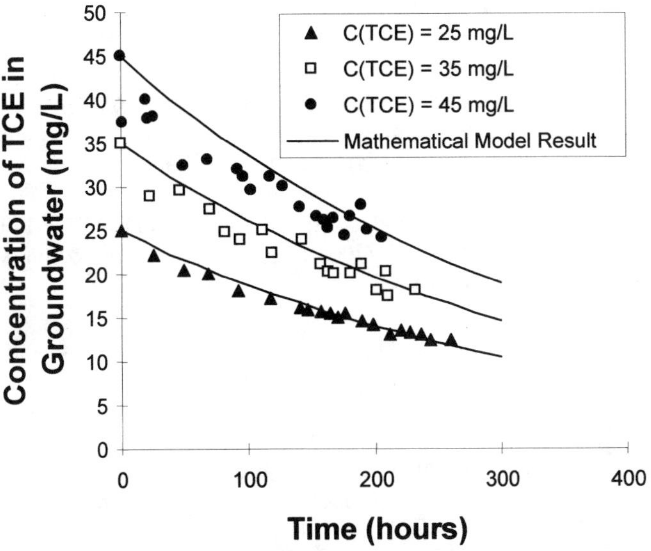

C(TCE) = 25 mg/L
C(TCE) = 35 mg/L
C(TCE) = 45 mg/L
Mathematical Model Result
Concentration of TCE in Groundwater (mg/L)
0
5
10
15
20
25
30
35
40
45
50
0
100
200
300
400
Time (hours)

FIGURE 4. Anaerobic-aerobic degradation of TCE in the gel bead with no organic substrate present in water.

CONCLUSIONS

Results presented in this paper show that active microorganisms encapsulated in gel beads can be used to biodegrade contaminants, such as TCE, without any organic substrates present. Most studies on biodegradation of chlorinated hydrocarbons, such as TCE have involved either anaerobic degradation or cometabolic transformation in the presence of a suitable organic substrate. The mechanism for TCE degradation inside the gel bead is mainly due to anaerobic transformation in the interior of the gel bead by the encapsulated microorganisms, followed by aerobic degradation of the anaerobic by-products near the outer section of the bead. Oxygen profiles, measured by other researchers, have shown that anaerobic conditions prevail inside the gel beads (Zhou and Bishop, 1997). This synchronous anaerobic-aerobic degradation process, which occurs inside the bead, is responsible for the degradation of TCE and can be used to bioremediate other contaminants, such as perchloroethylene, polycyclic aromatic hydrocarbons, polychlorinated biphenyls, etc. Further studies on the application of these gel beads for other contaminants and contaminated phases, such as soil, sediments, etc., need to be conducted in the future.

REFERENCES

Bouwer, E.J. and P.L. McCarty. 1983. "Transformation of 1-and 2-carbon halogenated aliphalic organic compounds under methanogenic conditions." *Applied and Environmental Microbiology*". 45: 1286-1294.

Fliermans, C.B., T.J. Phelps, D. Ringleberg, A.T. Mikell and D.C. White. 1988. "Mineralization of trichloroethylene by heterotrophic enrichment cultures." *Appl. Environ. Microbiol. 54:* 1709-1714.

Fogel, M.M., A.R. Taddeo, and S. Fogel. 1986. "Biodegradation of chlorinated ethenes by a methane-utilizing mixed culture." *Appl. Environ. Microbiol. 51*: 720-724.

Little, C.D., A.V. Palumbo, S.E. Herbes, M.E. Lidstrom, R.L. Tyndall, and P.J. Gilmer. 1988. "Trichloroethylene biodegradation by a methane-oxidizing bacterium." *Appl. Environ. Microbiol. 54*: 951-956.

Nelson, M.J.K., S.O. Montgomery, W.R. Mahaffey, and P.H. Pritchard. 1987. Biodegradation of trichloroethylene and involvement of an aromatic biodegradative pathway." *Appl. Environ. Microbiol. 53*: 949-954.

Parvatiyar, M.G., R. Govind, and D.F. Bishop. 1996. "Treatment of trichloroethylene (TCE) in a membrane biofilter." *Biotechnol. and Bioeng. 50* (1*)*: 58-64.

Phelps, T.J., D. Ringleberg, J. Davis, C.B. Fliermans, and D.C. White. 1989. "Microbial biomass and activities associated with subsurface environments contaminated with chlorinated hydrocarbons." *Geomicrobiol. J. 6*: 157-170.

Strandberg, G.W., T.L. Donaldson, and L.L. Farr. 1989. "Degradation of trichloethylene and trans-1,2-dichloroethylene by a methanotrophic consortium in a fixed-film, packed-bed bioreactor." *Environ. Sci. Technol. 23*: 1422-1424.

Wilson, J.T. , L.E. Leach, M. Hensen and J.N. Jones. 1986. "In situ biorestoration as a groundwater remediation technique", *Ground Water Monit. Rev., 6:* 56-67.

Wilson, J.T., and B.H. Wilson. 1985. "Biotransformation of trichloroethylene in soil." *Appl. Environ. Microbiol. 49*: 242-243.

Zhou, Q. and Bishop, P.L. 1997. "Determination of oxygen profiles and diffusivity in encapsulated biomass K-carrageenan gel beads." *Wat. Sci. Tech. 36* (1): 271-277.

USING A MOLECULAR APPROACH TO MONITOR A BIOAUGMENTATION PILOT

Edwin R. Hendrickson, Mark G. Starr, Margaret A. Elberson, Jo Ann. Payne, E. Erin Mack, Hui-Bin Huang (DuPont Co. Newark, DE), Michaye L. McMaster (GeoSyntec Consultants, Guelph, Ontario, Canada) and David E. Ellis (DuPont Co., Wilmington, DE.)

ABSTRACT: The Bioremediation Consortium of the Remediation Technologies Development Forum (RTDF) carried out a successful anaerobic bioaugmentation pilot to bioremediate a chloroethene Tetrachloroethene (PCE), Trichloroethene (TCE) and 1,2 *cis-dichloroethene* (cDCE) contaminated aquifer at Kelly Air Force Base near San Antonio, Texas (TX). An anaerobic dechlorinating enrichment culture, KB-1, was injected into the ground to duplicate the successful bioaugmentation pilot at Dover AFB, Delaware (DE) (1997-1999). The KB-1 community structure has been analyzed and shown to have a *Dehalococcoides ethenogenes*-like species present in its community structure. *Dehalococcoides ethenogenes* (DHE) is an organism described by Maymo-Gatell et al. (Science 276, 1568-1571, 1997). DHE was shown to dechlorinate PCE and TCE by removing all the chlorine atoms to form ethene, through a process known as dehalorespiration. *Dehalococcoides ethenogenes*-like (DHE-like) organisms detected in samples from approximately 30 different sites in North America and Europe have shown 16S rRNA gene sequences (rDNA) with signature sequences that are unique to the sampling site. This was found true for the KB-1 DHE-like organism. We have developed a specific 16S rRNA polymerase chain reaction (PCR) assay to detect DHE-like organisms. Using the PCR assay and 16S rDNA sequence information, groundwater samples were monitored during the course of the Kelly pilot study. The DHE-like organism was not detected in the control groundwater that had been given electron donors (methanol and acetate). After bioaugmentation, PCR analysis of groundwater from monitoring wells detected the DHE-like organism. Detection first appeared in the injection well and then in down gradient monitoring wells (first in the nearest well and then in the well wells further down gradient). The DHE-like organism was detected in the extraction wells, two months after bioaugmentation. Together, with field data from monitoring wells that have demonstrated dechlorination of PCE to ethene, the PCR and sequence data suggest that the bioaugmentation culture, KB-1, had colonized the test plot in the Kelly AFB chloroethene contaminated aquifer.

INTRODUCTION

Chloroethene solvents, perchloroethene (PCE) and trichloroethene (TCE), are widely used as solvents, degreasing, and cleaning agents. Because of past disposal practices and spills, these agents are contaminants in groundwater, sediments and soil. Standard remedial approaches have proven to be ineffective and costly for removing these substances from the environment. Within the past 15 years, dechlorination mechanisms by natural microbial populations have suggested that the destruction of chlorinated compounds can be practically achieved by stimulating bacterial reductive dechlorination in the field (Beeman et al., 1994; McCarty, 1997). The Remediation Technologies Development Forum (RTDF), a government and industry research consortium, conducted a successful pilot field dechlorination study at Dover Air Force Base (AFB), Dover, Delaware (DE) (1997 to 1999) (Ellis et al., 2000). The study demonstrated that biostimulation (the addition of nutrients and electron donors into the subsurface) and bioaugmentation (the injection of a dechlorinating culture into the subsurface) could be used to stimulate the dechlorination of TCE beyond 1,2-*cis*-dichloroethene (cDCE) to ethene. The biostimulation stimulated the formation of anaerobic environment needed for dechlorination and the bioaugmentation culture provided the organisms to complete the process (Ellis et al., 2000).

The culture used to bioaugment the pilot study plot was the "Pinellas Culture"; an enrichment culture developed from groundwater and soil taken from Department of Energy's Pinellas site in Largo, Florida. Its microbial population was enriched to dechlorinate PCE or TCE to ethene (DeWeerd et al., 1998; Harkness et al., 1999). We have developed a detection strategy that uses the Polymerase Chain Reaction (PCR) assay that uses species-specific primers to test for dechlorinating bacteria in the environment. The species-specific primers were designed using the variable region sequences in 16S rRNA gene sequences (rDNA) sequences from known dechlorinating bacteria found in GenBank. This assay was used to investigate the presence of dechlorinating bacteria in the Pinellas culture and their ability to survive in the environment (Elberson et al., 1999; Hendrickson et al., 2000).

Examination of 16S rDNA sequences from DNA extracted from dechlorinating organisms showed that the 16S rRNA variable sequences could be used to identify the organisms. By analyzing culture samples, the Pinellas culture's population has been partially characterized and shown to have three dehalorespirating-strains, *Dehalobacter restrictus*, *Dehalospirillum multivorans,* and *Dehalococcoides ethenogenes* (Elberson et al., 1999; Hendrickson et al., 2000). The latter strain *Dehalococcoides ethenogenes* is the only laboratory strain shown to completely dechlorinate PCE to ethene (Maymó-Gatell et al., 1997). These methods were used analyze pilot field samples and demonstrated that some of dechlorinating population in the Pinellas culture survived in the pilot's anaerobic subsurface and could be found throughout the pilot study area. Dehalorespirating organisms were not detected in samples taken outside the pilot area. It was concluded that some of the culture had indeed survived and colonized the pilot subsurface. The detection and sequencing assays were performed one, two and three years after the Pinellas culture had been injected into the ground. Results demonstrated that the organisms can survive for a long period after injection into the subsurface and continue to dechlorinate as long as the required anaerobic environment is maintained.

This paper discusses analysis of the microbiology component of the RTDF's second pilot to demonstrate microbiological dechlorination of chloroethene solvents in a contaminated aquifer at Kelly AFB, San Antonio Texas (TX). The site chosen was next to Building P360, which was historically used for the maintenance and repair of jet engines. These activities for more than 25 years resulted in solvent spills and subsequent contamination of the groundwater. Like the Dover AFB bioaugmentation pilot, the subsurface was enriched with electron donors and nutrients to develop anaerobic conditions in the subsurface to stimulate indigenous organisms (if they were present) to reductively dechlorinate the chloroethenes to ethene. Microcosms inoculated with Kelly pilot plot soil to study the degree of microbial dechlorination demonstrated that the indigenous organisms only partially dechlorinated PCE to cDCE (personal communication, Elizabeth Edwards, University of Toronto). To bioaugment the Kelly pilot, a new dechlorinating culture was chosen. This culture, known as KB-1, was a dechlorinating enrichment culture developed from groundwater and soil taken from a field site contaminated with chloroethenes near Kitchener, Ontario, Canada (personal communication, Elizabeth Edwards, University of Toronto). The KB-1 culture was found to have a dechlorinating strain in its population that was related to *Dehalococcoides ethenogenes*. The KB-1 strain had sequences in 16S rDNA that were unique and could be used to track the organism in the pilot study.

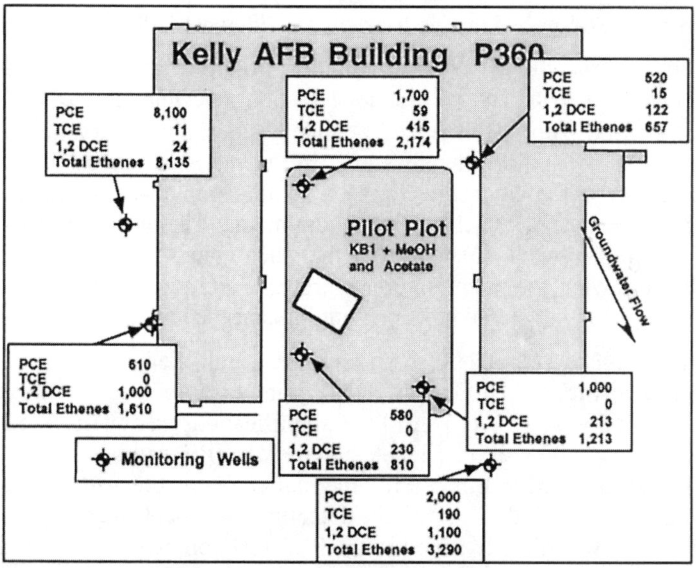
Kelly AFB Building P360
PCE 8,100
TCE 11
1,2 DCE 24
Total Ethenes 8,135
PCE 1,700
TCE 59
1,2 DCE 415
Total Ethenes 2,174
PCE 520
TCE 15
1,2 DCE 122
Total Ethenes 657
Pilot Plot
KB1 + MeOH and Acetate
Groundwater Flow
PCE 610
TCE 0
1,2 DCE 1,000
Total Ethenes 1,610
PCE 1,000
TCE 0
1,2 DCE 213
Total Ethenes 1,213
Monitoring Wells
PCE 580
TCE 0
1,2 DCE 230
Total Ethenes 810
PCE 2,000
TCE 190
1,2 DCE 1,100
Total Ethenes 3,290

FIGURE 1. Chloroethene Concentrations in Kelly AFB Pilot Test Area, Building P360

Objective. The objective of the microbiology component was to determine if dechlorinating bacteria and specifically one that is related to *Dehalococcoides ethenogenes* (DHE) could colonize the Kelly PCE contaminated aquifer. Could this occur either when the pilot aquifer is biostimulated or when it is bioaugmented with a dechlorinating culture? To accomplish this objective we asked the following questions: (i) Are dechlorinating bacteria such as DHE present in the pilot area? (ii) If a DHE-related organism is present, does it have signature sequences that can be used to differentiate it from those of the KB-1 strain? (iii) Can the KB-1 dechlorinating culture colonize the subsurface aquifer?

(iv) If it does colonize the subsurface, can we track its progress with by tracking the DHE-like strain in the subsurface with the PCR detection assay? (v) Does the apparent colonization of the aquifer subsurface correlate with complete dechlorination of PCE to ethene (if this could be stimulated to occur)?

MATERIALS AND METHODS

Groundwater Samples. One thousand-milliliter groundwater samples were taken from the injection, extraction, and monitoring wells of the chloroethene contaminated pilot using plastic sampling bottles, filled-up to the top, sealed, doubled bagged and shipped to the lab on ice. Upon arrival, the samples were either stored overnight at 4° C or immediately centrifuged using a GSA rotor in a RC5B Sorvall Superspeed centrifuge. The resulting cell/soil pellets were resuspended in 2 mL of 1x PBS (10 mM Na phosphate, 150 mM Na chloride, pH 7. 6) and either stored at -20°C or extracted for genomic DNA

Design of *Dehalococcoides ethenogenes*-specific Primers for the PCR Assay. The PCR primers were designed using unique sequence from variable and hypervariable regions of the *Dehalococcoides ethenogenes* 16S rDNA sequence using the procedure described in Hendrickson and Ebersole (2000).

DNA Extraction Procedure. DNA was extracted from the microcosm cultures or groundwater samples by a bead mill homogenization procedure, FastDNA® SPIN Kit Spin Kit for Soil (Bio 101, Vista, CA), that was designed to isolate genomic DNA from all cell types. For groundwater, 1 mL of the resuspended pellet in 1x PBS was used. For soil microcosm cultures, a 10 mL sample was centrifuged to a pellet and resuspended in 500 μL of the culture media. The genomic DNA from both types of samples were isolated using the silica matrix system of FastDNA SPIN Kit by following the manufacturer's protocol and recommendations. The isolated DNA sample was stored at -20° C until it is needed for the PCR assay.

PCR Procedure. The 16S rDNA sequences were amplified from the samples using *Dehalococcoides*-specific 16S rDNA primers as previously described (Hendrickson and Ebersole, 2000). All PCR amplifications were performed using the GeneAmp PCR kit with Taq DNA polymerase (PE Applied Biosystems, Branchburg, NJ) in a Perkin Elmer 9600 thermal cycler as previously described by Hendrickson et al. (2001). A "direct detection" protocol used 1 μL of the microcosm culture was directly added to the PCR reaction, which was conducted as previously described above (Hendrickson and Ebersole, 2000).

Analysis of *Dehalococcoides ethenogenes* Related Sequences. After the *Dehalococcoides ethenogenes*-like sequences are detected in the groundwater or microcosms developed from contaminated soil, they were amplified, cloned, sequenced and analyzed as previously described by Hendrickson and Ebersole. (2000). .

Groundwater Modeling and Design. Groundwater modeling and design was previously discussed by Major et al. (2001). A layout of the monitoring wells is shown in Figure 2. Monitor wells (B1, B2 and B3) were spaced along the central plot's downgradient flow path from the injection well (I1) to the ejection well (E1) at distances of 8, 12 and 23 feet. Groundwater was extracted from three downgradient wells (E2, E1 and E3) spaced 5 feet apart where E1 is on the plot centerline. Two transgradient (T1 and T2) wells were place in a perpendicular line to the centerline of the plot, 10 feet from either side of monitoring well B2. These were used to measure pilot test area pore volume and estimating the area of influence. Water extracted was combined, amended, and re-injected into I1. The pumping rate, subsurface residence time, electron donors/nutrients delivery, sampling schedule and analysis of parameters were previously discussed (Major et al., 2001). Additional samples outside the pilot area were sampled to assess the recalculation system's impact outside of the pilot area.

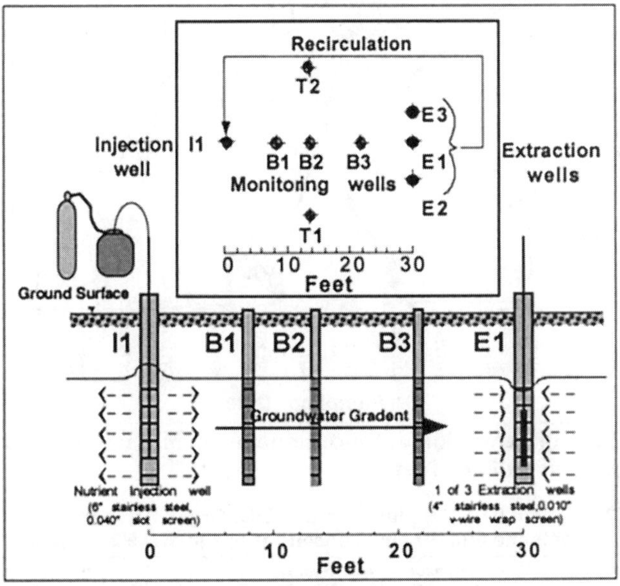
Recirculation
T2
E3
Injection well
I1
B1 B2 B3
Monitoring wells
E1
E2
Extraction wells
T1
0 10 20 30
Feet
Ground Surface
I1
B1
B2
B3
E1
Groundwater Gradient
Nutrient Injection well
(6" stainless steel, 0.040" slot screen)
1 of 3 Extraction wells
(4" stainless steel,0.010" v-wire wrap screen)
0 10 20 30
Feet

FIGURE 2. Schematic of Pilot test Area's Well Systems and Locations.

RESULTS AND DISCUSSION

The injection methanol and nutirents began on Day 1. The effect of the addition of methanol in monitoring well (MW) B1 are shown in Figure 3 (Major et al., 2001). By Day 89, analytical data revealed that that the dissolved oxygen had not declined and redox potential had not been reduced. Further, the PCE was not being converted to cDCE. Acetate was added with methanol as a second electron donor to induce anaerobic conditions. Consequently, on Day 125, the dissolved oxygen concentrations (1.1 mg/L) and redox potential (-233 mV) became indicative of anaerobic conditions. Concentration of cDCE had begun to increase. Similar results were found on Days 164 and 173. By this time, near stoichiometric conversion of PCR to cDCE had occurred; however, no evidence for dechlorination of PCE beyond cDCE was observed (Major et al., 2001).

Dehalococcoides-like organisms had not been detected in the original sample at Day 1 and were not stimulated by the addition of nutrients and electron donors. As shown in Table 1 and in Figure 4, DHE-like strains could not be detected by the PCR assay prior to bioaugmentation with culture KB-1. In Figure 4A, the PCR assay did not yield a detectable product in the electrophoresis gel for MWs B1 and B2 and the injection well I1 at Day 173 (3 days before bioaugmentation). Bioaugmentation was conducted on Day 176 with injection of 14 Liters of the KB-1 culture into the subsurface, which was characterized as dechlorinating culture (able to convert TCE to ethene) at the time of injection.

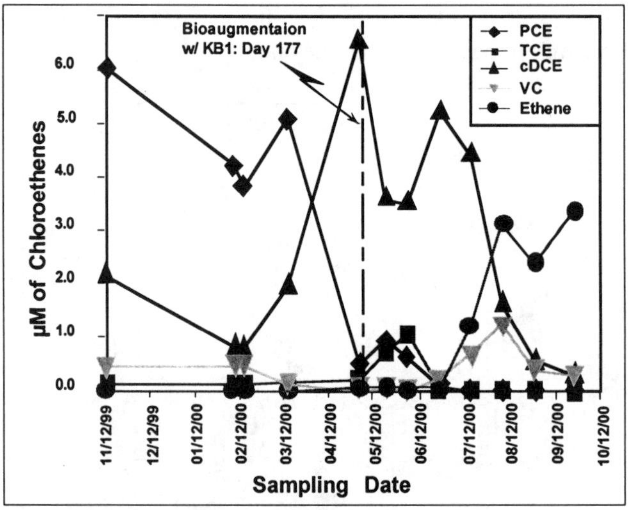
Bioaugmentaion
w/ KB1: Day 177
PCE
TCE
cDCE
VC
Ethene
µM of Chloroethenes
6.0
5.0
4.0
3.0
2.0
1.0
0.0
11/12/99
12/12/99
01/12/00
02/12/00
03/12/00
04/12/00
05/12/00
06/12/00
07/12/00
08/12/00
09/12/00
10/12/00
Sampling Date

FIGURE 3. Concentration of Chloroethenes in MW B1 during the Bioaugmentation Pilot Test

TABLE 1. Detection of the KB1 Dehalococcoides-Like Organism in the Bioaugmentation Pilot Plot and Chloroethene and Ethene Concentrations in MW B2 on Sampling Dates

Date	Exp. #Day	Bgmnt. #Day	Monitoring Well Number									[Chloroethenes] in MW B2			
			I1	B1	B2	B3	E1	T1	T2	E2	E3	µM PCE	µM cDCE	µM VC	µM Ethene
3-May-00	173	-3	–	–	–	NT	–	NT	NT	NT	NT	0.17	6.40	nd	nd
7-May-00	177	1	++	–	NT	NT	–	NT	NT	NT	NT	NT	NT	NT	NT
22-May-00	192	17	++	–	NT	NT	–	NT	NT	NT	NT	0.58	4.33	nd	nd
5-Jun-00	206	30	++	+/–	NT	NT	–	NT	NT	NT	NT	0.23	3.40	nd	nd
27-Jun-00	228	51	++	++	NT	NT	–	NT	NT	NT	NT	0.16	4.75	0.13	nd
18-Jul-00	249	72	++	++	++	+/–	–	NT	NT	NT	NT	0.03	4.54	0.93	1.25
7-Aug-00	269	93	++	++	++	+/–	+/–	NT	NT	NT	NT	0.05	1.65	0.74	3.14
29-Aug-00	291	115	++	++	++	++	++	++	++	–	NT	0.03	0.74	0.38	2.66
25-Sep-00	318	143	+/–	++	++	++	++	++	++	+/–	NT	0.1	0.24	0.19	3.07

Exp. = RTDF pilot experiment; nd = non-detect; – = DHE not detected
Bgmnt. = bioaugmentation w/ KB1; NT= No test was done; ++ = DHE Detected
#Day = number of days; +/– = DHE detected in a trace band

The first indication of dechlorination beyond cDCE was the occurrence of trace amounts of VC on Day 228 or 51 days after bioaugmentation. Ethene was detected at 21 days later on Day 249. Vinyl chloride was now readily detected. By Day 318, all of the PCE and most of the cDCE and VC had been converted to Ethene.

After the injection of KB-1 for bioaugmentation, 1 liter samples of groundwater were taken for "*Dehalococcoides* analysis" as part of the sampling regime of monitoring wells for measuring the changes in the concentration of chloroethenes and other geochemical parameters. As shown in Table 1 and Figure 4 A, DHE-like organism was detected in the in the injection well (I1) on the first day after bioaugmentation. It was not detected in MW B1 or the extraction well E1.

Examination of the data in Table 1 demonstrates that *Dehalococcoides*-like organism was gradually detected, with respect to time, down gradient from the injection well after bioaugmentation. It was first detected in MW B1, 30 days after the KB-1 injection. DHE was later detected in samples from wells, simultaneously in B2 and B3 after 72 days, in E1 after 93 days, simultaneously in T1 and T2 after 115 days and in E2, 143 days after bioaugmentation. Figure 4 shows an example of a gel, detecting PCR products obtained with one set of DHE-specific PCR primers. Sequences amplified with DHE-specific 16S rRNA primers were sequenced and found to have the same signature sequences in the

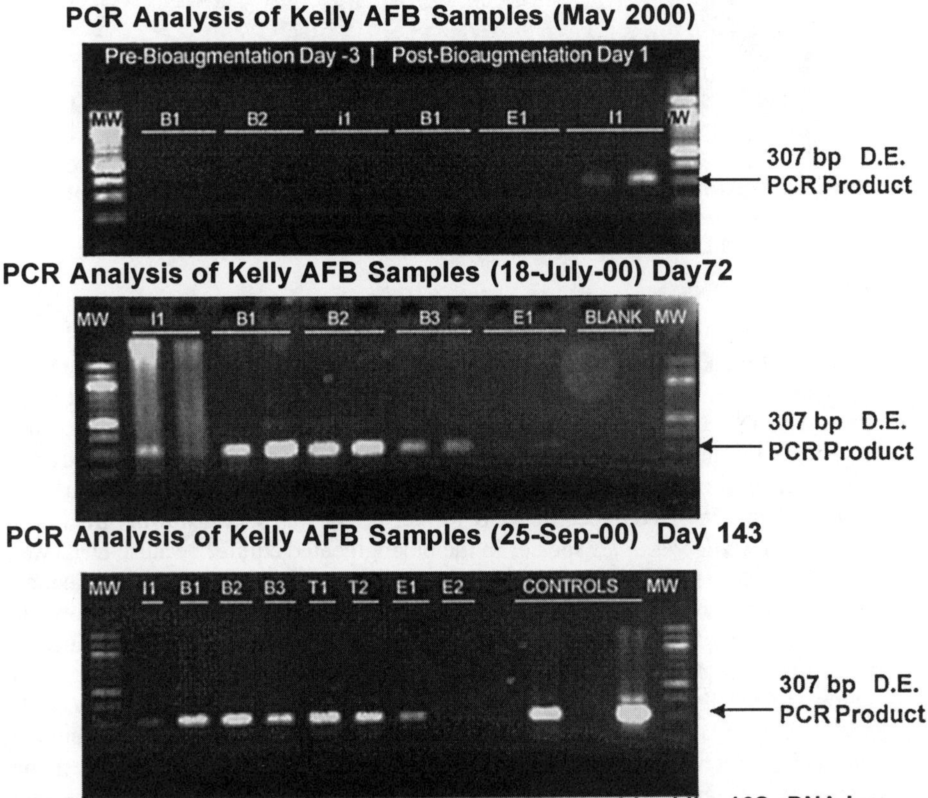
PCR Analysis of Kelly AFB Samples (May 2000)
Pre-Bioaugmentation Day -3 | Post-Bioaugmentation Day 1
MW B1 B2 I1 B1 E1 I1 MW
307 bp D.E. PCR Product
PCR Analysis of Kelly AFB Samples (18-July-00) Day72
MW I1 B1 B2 B3 E1 BLANK MW
307 bp D.E. PCR Product
PCR Analysis of Kelly AFB Samples (25-Sep-00) Day 143
MW I1 B1 B2 B3 T1 T2 E1 E2 CONTROLS MW
307 bp D.E. PCR Product

FIGURE 4. PCR Assay for the Detection *Dehalococcoides*-Like 16S rDNA in Groundwater Samples Taken from the Wells in the Pilot Test Area during Bioaugmentation Study.

variable 16S rDNA regions as the KB-1 DHE-like organisms. This indicted that the Kelly DHE-like sequences originated from the KB-1 culture. This is similar to data from the Dover AFB bioaugmentation study, where the signature sequences 16S rDNA of the Dover DHE-like test pilot organisms were found to be the same as those found in the Pinellas bioaugmentation culture. One hundred-twenty nine days after bioaugmentation, background control groundwater samples, SB235 and SB236, were taken from control two wells approximately 40 yards, north and south, perpendicular to the bioaugmentation *Dehalococcoides*-like organism. These controls indicated that the plot. In addition, a third groundwater sample, SB237, was taken 6 feet down gradient from E1. As shown in Figure 5, all samples were negative for the *Dehalococcoides*-like organism was not naturally occurring in the aquifer as a separate coincidence from the pilot experiment.

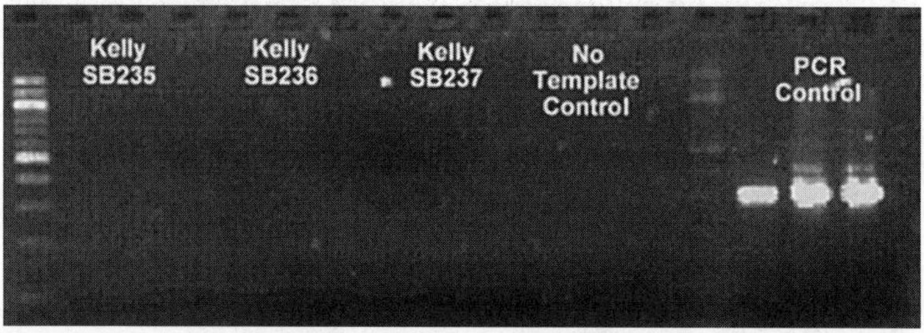
Kelly SB235
Kelly SB236
Kelly SB237
No Template Control
PCR Control

FIGURE 5. PCR Assay Results for DHE in Pilot Back ground Control Samples taken from Area outside the Plot Area. The PCR control is a Cloned *Dehalococcoides*-Like 16S rDNA Isolated from an Industrial Chloroethene Dechlorinating Site

CONCLUSIONS.

Dehalococcoides-like organisms were not detected in soil and groundwater taken from the Kelly AFB Pilot, suggesting that DHE-like organisms do not naturally inhabit the Kelly aquifer. The DHE-like 16S rRNA isolated from groundwater samples from the Kelly AFB bioaugmented pilot had the same signature sequences as the *Dehalococcoides*-Like species found in the KB-1 bioaugmentation culture indicating that its origin was from KB-1. This data suggest that 16S rDNA specific PCR primer assay was successful in "tracking" the KB-1 *Dehalococcoides* species in the bioaugmented aquifer because DHE was detected with respect to time, down gradient from the bioaugmentation injection well. This observation occurred simultaneously with demonstration of complete dechlorination of PCE to ethene in the pilot test plot. Together, the dechlorinating data, the PCR data and the sequence data suggests that the KB-1 culture had colonized the Kelly AFB PCE-contaminated aquifer. The RTDF's Kelly AFB and Dover AFB bioaugmentation pilot's data provides evidence to substantiate the use of dechlorinating consortia for the bioaugmentation of chloroethene contaminated aquifers that are not naturally attenuating.

REFERENCES

Beeman, R. E., J. E. Howell, S. H. Shoemaker, E. A. Salazar, and J. R. Buttram. 1994. "A field evaluation on *in situ* microbial reductive dehalogenation by biotransformation of chlorinated ethylenes." p. 14-27. *In* R. E. Hinchee, A.

Leeson, L. Semprini, and S. K. Ong (ed.), *Bioremediation of chlorinated and polycyclic aromatic hydrocarbon compounds*. Lewis Publishers, Boca Raton, FL.

DeWeerd, K. A., W. P. Flanagan, M. J. Brennan, J. M. Principe, and J. L. Spivack. 1998. "Biodegradation of Trichloroethylene and Dichloroethylene in Contaminated Soil and Groundwater." *Biorem. J.* 2:29-42.

Elberson, M. A., E. R. Hendrickson, J. A. Tabinowski, and D. E. Ellis. 1999. "Detection of Dechlorinating Bacteria in Groundwater and Soils from Waste Sites Contaminated with PCE and TCE." *Abstract of the Proceedings of the 99th General Meeting Of the American Society for Microbiology*, Chicago, Illinois, USA, May 3-June 3, 1999.

Ellis, D. E., E. J. Lutz, J. M. Odom, R. J. Buchanan, Jr., M. D. Lee, C. L. Bartlett, M. R. Harkness, and K. A. DeWeerd. 2000. "Bioaugmentation for Accelerated In Situ Anaerobic Bioremediation." *Environ. Sci. Technol.* 34(11): 2254-2260.

Harkness, M. R., A. A. Bracco, J. Brennan, M. J., K. A. DeWeerd, and J. L. Spivack. 1999. "Use of Bioaugmentation to Stimulate Complete Reductive Dechlorination of Trichloroethene in Dover Soil Columns." *Environ. Sci. Technol. 33*:1100 -1109.

Hendrickson, E. R., M. A. Elberson, J. A. Tabinowski, and D. E. Ellis. 2000. "Molecular Confirmation of Aquifer Colonization by an In Situ Bioaugmentation Culture." *Abstract. Bioaugmentation and Biomonitoring* The Second International Conference on Remediation of Chlorinated and Recalcitrant Compounds, Monterey, CA, USA, May 22-25, 2000:E10.

Hendrickson, E. R., and R. C. Ebersole. October 2000. Nucleic acid fragments for the identification of dechlorinating bacteria. PCT: WO 0063443 A 32. pp 1-55

Major, D.W., M.L. McMaster, E.E. Cox, B.J. Lee and E.E. Gentry, E.R. Hendrickson, E.A. Edwards and S. Dworatzek. "Successful Field Demonstration of Bioaugmentation To Degrade PCE And TCE to Ethene" *Abstract. The Sixth International Symposium on In Situ and On-Site Bioremediation.* , San Diego, CA, USA, June 4-7, 2000.

Maymó-Gatell, X., Y. T. Chien, J. M. Gossett, and S. H. Zinder. 1997. "Isolation of a bacterium that reductively dechlorinates tetrachloroethene to ethene." *Science 276*:1568-1571.

McCarty, P. L. 1997. "Breathing with Chlorinated Solvents." *Science 276*:1521-1522.

BIOAUGMENTATION WITH *BURKHOLDERIA CEPACIA* PR1$_{301}$: IMMOBILIZATION FOR ACTIVITY RETENTION ENHANCEMENT

Daniel J. Adams (Camp Dresser & McKee Inc., Wichita, Kansas)
Kenneth F. Reardon (Colorado State University Fort Collins, Colorado)

ABSTRACT: Bioaugmentation with constitutive trichloroethylene (TCE)-degrading bacteria has been proposed as one method to remediate TCE-contaminated aquifers. One factor in the success of such a strategy is the survival and TCE-degrading activity of injected cells within the aquifer. In laboratory experiments, two methods of cellular immobilization were studied to enhance the TCE-degrading activity retention of *Burkholderia cepacia* PR1$_{301c}$ (PR1) added to soil. PR1 constitutively expresses toluene *ortho*-monooxygenase (TOM), a non-specific enzyme that catalyzes the oxidation of TCE. PR1 was immobilized on small porous ceramic pellets (Isolite) and microencapsulated in an alginate gel. Soil microcosms were used to study the TOM activity retention of PR1 in sterile and non-sterile soil, and in non-sterile soil in the presence of TCE. Methods of immobilization were developed that increased activity retention for PR1. The Isolite-immobilized cells retained their TCE-degrading activity 4 to 5 times longer than freely suspended cells in non-sterile soil in the presence of TCE. The microencapsulated cells retained their activity 2 to 3 times longer than freely suspended cells in non-sterile soil in the presence of TCE. Both of these immobilization methods provide opportunities for successful field application of an aerobic co-metabolic system using bioaugmentation, without the need to supply an inducer.

INTRODUCTION

Background. Trichloroethylene (TCE)-contaminated sites that lack the metabolic capabilities and proper geochemical conditions for the biodegradation of TCE can be treated using bioaugmentation, the addition of microorganisms to the subsurface to promote bioremediation. Bioaugmentation with an organism such as *Burkholderia cepacia* PR1$_{301c}$ (PR1) avoids some of the problems associated with anaerobic reductive dechlorination and conventional aerobic cometabolism of TCE. PR1 is a chemical mutant that constitutively expresses toluene *ortho*-monooxygenase (TOM), a non-specific enzyme that catalyzes the oxidation of TCE (Munakata-Marr et al., 1996), thus removing the need to supply an inducer for aerobic biodegradation and avoiding the accumulation of vinyl chloride, which may occur during anaerobic reductive dechlorination.

One of the primary factors that limit the survival and activity of added cells is the environment into which the cells are added. The cells must overcome the stress of being placed into a nutrient-limited system where natural organisms prey upon them and compete with them for the few nutrients available.

Immobilization of cells can enhance the viability and activity of cells added to the environment by protecting the added cells from predation and allowing for the coimmobilization of nutrients that would be less available to the native bacteria. The inclusion of nutritional amendments in the immobilization matrix may at least delay the undesirable changes in metabolism resulting from starvation. Lin et al. (1995) increased the long-term viability of two strains of bacteria by immobilizing them on vermiculite. O'Reilly and Crawford (1989) showed that cells entrapped in or in the presence of polyurethane foam effectively degraded high levels of pentachlorophenol (PCP), while free cells did not. Alginate-encapsulated *Pseudomonas fluorescens* exhibited better survival and root colonization rates than free cells added to the same non-sterile soil (van Elsas et al., 1992). And Cassidy et al. (1997) showed that encapsulating cells in κ-carrageenan enhanced PCP degradation.

Objectives. The overall goal of this research is to improve bioaugmentation techniques to increase the chances for success as a field application. The specific objective of this study in reaching this goal is to increase the activity retention of PR1 added to non-sterile soil containing TCE through cellular immobilization and encapsulation.

MATERIALS AND METHODS

Microorganism and Medium. PR1 was grown in a basal salts medium consisting of 8.9 g/L $(NH_4)_2SO_4$, 3.24 g/L K_3PO_4, 1.0 g/L $NaPO_4$, and 10 mL/L of a trace metals solution containing 12 g/L nitrolotriacetic acid, 20 g/L $MgSO_4{\bullet}7H_2O$, 0.3 g/L $ZnSO_4{\bullet}7H_2O$, 0.3 g/L $MnSO_4{\bullet}7H_2O$, and 1.2 g/L $FeSO_4{\bullet}7H_2O$. The basal salts medium was supplemented with 6.0 g/L glucose. Cells were grown at 30 °C and 150 rpm. Cells were harvested after 18-21 hours of growth in a batch culture, corresponding to an optical density of 1.5-2.0, as measured on a Beckman DU 640 spectrophotometer at a wavelength of 600 nm. Cells were harvested by centrifugation at 8000 rpm (7625 x g) for 8 minutes in a Sorvall RC-5B refrigerated centrifuge.

Soil and Groundwater. The aquifer soil was a well-graded sand excavated from the saturated zone at the Gilbert-Mosley site in Wichita, Kansas contaminated with tetrachloroethylene (PCE), TCE, and low levels of anaerobic degradation byproducts. The soil contained approximately three percent fines (i.e., passing the 0.075 mm sieve) and was classified as "SW" based on the Unified Soil Classification System (USCS). The soil was passed through a No. 8 sieve prior to use. For experiments using sterile soil, soil was autoclaved at 120 °C for 30 minutes on three consecutive days. All experiments used groundwater from the same Wichita site. All sterile groundwater was filter-sterilized with a Nalgene 0.2 μm bottle top vacuum filter.

Microcosms. The microcosms were composed of 250-mL bottles filled with soil and groundwater to reach a total solids mass of 30 g and aqueous volume of 30 mL, taking into account both the moisture content of the soil and the volume or mass of any added material (i.e. Isolite, microbeads, and cell suspension). Bottles were filled with sterile soil and groundwater, non-sterile soil and groundwater, or non-sterile soil and groundwater with 10 mg/L TCE, in duplicate, and wrapped in aluminum foil. After inoculation with cells, microcosms were placed in an Innova 4230 (New Brunswick Scientific) refrigerated incubator at 18 °C and 100 rpm for 2 hours before TOM activity sampling began.

Enzyme Activity Measurement. The assay used to measure TOM enzyme activity in the microcosms was adapted from Shields and Reagin (1992). A 1-g sample from each microcosm (0.5 mL of groundwater and 0.5 g of soil) was placed into a 2-mL microcentrifuge tube. The samples were centrifuged at 15,000 rpm (15,600 x g) (Eppendorf 5414 microcentrifuge) for 3 minutes and the supernatant was removed. After addition of the triflouromethyl phenol (TFMP) solution, the tubes were vortexed to achieve adequate mixing of the soil slurry. The slurry was then transferred to 25-mL Erlenmeyer flasks for incubation at 30 °C and 150 rpm for 20 minutes. The absorbance of this solution was then measured at 386 and 600 nm and specific activity was determined as prescribed by Shields and Reagin (1992). The incubation time during the enzyme activity assay was increased from 20 minutes to 60 minutes for experiments with low expected activity to insure a proper background level that did not decrease to below zero. The correction factor, C, was 2.1 for samples containing soil and groundwater. The TOM activity was measured in uninoculated microcosms and subtracted from the values obtained for each inoculated microcosm to obtain the data that are presented in the results section below.

Isolite Immobilization. The Isolite pellets (FOREMOST Solutions, Inc.) used in this study have a particle density of 2.27, are 1 mm in diameter, and have continuous, interconnected, and open-ended pores. The pore size varies between 0.1 and 2.0 μm with 30% of the pores greater than 1.0 μm. The specific surface area of 1 g of Isolite is approximately 4.6 m^2 and it is composed of 78% SiO_2, 12% Al_2O_3, and 5% Fe_2O_3. All Isolite used in this research was sterilized by autoclaving for 20 minutes at 120 °C on the dry cycle.

A volume of 7.5 mL of cell suspension at an absorbance of 1.5 at 600 nm was resuspended in 10 mL of 1% skim milk in sterile groundwater. This cell mixture was then added to 5 g Isolite and incubated at 30 °C and 150 rpm with manual stirring every hour. After 6 hours, 10 g of this slurry was added to the microcosm. The final cell concentration of added cells in each microcosm was approximately 4 x 10^8 cells/mL.

Microencapsulation. The microencapsulation method was adapted from Stormo and Crawford (1992). This method involved pumping a cell suspension/sodium alginate solution (Sigma: sodium salt of alginic acid, low viscosity, # A2158) through an atomizing nozzle (Turbosonic Technologies Inc., SESNZ-052HA00, with adapter SESNA-HA01) under pressure into a receiving bath of $CaCl_2$ that was constantly stirred. The receiving bath was a 12-L glass carboy filled with 3 L of $CaCl_2$. The atomizing nozzle was suspended approximately 10 cm from the surface of the $CaCl_2$ solution. The solution was pumped through the nozzle at 15 mL/min under a pressure of 14 psig.

Cells were encapsulated with skim milk and successively filtered through 105 and 53 μm nylon mesh screens in an attempt to produce microencapsulated cells that are capable of mechanical transport through the soil based on the filter criteria of the soil. The microencapsulated PR1 had a final alginate and skim milk concentration of 3% and 1%, respectively. After microencapsulation, the microbeads were filtered and incubated in 2% skim milk in HEPES buffer at 30 °C and 150 rpm for 2 hours. Each microcosm was inoculated with 4.8 g of filtered beads, resulting in an added cell concentration of approximately 3 x 10^8 cells/mL in each microcosm.

Size distributions of the microencapsulated cells and the filter criteria of the soil are presented in Figure 1. The size distribution of the soil was determined through mechanical sieve analysis and filter criteria according to a published definition (Lambe and Whitman, 1969). The particle size distribution of the microencapsulated cells was determined using a Coulter LS Particle Size Analyzer with the Fraunhofer optical model.

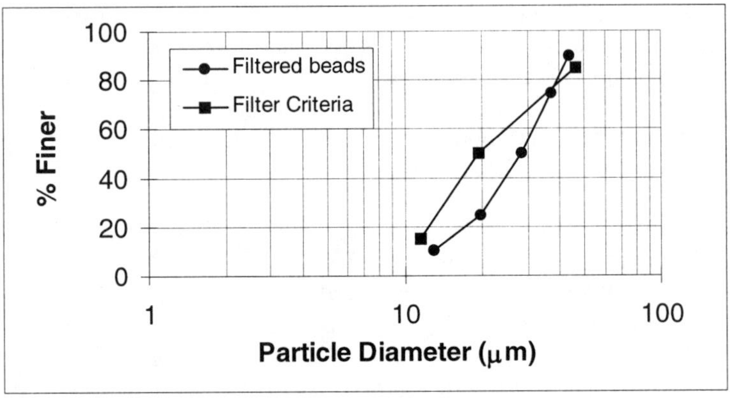
100
80
60
40
20
0
% Finer
Filtered beads
Filter Criteria
1
10
100
Particle Diameter (μm)

FIGURE 1. Size distribution of microencapsulated PR1.

The enumeration of viable microencapsulated cells in each microcosm was accomplished by dissolving a known mass of microbeads in a known volume of 200 mM phosphate buffer and a sample of this volume was spread on plates of ¼ -strength tryptic soy agar (7.5 g/L tryptic soy broth). Colony forming units (CFUs) were counted after 48 hours incubation at 30 °C.

RESULTS AND DISCUSSION

TOM Activity Retention of Freely Suspended PR1. There was a significant difference between activity retention in sterile and non-sterile soil, for freely suspended PR1 added to microcosms at a final concentration of 1 x 10^8 cells/mL

(Figure 2). The activity for cells added to sterile soil remained approximately 0.02 nmol TFHA/min/g soil or higher for 15 days, while the activity in non-sterile soil dropped to zero after 5 days (Figure 2). This verified previous research indicating that predation and/or competition has a significant effect on the survival and activity retention of bacteria added to soil. The addition of TCE to non-sterile soil further decreased the ability of PR1 to retain TOM activity, which dropped to zero after only 2 days. The TCE most likely acts as an energy sink quickening the onset of a starvation mode. The oxidation of TCE can also potentially produce toxic oxidation byproducts.

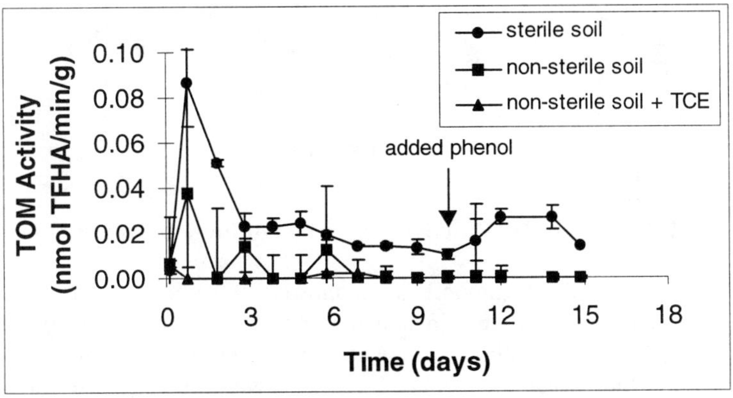
sterile soil
non-sterile soil
non-sterile soil + TCE
added phenol
TOM Activity
(nmol TFHA/min/g)
0.10
0.08
0.06
0.04
0.02
0.00
0
3
6
9
12
15
18
Time (days)

FIGURE 2. TOM activity retention of freely suspended PR1.

The initial decrease in TOM activity in the sterile microcosm and the increase after the addition of phenol indicates that TOM activity retention is not only dependent on predation and cell death, but also on the oligotrophic environment and cell starvation. The decrease in activity of the cells added to sterile microcosms cannot be attributed to predation. The increase in activity after substrate (phenol) addition suggests activity loss due to the carbon-deficient environment of the soil and groundwater (which was reversed after increasing available carbon).

TOM Activity Retention of Immobilized PR1. A comparison of activity retention of Isolite-immobilized cells and microencapsulated cells with freely suspended cells is presented in Figure 3. The Isolite-immobilized cells retained a TOM activity of approximately 0.01 nmol TFHA/min/g in the non-sterile microcosms with TCE for up to 17 days. The microencapsulated cells retained TOM activity above 0.02 nmol TFHA/min/g for 5 days and maintained a TOM activity above 0.01 nmol TFHA/min/g up to 9 days before dropping to background levels on day 10. The TOM activity of the freely suspended cells dropped to nearly background within 2 days. The cell concentration of the freely suspended cells in this experiment was 1 x 10^8 cells/mL, compared to 4 x 10^8

cells/mL for the Isolite-immobilized cells and 3 x 10^8 cells/mL for the encapsulated cells.

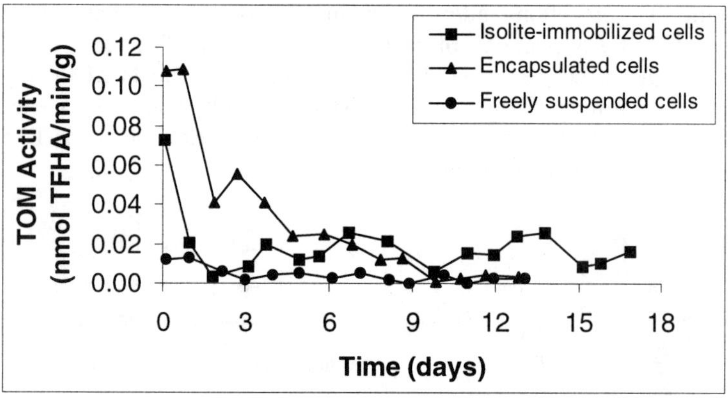
Isolite-immobilized cells
Encapsulated cells
Freely suspended cells
TOM Activity
(nmol TFHA/min/g)
0.12
0.10
0.08
0.06
0.04
0.02
0.00
0
3
6
9
12
15
18
Time (days)

FIGURE 3. TOM activity retention of immobilized PR1.

These methods of immobilization allowed PR1 to retain TOM activity for a much greater period of time in non-sterile soil in the presence of TCE than freely suspended cells. Although there appeared to be an initial rapid decrease in activity, other measurements indicated that TCE was still degraded after 3 hours (results not shown). Not only did the immobilization provide some protection from predation but it also provided a means to make a carbon source available to PR1. In addition, the acclimation period prior to the cells' addition to the microcosms allowed the PR1 time to adjust to the groundwater conditions before being placed into the competitive non-sterile environment. This acclimation period also allowed for the adjustment to a new substrate and maintenance of the enzymatic systems required for TCE degradation that might otherwise have been shut down if added immediately to the carbon deficient microcosms containing TCE.

CONCLUSIONS

The immobilization of PR1 in addition to the coimmobilization with skim milk increased the TCE-degrading enzyme activity retention of PR1 added to non-sterile TCE-contaminated soil. These methods of immobilization address both the needs to alleviate the carbon and nutrient deficient conditions found in the subsurface and the need for protection from predation. Laboratory grown cells added to a harsh environment undergo physiological changes that may result in their inability to degrade TCE for long periods of time. Immobilization methods such as the ones developed in this work may improve bioaugmentation strategies by lengthening the time between required cell culture additions. The Isolite immobilization may be used in a permeable reactive biowall as a passive bioremediation approach and the microencapsulated cells may be injected into the subsurface for a more aggressive approach when contamination is too deep for a permeable reactive wall or at sites where contamination occurs under existing structures.

ACKNOWLEDGEMENTS

We would like to acknowledge the City of Wichita for providing soil and groundwater for this research. We also acknowledge and offer thanks to Malcolm Shields from the Center for Environmental Diagnostics and Bioremediation at the University of South Florida for providing *B. cepacia* PR1 and William Mahaffey of Pelorus Environmental in Golden, Colorado for providing the Isolite used in this study.

REFERENCES

Cassidy, M.B., H. Mullineers, H. Lee, and J.T. Trevors 1997. "Mineralization of Pentachlorophenol in a Contaminated Soil by *Pseudomonas* sp UG 30 cells Encapsulated in κ-carrageenan." *Journal of Industrial Microbiology. 19*: 43-48.

Lambe, W.T. and Whitman, R.V. 1969. *Soil Mechanics.* Wiley Press, New York, NY.

Lin, J., S. Lantz, W.W. Schultz, J.G. Mueller, and P.H Pritchard. 1995. "Use of Microbial Encapsulation/Immobilization for Biodegradation of PAHs." In R.E. Hinchee, J. Frederickson, and B.C. Alleman (Eds.), *Bioaugmentation for Site Remediation.* Battelle Press, Columbus, OH.

Munakata-Marr, J., P.L McCarty, M. S. Shields, M. Reagin, and S.C. Francesconi 1996. "Enhancement of Trichloroethylene Degradation in Aquifer Microcosms Bioaugmented with Wild Type and Genetically Altered *Burkholderia (Pseudomonas) cepacia* G4 and PR1." *Environmental Science and Technology. 30*(6): 2045-2052.

O'Reilly, K.T. and R.L. Crawford. 1989. "Degradation of Pentachlorophenol by Polyurethane-Immobilized Flavobacterium Cells." *Applied and Environmental Microbiology. 55*(9): 2113-2118.

Shields, M.S. and M.J. Reagin. 1992. "Selection of a *Pseudomonas cepacia* Strain Constitutive for the Degradation of Trichloroethylene." *Applied and Environmental Microbiology. 58*(12): 3977-3983.

Stormo, K.E. and R.L. Crawford. 1992. "Preparation of Encapsulated Microbial Cells for Environmental Applications." *Applied and Environmental Microbiology. 58*(2): 727-730.

van Elsas, J.D., J.T. Trevors, D. Jain, A.C. Wolters, C.E. Heijnen, and L.S. van Overbeek. 1992. "Survival of, and Root Colonization by, Alginate-Encapsulated *Pseudomonas fluorescens* Cells Following Introduction into Soil." *Biology and Fertility of Soils. 14*: 14-22.

IN SITU BIOREMEDIATION OF MTBE USING BIOBARRIERS OF SINGLE OR MIXED CULTURES

J.P. Salanitro, G.E. Spinnler, P.M. Maner, D.L. Tharpe, D.W. Pickle, and H.L. Wisniewski (Equilon Enterprises, LLC, Houston, Texas)
P.C. Johnson and C. Bruce (Arizona State University, Tempe, Arizona)

ABSTRACT: We previously conducted in situ enhanced bioremediation pilot tests in the MTBE (methyl t-butyl ether) only portion of the ground water plume at the USN (U.S. Navy) Environmental Test Site at Port Hueneme (PH), CA. These experiments showed that an oxygenated and bioaugmented barrier containing a high activity ether oxygenate-degrading mixed culture (MC-100) could consistently remove ground water levels of MTBE and TBA (t-butyl alcohol) to or below their detection limits. Our current experiments at PH are focused on comparing the bioremediation potential of a new single culture MTBE-degrading isolate, SC-100 (Rhodococcus sp. nov.) and the mixed consortium MC-100 in aerated or oxygenated bioactive zones. Microcosm experiments of PH soils amended with SC-100 or MC-100 cultures degraded MTBE at rates much faster than uninoculated samples. Test plots consisted of 1) MC-100+air, 2) SC-100+ air, and 3) SC-100+O_2 . Biobarriers (ca. 20ft (6 m) wide) developed in the aquifer, 10-20 ft (3-6 m) BGS, were intermittently sparged with air or O_2. Monitoring wells (MW) upgradient, within and downstream of the biobarriers were installed to assess removal of MTBE and TBA. MTBE and TBA concentrations in MW samples upgradient of the biobarrier were consistently about 1000-2000μg/L (range, 500-3000μg/L) and 35μg/L (range, 5-150μg/L), respectively. After 275 days MTBE levels in shallow and deep MW samples within the biobarriers declined to 1-10μg/L and were similar among the three plots with air or O_2 and MC-100 or SC-100. TBA concentrations within the biobarriers also decreased to the detection limit (5-25μg/L). These field data indicate that the implementation of oxygenated bioactive zones containing microbes with high MTBE-degrading activity (e.g. MC-100 and SC-100) may be used to control the migration of oxygenates (MTBE and TBA) in ground water plumes.

INTRODUCTION

Bioaugmentation is a remediation technology in which wastes, sludges, soils, and ground water are amended with cultures of microorganisms to accelerate the rate of biodegradation/biotransformation of contaminant compounds. Such specialized microbes derived from indigenous or non-indigenous sources, when properly used over existing non-specific heterogeneous populations, have the ecological advantage of metabolizing difficult-to-degrade chemicals. Biotic factors which could influence the remediation of aquifers by microbial inocula are many, including long-term viability and survival, relative immobilization/detachment within soil particles, inoculant dose, ability to grow on contaminant organics and availability of suitable electron acceptors; environmental conditions such as ground water pH and redox potential, hydraulic conductivity and soil composition and properties could also affect inoculant activity. Although bioaugmentation for many recalcitrant chemicals in ground water has not been entirely optimized, several laboratory and pilot and field-scale studies suggest aquifer seeding may be a viable remedial option for controlling plume migration and stimulating contaminant degradation. This is especially true for chemicals that are poor microbial growth substrates and for cases where indigenous degrader populations are limited in soil.

In experiments reported in the literature, the addition of aerobic bacterial cultures to aquifer microcosms or soil columns have shown that trichloroethylene (TCE) (Munakata – Marr et al., 1996; 1997), CCl_4 (Mayotte et. al., 1996), trichlorobenzene (Tchelet et al., 1999), chlorinated phenols (Steinle et al., 2000; Barbeau, et al., 1997), atrazine (Franzmann et al., 2000) and benzene, toluene, and xylene (BTX) (Weber and Corseuil, 1994) degradation was markedly enhanced (10 to 100-fold) when high cell concentrations of specific aerobic indigenous or non-indigenous microbes (pure or mixed cultures) and adequate O_2, and in some cases, small amounts of co-substrate were employed. The reductive dechlorination of TCE to ethene was demonstrated by Harkness et al., (1999) in anaerobic soil columns amended with a non-indigenous soil enrichment culture that may have used cis-dichloroethylene as an electron acceptor. Pilot-scale tests in small sections (few meters) of an aquifer have demonstrated the enhanced breakdown of TCE using methanotrophs which cometabolize TCE when grown on CH_4 (Duba et al., 1996) or anaerobic consortia (Ellis et al., 2000). Bioaugmentation has also been evaluated for CCl_4 in a pilot field test in which a specific denitrifying culture (Ps. stutzeri KC), acetate, and NO_3^- and PO_4^{-3} nutrients were injected into test sections of an aquifer (Dybas et al., 1998).

The gasoline oxygenate, MTBE (methyl-t-butyl ether), is another recalcitrant groundwater contaminant that may be remediated by a bioaugmentation approach. We have previously reported on specific mixed (Salanitro et al., 1994; Salanitro et al., 1998) and single cultures (Salanitro et al., 1999) which can completely degrade MTBE but grow poorly on the ether. The inability to use MTBE as a carbon source for growth has also been shown for PM1 (Rubrivivax sp.) by Hanson et al. (1999) and for propanotrophs by Steffan et al. (1997). The inability to grow well on MTBE and the presence of very low numbers of alkyl-ether degraders observed in most aquifers containing MTBE from UST releases of reformulated gasoline has been reported (Salanitro et al., 1998). Our recent pilot studies in the Port Hueneme aquifer have shown that soil inoculation with high cell concentrations of MTBE-degraders (MC-100) in an oxygenated biobarrier zone can control the transport of the MTBE plume (Salanitro et al., 2000). These initial field experiments included 20ft biobarriers containing O_2 only, O_2 and MC-100 (mixed bacterial consortium degrading MTBE) and an unsupplemented control plot. In the field experiments of the present study, we constructed air and O_2 sparged biobarriers seeded with a single culture MTBE-degrader (SC-100, Rhodococcus sp. nov.) or a mixed culture (MC-100) MTBE-degrading consortium, and compared the degradation of MTBE across the bioactive zones. Our results clearly demonstrate the enhanced destruction of MTBE (and t-butyl alcohol) when an aquifer is supplemented with high concentrations of specialized ether-degrading cultures.

MATERIALS AND METHODS

Experimental Site and Field Test Plots. The Port Hueneme (PH) site hydrogeological features and MTBE plume characteristics were previously described (Salanitro et al., 2000). MTBE from the original underground storage tank gasoline release has been transported in the aquifer to over 5000 ft (1524 m) The current biobarrier experiments were located approximately midway down the advancing plume where MTBE (and low concentrations of TBA) is the predominant soluble contaminant. Initial MTBE and TBA concentrations varied across the plots from 500-3000µg/L and 5-150µg/L, respectively and the DO (dissolved oxygen) levels were ≤ 1mg/L. Each seeded biobarrier plot (shown in Figure 1) was aligned perpendicular to the ground water flow and measured approximately 20ft (6 m) wide x 45ft (14 m) long. The transverse spacing between biobarriers was 10ft (3 m). The microbial injection and gas sparging areas of each plot were located 25-30 ft (8-9 m) downgradient from the top of the plot and within the 10-20 ft (3-6 m) BGS thick aquifer. The experimental plot layouts were similar to those given in a previous study (Salanitro et al., 2000). The MC-100 + air plot consisted of an air sparged zone and inoculation of MC-100 mixed bacterial MTBE-degrading consortium. The SC-100 + air and SC-100+O_2 plots were sparged with air or oxygen and seeded with SC-100, a pure culture of a Rhodococcus (sp. nov.) isolated from the MC-100 culture which also degrades MTBE and TBA. A previously constructed plot containing O_2 only served as the oxygenated, unseeded control. The gas delivery wells were installed and screened in the upper (10-15 ft) and lower (15-20ft) portions below the water table (Salanitro et al., 2000). Air or O_2 sparging was supplied intermittently by an air compressor and O_2 generating Air Sep AS 80 pressure swing system, respectively. Ground water monitoring wells (1 inch PVC pipe) were distributed upgradient and downgradient of the seeded zone and within the gas delivery area. The wells were installed by a direct-push soil coring system and screened for composite shallow or deep ground water sample points.

Microbial cultures MC-100 and SC-100 were injected (~5 gal amounts) under pressure on 1 ft centers horizontally and vertically into the aquifer using direct push tools and a high-pressure pump. The cultures were diluted with local ground water to a cell concentration of 2.5-5g dry wt/L. About 9-10 kg dry wt of SC or MC biomass were injected into 20ft of inoculated zone.

MC-100 and SC-100 Cultures and Sampling and Analysis. MC-100 culture was a mixed bacterial culture grown on MTBE as previously described (Salanitro et al., 1994; 2000). The MTBE removal activity in a biodegradation ("die-away") assay of MC-100 varied from 4-8mg MTBE/g cells/hr. The SC-100 single culture was grown at 30C in a 100 gal fermenter at the University of Georgia on NH_4^+ and PO_4^{-3} salts medium with 25g/L cerelose (polyglucose) as sole carbon source. The SC-100 culture was harvested by centrifugation to a 25% dry wt paste and stored at –70C until diluted with ground water at the site. The MC-100 culture was saturated with oxygen and transported under refrigeration (4-6C) in 300gal (1134 L) tote (tank) vessels and also diluted with local ground water.

Ground water samples for MTBE and TBA analysis were taken from monitoring and gas injection wells in test plots (Figure 1) using a peristaltic pump after removal of one well volume of water. DO was measured in a flow cell assembly using a YSI DO meter. MTBE and TBA in ground water samples were analyzed by headspace/PID or direct injection/FID gas chromatographic methods as given previously (Salanitro et al., 2000).

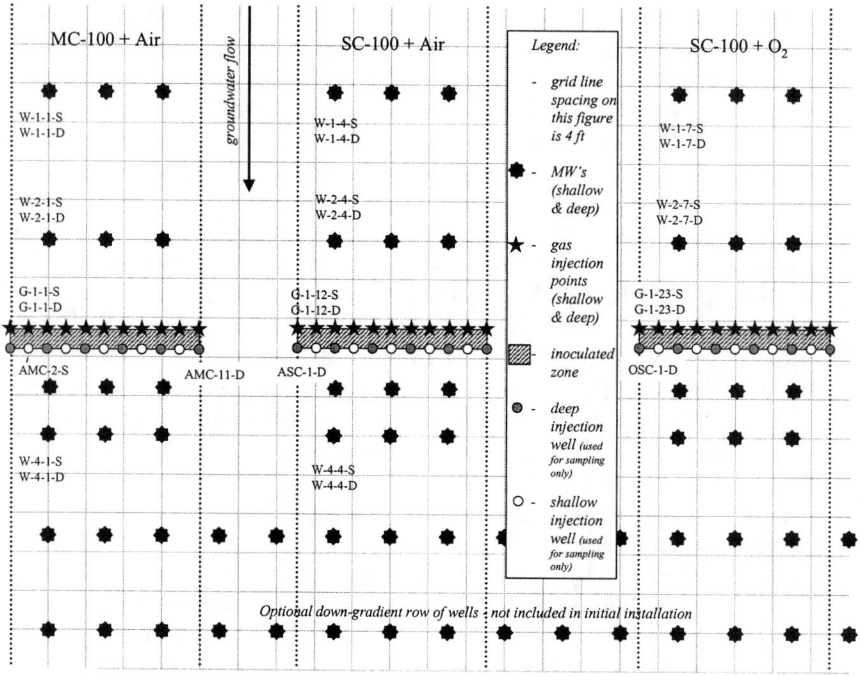
MC-100 + Air
groundwater flow
SC-100 + Air
SC-100 + O_2
Legend:
- grid line spacing on this figure is 4 ft
- MW's (shallow & deep)
- gas injection points (shallow & deep)
- inoculated zone
- deep injection well (used for sampling only)
- shallow injection well (used for sampling only)
W-1-1-S
W-1-1-D
W-2-1-S
W-2-1-D
G-1-1-S
G-1-1-D
AMC-2-S
AMC-11-D
W-4-1-S
W-4-1-D
W-1-4-S
W-1-4-D
W-2-4-S
W-2-4-D
G-1-12-S
G-1-12-D
ASC-1-D
W-4-4-S
W-4-4-D
W-1-7-S
W-1-7-D
W-2-7-S
W-2-7-D
G-1-23-S
G-1-23-D
OSC-1-D
Optional down-gradient row of wells : not included in initial installation

Figure 1. Layout of experimental plots and biobarriers showing shallow and deep monitoring and gas injection wells and culture (MC-100 and SC-100) inoculated zones.

RESULTS AND DISCUSSION

Distribution of DO in the Air and O_2 Sparged Biobarriers. Data on the DO levels in shallow and deep wells of the air and O_2 sparged seeded zones are in Figures 2 and 3. Prior to air or O_2 sparging (-40 days), groundwater levels of DO were usually ≤1mg/L across the plots.

The gas injection system was initiated several days prior to biomass injections. At 0 days (time of culture inoculation) the DO in shallow wells measured 5 mg/L and 13-17 mg/L in the air and O_2 sparged biobarriers, respectively. The DO concentrations stabilized in shallow wells from 64 days to

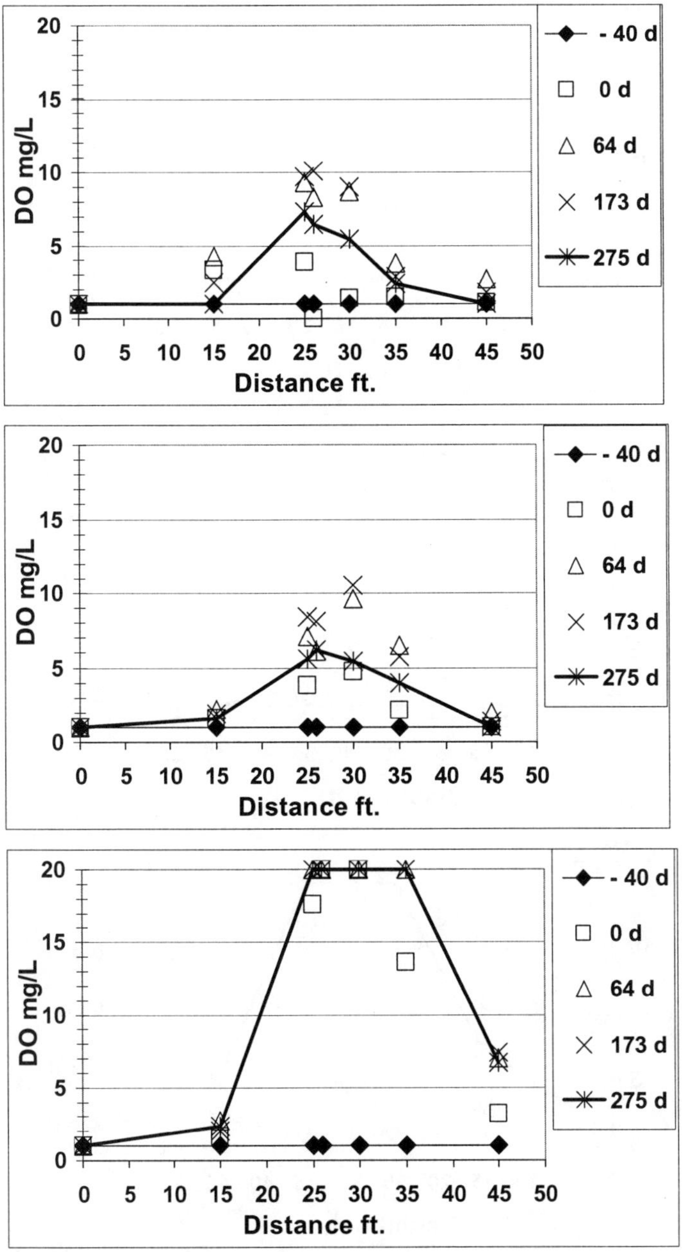
DO mg/L
Distance ft.
0 5 10 15 20 25 30 35 40 45 50
0 5 10 15 20
- 40 d
0 d
64 d
173 d
275 d
(a)
(b)
(c)

FIGURE 2. Dissolved oxygen concentrations in ground water samples from the shallow monitoring wells within the biobarrier zones of plots MC-100+Air (a), SC-100+Air (b), and SC-100+O2 (c). The seeded zone is located at 25-27ft.

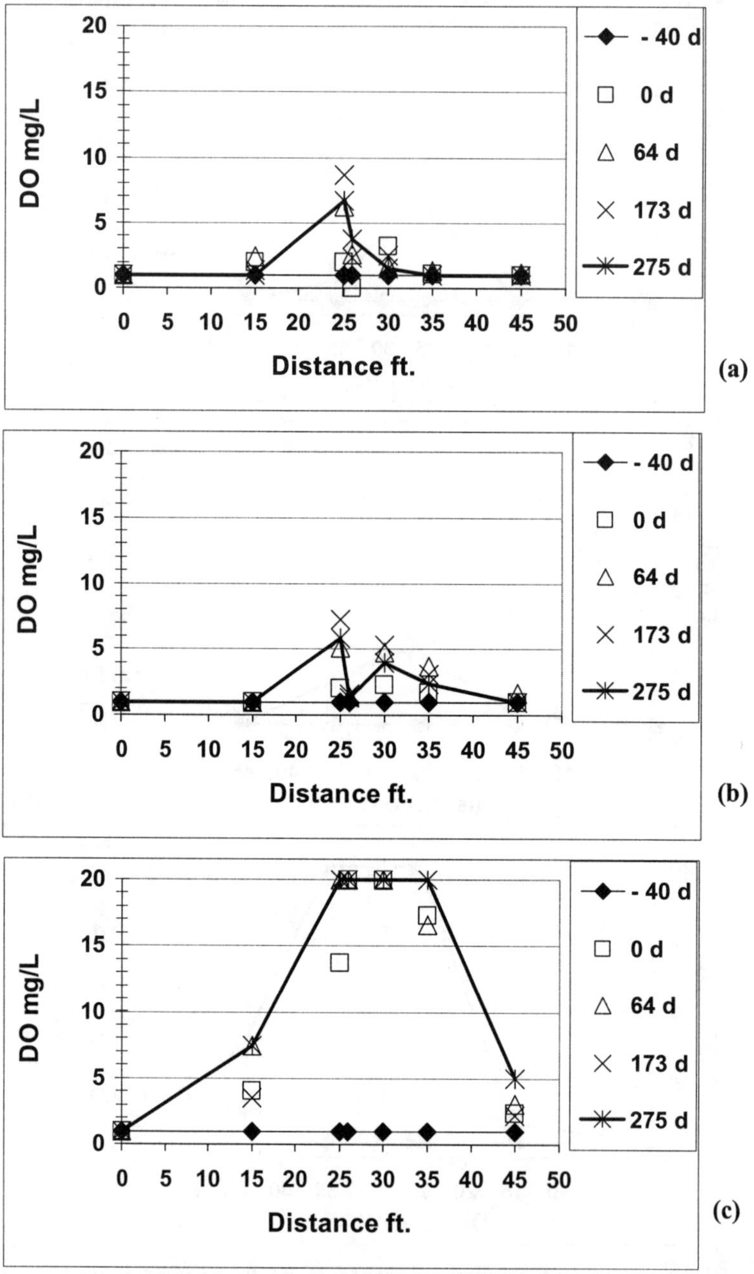
DO mg/L
Distance ft.
- 40 d
0 d
64 d
173 d
275 d
(a)
(b)
(c)

FIGURE 3. Dissolved oxygen concentrations in ground water samples from the deep monitoring wells within the biobarrier zones of plots MC-100+Air (a), SC-100+Air (b), and SC-100+O2 (c). The seeded zone is located at 25-27ft.

275 days at 5-10 mg/L in the air and ≥20mg/L in the O_2 – sparged areas (Figure 2). In the air-sparged plot, DO levels declined sharply to 1-3 mg/L in monitoring wells 10-15 ft downstream of the seeded zone. The increase in DO in groundwater from deep wells show very similar trends as the shallow plots (Figure 3). The DO varied from 5-9 mg/L and ≥20mg/L in the air and O_2 sparged zones, respectively. The lower DO levels in deep wells sampled at day 0 in the air injected zones was due to an error in the timer relays for the gas delivery systems. Adequate DO levels were present after one to two months (day 64, Figure 3).

Effects of MC-100 and SC-100 Seeded Biobarriers Plots on MTBE Biodegradation. Data on the decline in MTBE concentrations across the biobarriers in shallow and deep well groundwater samples are shown in Figures 4 and 5, respectively. MTBE concentrations are given as the geometric mean of monitoring well rows spaced along each plot. Groundwater concentrations of MTBE upstream of the biobarriers varied from 1-3 mg/L in shallow and deep well samples. Rapid decline in MTBE concentration occurred at 25-30 ft (8-9 m), the location of the bioactive zone. At 64 days, MTBE in shallow wells decreased to ≤10μg/L in the MC-100 + air, SC-100 + air and SC-100 + O_2 plots (Figure 4). After 173 and 275 days, MTBE in these same shallow wells decreased to 1 μg/L MTBE. In the deep wells, the decrease in MTBE was slower that that observed in shallow wells at day 64 and 173. This may have been due to the loss in aeration for one month as mentioned previously. MTBE from deep well samples decreased in all plots to 10-500 μg/L. In our previous biobarrier study (Salanitro et al., 2000) we showed that in the O_2 only plot, MTBE did not appreciably decrease for 186 days; after 261 days, however, MTBE decreased to 10-100 μg/L (from initial concentrations of 2-8 mg/L). These observations suggested that indigenous MTBE degraders were responsible for the decrease in this oxygenated zone.

Concentrations of TBA within the seeded zones were similar and varied from ≤5-25 μg/L. TBA levels in the aquifer upstream of the biobarriers varied from 5-150 μg/L.

CONCLUSIONS

In summary, pilot scale studies on the bioagumentation of the Port Hueneme aquifer with specialized aerobic MTBE-degrading bacterial cultures (MC-100 and SC-100) demonstrated that the oxygenates MTBE and TBA are rapidly degraded in bioactive zones enriched with air or O_2 sparging. Such declines in MTBE concentrations (without the accumulation of TBA as a byproduct) after two months of monitoring shallow and deep wells could only be attributed to the presence of the added MTBE-degrading cultures. High concentrations of bacterial cultures inoculated directly into aquifers appears to be a suitable in situ remedation technology for difficult-to-degrade ether oxygenates like MTBE.

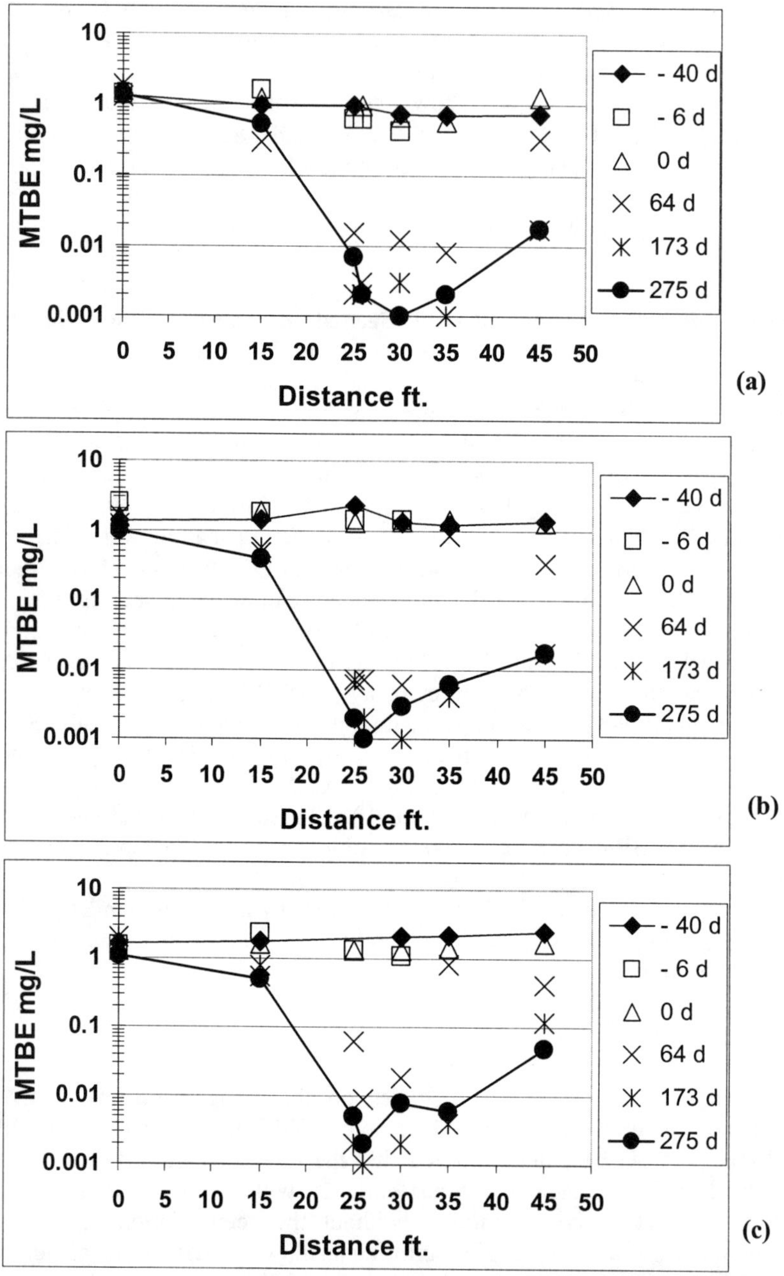
MTBE mg/L
10
1
0.1
0.01
0.001
0 5 10 15 20 25 30 35 40 45 50
Distance ft.
- 40 d
- 6 d
0 d
64 d
173 d
275 d
(a)
(b)
(c)

FIGURE 4. Decline in MTBE concentrations in shallow monitoring wells traversing the biobarriers MC-100 + Air(a), SC-100+Air (b), and SC-100+O2(c). The seeded zone is located at 25-27 ft.

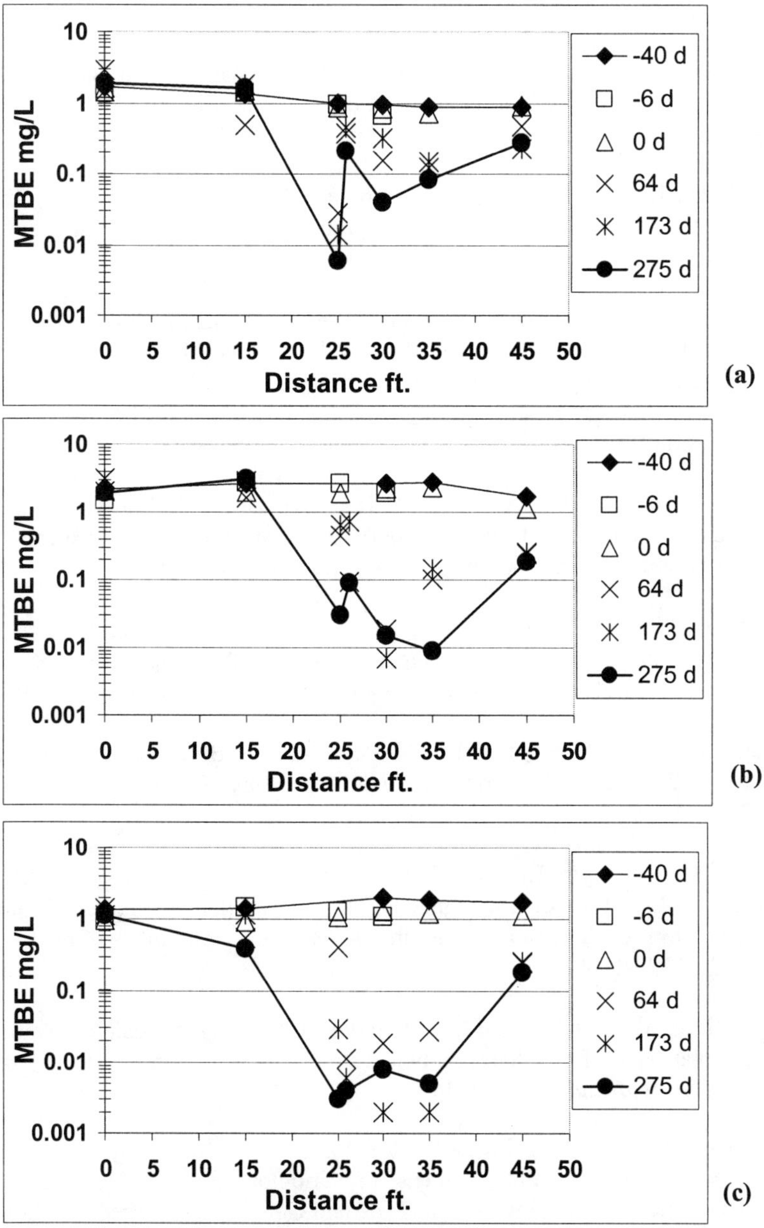
10
1
0.1
0.01
0.001
MTBE mg/L
0 5 10 15 20 25 30 35 40 45 50
Distance ft.
-40 d
-6 d
0 d
64 d
173 d
275 d
(a)
(b)
(c)

FIGURE 5. Decline in MTBE concentration in deep monitoring wells traversing the biobarriers MC-100+Air (a), SC-100+Air (b), and SC-100+O2 (c). The seeded zone is located at 25-27 ft.

REFERENCES

Barbeau, C., L. Deschenes, D. Karamonev, Y. Comeau, and R. Samson. 1997. Bioremediation of Pentachlorophenol – Contaminated Soil by Bioaugmentation Using Activated Soil. *Appl. Microbiol. Biotechnol. 48*:745-752.

Duba, A. G., K. J. Jackson, M. C. Jovanovich, R. B. Knapp, and R. T. Taylor. 1996. TCE Remediation Using In Situ, Resting-State Bioaugmentaion. *Environment Sci. Technol. 30*:1982-1989.

Dybas, M. L, M. Bareclona; S. Bezborodnikov, S. Davies, L. Forney, H. Heuer, O. Kawka, T. Mayotte, L. Sepulveda-Torres, K. Smalla, M. Sneathen, J. Tredge, T. Voice, D. C. Wiggert, M. E. Witt, and C. S. Criddle. 1998. Pilot-Scale Evaluation of Bioaugmentation for In Situ Remedation of a Carbon Tetrachloride – Contamination Aquifer. *Environ. Sci. Technol. 32*:3598-3611.

Ellis, D. E., E. J. Lutz, J. M. Odom, R. L. Buchanan, Jr., C. L. Bartlett, M. D. Lee, M. R. Harkness, and K. A. Deweerd. 2000. Bioaugmentation for Accelerated In Situ Anaerobic Bioremediation. *Environ. Sci. Technol. 34*:2254-2260.

Franzmann, P. D., L. R. Zappia, A. L. Tilbury, B. M. Patterson, G. B. Davis, and R. T. Mandelbaum. 2000. Bioaugmentation of Atrazine and Fenamiphos Impacted Groundwater: Laboratory Evaluation. *Bioremediation J. 4*:237-248.

Hanson, J. R., C. E. Akerman and K. M. Scow. 1999. Biodegradation of Methyl tert - Butyl Ether by a Bacterial Pure Culture. *Appl. Environ. Microbial. 65*: 4788-4792.

Harkness, M. R., A. A. Bracco, M. J. Brennan, Jr., K.A. Deweerd, And J. L. Spivack. 1999. Use of Bioaugmentation to Stimulate Complete Reductive Dechloroination of Trichloroethene in Dover Soil Columns. *Environ. Sci. Technol.* 33:1100-1109.

Mayotte, T. J., M. J. Dybas, and C. S. Criddle. 1996. Bench-Scale Evaluation of Bioaugmentation to Remediate Carbon Tetrachloride – Contaminated Aquifer Materials. *Groundwater. 34*:358-367.

Munakata-Marr, J., P. L. McCarty, M. S. Shields, M. Reagin, and S. C. Francesconi. 1996. Enhancement or Trichloroethylene Degradation in Microcosms Bioaugmented with Wild Type and Genetically Altered Burkholderia (Pseudomonas). Cepacia G4 and PR1. *Environ. Sci. Technol. 30*:2045-2052.

Munkata-Marr, J., V. G. Matheson, L. J. Forney, J. M. Tiedje , and P. L. McCarty. 1997. Long-Term Biodegration of Trichloroethylene Influenced by Bioaugmentation and Dissolved Oxygen in Aquifer Microcosms. *Environ. Sci. Technol. 31*:786-791.

Salanitro, J. P., L. A. Diaz, M. P. Williams, and H. L. Wisniewski. 1994. Isolation of a Bacterial Culture that Degrades Methyl T-Butyl Ether. *Appl. Environ. Microbial. 60*:2593-2596.

Salanitro, J. P., C. C. Chou, H. L. Wisniewski, and R. E. Vipond. 1998. Perspectives on MTBE Biodegradation and the Potential for In Situ Aquifer Remediation. Southwest Regional Conference of the National Groundwater Association, June 4-8, Anaheim, CA.

Salanitro, J., G. Spinnler, P. Maner, H. Wisneiwski, and P. C. Johnson. 1999. Potential for MTBE Bioremediation-In Situ Inoculation of Specialized Cultures. Proceedings Petrol. Hydrocarbons and Org. Chem in Groundwater: Prevention, Detection and Remediaton Conference, Nov. 17-20, Houston, TX.

Salanitro, J. P., P. C. Johnson, G. E. Spinnler, P. M. Maner, H. L. Wisniewski, and C. Bruce. 2000. Field-Scale Demonstration of Enhanced MTBE Bioremediation through Aquifer Bioaugmentation and Oxygenation. *Environ. Sci. Technol. 34*:4152-4162.

Steffan, R. J., K. McClay, S. Vainberg, C. W. Condie and D. Zhang. 1997. Biodegradation of the Gasoline Oxygenates Methyl tert – Butyl Ether, Methyl tert – Butyl Ether and tert – Amyl Methyl Ether by Propane – Oxidizing Bacteria. *Appl. Environ. Microbial. 63*:4216-4222.

Steinle, P., P. Thalmann, P. Hohener, K. W. Hanselmann, and G. Stucki. 2000. Effect of Environmental Factors on the Degradation of 2,6-Dichlorophenol in Soil. *Environ. Sci. Technol. 34*:771-775.

Tchelet, R., R. Meckenstock, P. Steinle, and J. R. van der Meer. 1999. Population Dynamics of an Introduced Bacterium Degrading Chlorinated Benzenes in a Soil Column and in Sewage Sludge. *Biodegradation. 10*:113-125.

Weber, Jr., W. J., and H. S. Corseuil. 1994. Inoculation of Contaminated Subsurface Soils with Enriched Microbes to Enhance Bioremediation Rates. *Water Res. 78*:1407-1414.

ORGANIC MULCH BIOWALL TREATMENT OF CHLORINATED SOLVENT-IMPACTED GROUNDWATER

Carol E. Aziz, Mark M. Hampton, and Mark Schipper (Groundwater Services, Houston, TX)
Patrick Haas (Air Force Center for Environmental Excellence, San Antonio, TX)

ABSTRACT: To enhance biological reductive dechlorination of trichloroethylene (TCE)-impacted groundwater at Offutt AFB, Nebraska, an *in situ* pilot-scale permeable reactive treatment wall (100 ft [L] x 1 ft [W] x 23 ft [D]) was constructed under a contract with the Air Force Center for Environmental Excellence. The biowall was installed using a continuous trenching and backfill method and filled with a 50:50 mixture of organic mulch and concrete sand. The mulch served as an inexpensive source of fermentable organic matter, which produces hydrogen to promote reductive dechlorination. A comparison of the aquifer conditions before the emplacement of the biowall and during the 18-month test showed a decrease in the oxidation-reduction potential, dissolved oxygen, and pH of the groundwater. After the first six months, there was significant production of *cis*-1,2-dichloroethene (*cis*-DCE) downgradient of the biowall, indicating that the mulch was promoting reductive dechlorination. During the 18 months of operation, TCE removals as high as approximately 74-89% were achieved. Low concentrations of vinyl chloride were produced downgradient (<0.0025 mg/L). Production of ethene and ethane, final reductive dechlorination products, was observed. In summary, the mulch biowall was effective in removing chlorinated ethenes from groundwater without significant vinyl chloride production.

INTRODUCTION

In the "Technical Protocol for Evaluating Natural Attenuation of Chlorinated Solvents In Ground Water" (Wiedemeier et al., 1998), a Type 2 plume is one where reductive dechlorination is supported by the utilization of a naturally occurring organic carbon source. Type 2 plume behavior has been documented in coastal regions and wetland environments where chlorinated solvents migrate into organic carbon-rich zones. To facilitate enhanced biodegradation, naturally occurring organic amendments can be put in contact with the contaminated groundwater to promote reduction, simulating Type 2 plume behavior. This approach has been used previously to promote nitrate attenuation by heterotrophic denitrification (Robertson et al., 2000; Schipper and Vojvodic-Vukovic, 1998). In this work, natural organic matter in the form of mulch is used in a permeable reactive biowall to promote the in situ reductive dechlorination of chlorinated solvents in groundwater. The mulch lowers the dissolved oxygen concentration and oxidation-reduction potential (ORP) in the aquifer by acting as a source of available carbon to aerobic bacteria. Once anaerobic conditions are created, fermentation of the organic matter generates hydrogen, which can be used to promote biological reductive dechlorination. Because mulch is inexpensive and permeable walls are passive, this technology has the potential to be a cost-effective solution for chlorinated solvent-impacted groundwater.

Objectives. The objectives of this work are as follows:

- To test the efficacy of organic mulch as an electron donor to promote the biological reductive dechlorination of trichloroethylene (TCE)-impacted groundwater; and,
- To develop a low maintenance, cost-effective, in situ treatment wall technology.

Site Description. The test was conducted at Offutt AFB, located five miles south of Omaha, Nebraska. The in situ mulch biowall and control plot were located near the base boundary in the vicinity of Building 301. The biowall was installed near MW-9S where the depth to groundwater was about 6 ft below ground surface (bgs). The hydraulic conductivity in the alluvial silt and clay near MW-9S averaged 3.5 ft/day. The groundwater flow was predominantly westward, with a hydraulic gradient of 0.01 ft/ft. Assuming an effective porosity of 0.15, the computed groundwater seepage velocity was 0.23 ft/day or 85 ft/year.

Groundwater quality data collected prior to the test indicated that TCE and *cis*-1,2-dichloroethene (*cis*-DCE) were the primary constituents of concern, with TCE being measured at 1.28 mg/L and *cis*-DCE being detected at concentrations of 0.014 mg/L near the test location. Relatively low levels of *trans*-1,2-dichloroethene and 1,1-dichloroethene were detected, indicating that *cis*-DCE was a product of reductive dechlorination. Vinyl chloride (VC), a reductive dechlorination product of *cis*-DCE, was detected at 0.002 mg/L in only one sample, and no ethene was detected at quantifiable concentrations. These data indicated that reductive dechlorination was generally not proceeding past the transformation of TCE to *cis*-DCE.

MATERIALS AND METHODS

Mulch was generated at Offutt AFB by shredding fallen trees and leaves using a tub grinder. Using a backhoe, the actively composting mulch was mixed with concrete sand (1:1 volumetric ratio) for stability. The biowall (100 ft long x 1 ft wide x 23 ft deep) was installed using a one-pass trencher. The trencher permitted simultaneous construction and backfilling of the wall with the mulch mixture. The trench was filled to 2 ft bgs with the mulch/sand mixture and was capped with the cuttings.

Two-inch diameter monitoring wells were installed to 20 ft bgs via hollow stem auger both upgradient and downgradient of the biowall and control plot. The upgradient wells and the wells in the control plot area served to measure the effects of natural attenuation. The wells were sampled for volatile organic compounds (perchloroethylene(PCE), TCE, DCE isomers, VC), natural attenuation parameters (dissolved oxygen, nitrate, sulfate, ferrous iron, hydrogen, redox potential, ethene, ethane, methane, and alkalinity) and water quality indicators (pH, total organic carbon, temperature, specific conductance, and chloride). Sampling occurred at start-up (1/99) and then after approximately 6, 12, and 18 months.

RESULTS AND DISCUSSION

Depression of Dissolved Oxygen. Within 6 months of the biowall installation, there was a marked decline in the dissolved oxygen (from 2.4 mg/L to <0.5 mg/L) as shown in Figure 1 and a depression in the oxidation-reduction potential and pH. Presumably, the consumption of the organic matter by aerobic bacteria served to lower the dissolved oxygen in the aquifer, creating conditions conducive to reductive dechlorination. Some methane was produced (0.5 to 3 mg/L); however, significant reductive dechlorination occurred indicating that the native dechlorinating bacteria were able to compete successfully for the available electron donor.

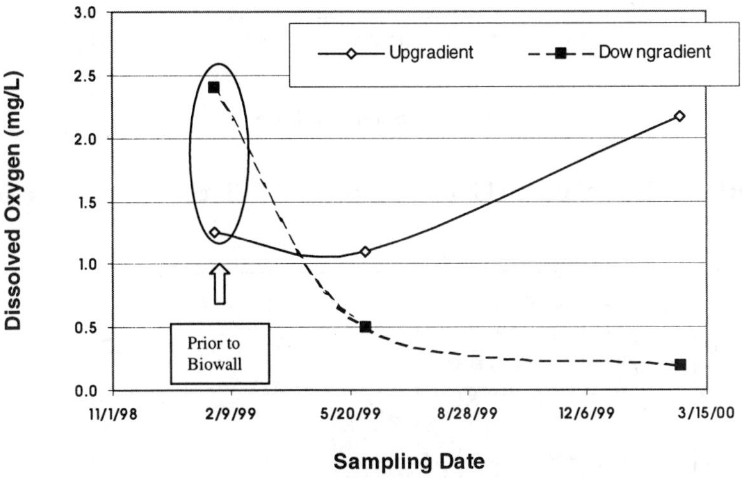

Dissolved Oxygen (mg/L)
3.0
2.5
2.0
1.5
1.0
0.5
0.0
Upgradient
Dow ngradient
Prior to
Biowall
11/1/98
2/9/99
5/20/99
8/28/99
12/6/99
3/15/00
Sampling Date

FIGURE 1. Dissolved oxygen concentration upgradient and downgradient of the mulch biowall.

Generation of Daughter Products. After 6 months of operation, a 460-fold increase in the ratio of *cis*-DCE:TCE downgradient of the biowall compared to upgradient was observed (Figure 2). Because *cis*-DCE is a reductive dechlorination by-product, this increase in the *cis*-DCE:TCE ratio was evidence of increased reductive dechlorination as a result of the addition of the mulch carbon source. Furthermore, the production of *cis*-DCE downgradient of the wall provided evidence that the contaminated groundwater was flowing through the wall and that sorption was not the predominant removal mechanism. From the 6- to 18-month mark, the *cis*-DCE:TCE ratio declined downgradient. This result corresponded to a 2-3 order of magnitude increase in the amount of ethene and ethane produced (Figure 3). Ethane concentrations as high as 0.010 mg/L were observed, but vinyl chloride concentrations remained around 0.002 mg/L. Ethene and ethane concentrations were non-detect prior to the biowall installation. These data suggest that decreased *cis*-DCE concentrations might be linked to more rapid degradation of *cis*-DCE and vinyl chloride to dechlorinated end-products.

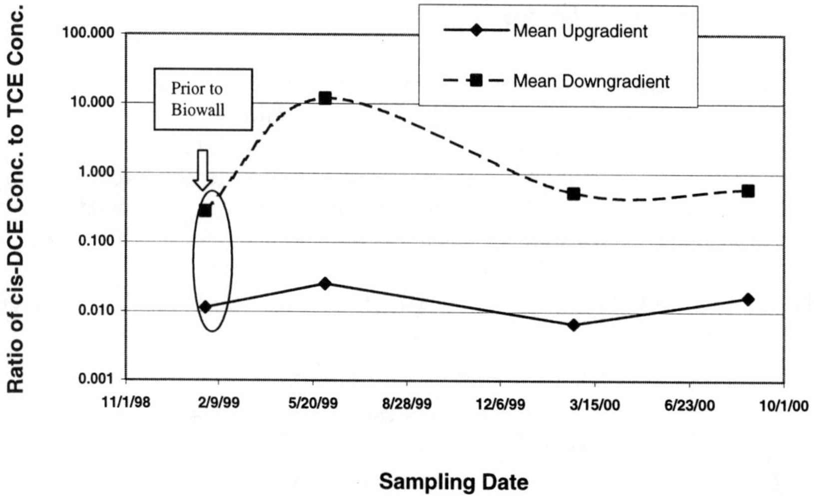
Ratio of cis-DCE Conc. to TCE Conc.
100.000
10.000
1.000
0.100
0.010
0.001
Mean Upgradient
Mean Downgradient
Prior to Biowall
11/1/98
2/9/99
5/20/99
8/28/99
12/6/99
3/15/00
6/23/00
10/1/00
Sampling Date

FIGURE 2. Effect of mulch biowall on *cis*-DCE:TCE concentration ratios.

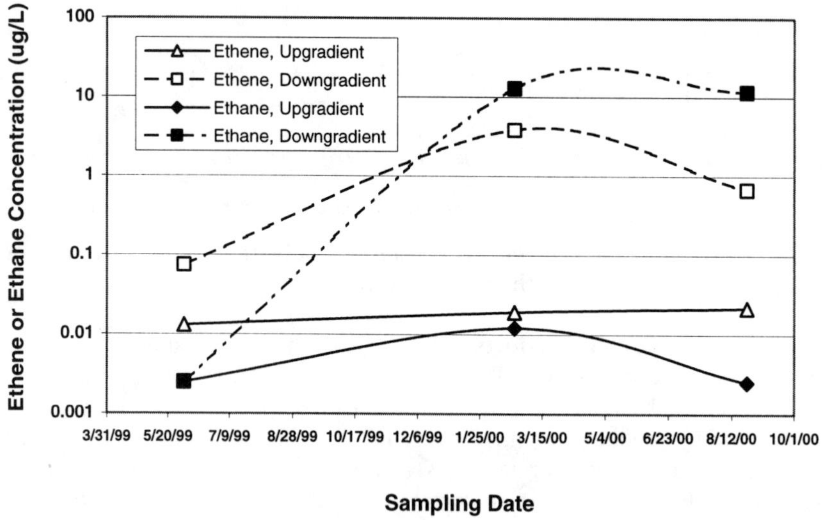
Ethene or Ethane Concentration (ug/L)
100
10
1
0.1
0.01
0.001
Ethene, Upgradient
Ethene, Downgradient
Ethane, Upgradient
Ethane, Downgradient
3/31/99
5/20/99
7/9/99
8/28/99
10/17/99
12/6/99
1/25/00
3/15/00
5/4/00
6/23/00
8/12/00
10/1/00
Sampling Date

FIGURE 3. Effect of mulch biowall on ethene and ethane concentrations.

Chlorinated Solvent Removal. Within 6 months and during the 18-month test, significant reductive dechlorination of TCE and *cis*-DCE was observed.

Approximately 74-89% of TCE was removed as a result of passing through the biowall (as shown in Figure 4). The effect of the biowall was much greater than the effect of natural attenuation, which averaged 6% TCE removal as determined by samples taken from the wells located in the control plot and prior to start-up. The TCE removal was calculated by subtracting the mean concentration in downgradient wells from the mean concentration in upgradient wells.

The February 1999 sampling event yielded unusual results. The upgradient TCE concentrations were unexpectedly low (i.e., 0.265 mg/L vs. 1.29-2.10 mg/L measured during the other sampling events), yielding an apparent increase in TCE when compared with downgradient TCE concentrations or a negative % removal. The upgradient TCE concentrations appeared low, but no sampling or analytical anomalies were noted. A large increase in *cis*-DCE was observed downgradient during the same sampling event (Figure 2), indicating that indeed biological reductive dechlorination was occurring and that TCE must be biodegrading.

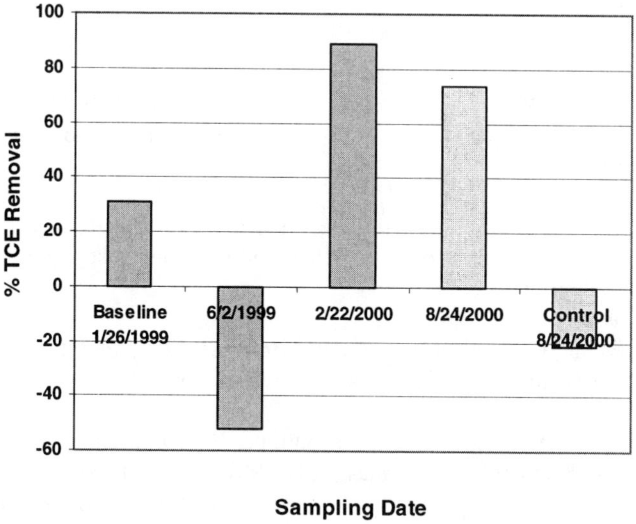
% TCE Removal
100
80
60
40
20
0
-20
-40
-60
Baseline
1/26/1999
6/2/1999
2/22/2000
8/24/2000
Control
8/24/2000
Sampling Date

FIGURE 4. Percent TCE removal effected by the mulch biowall.

Projected Cost of Technology. One distinct advantage of this passive technology is the low operating and maintenance costs. The capital expenditures include the installation of the biowall ($28,000 mobilization/demobilization, plus approximately $270 per linear foot) and the installation of a suitable number of monitoring wells. Mulch fill can be generated cheaply. In this case, the mulch was provided free of charge because of the excess mulch produced at Offutt AFB.

CONCLUSIONS

The installation of the mulch biowall was successful in turning the aquifer anaerobic, thereby creating conditions conducive to reductive dechlorination. The mulch provided sufficient organic matter to stimulate biological reductive dechlorination over a period of 18 months. TCE removals as high as approximately 74-89% were achieved, with minimal VC production. Complete dechlorination products, ethene and ethane, were also observed.

The current pilot-scale biowall will be expanded to 400 feet in 2001. The operation of this biowall will provide further performance data over a longer time period (on the order of 3-5 years).

This technology has the advantage of low operating and maintenance costs and is suitable for sites with shallow groundwater. Mulch biowalls can act as a stand-alone technology and may have applications as a polishing step following zero-valent iron walls.

ACKNOWLEDGMENTS

This work was funded by the Technology Transfer Division of the Air Force Center for Environmental Excellence under contract number F41624-97-C-8020.

REFERENCES

Robertson, W.D. D.W. Blowes, C.J. Ptacek, and J.A. Cherry. 2000. "Long-Term Performance of In Situ Reactive Barriers for Nitrate Remediation." *Ground Water. 38*(5):689-695.

Schipper, L., and M. Vojvodic-Vukovic. 1998. "Nitrate Removal from Groundwater Using a Denitrification Wall Amended with Sawdust: Field Trial." *J. Environ. Qual. 27*:664-668.

Vidic, R.D. and F.G. Pohland. 1996. *Treatment Walls*. Prepared for Ground-Water Remediation Technologies Analysis Center. Pittsburgh, PA.

Wiedemeier, T.H., M.A. Swanson, D.E. Moutoux, E.K. Gordon, J.T. Wilson, B.H. Wilson, D.H. Kampbell, P.E. Haas, R.N. Miller, J.E. Hansen, and F.H. Chapelle. 1998. *Technical Protocol for Evaluating Natural Attenuation of Chlorinated Solvents in Ground Water*. U.S. EPA, EPA/600/R-98/128.

BIOFILM BARRIERS FOR GROUNDWATER CONTAINMENT

Garth James (MSE Inc., Bozeman, MT)
Randy Hiebert (MSE Inc., Butte, MT)

ABSTRACT: Biofilm barriers are a promising new technology for the containment of groundwater contaminants and other applications that require manipulation of porous media hydraulic conductivity. These barriers are formed by the subsurface injection of bacteria and/or nutrients. The subsequent growth of the bacterial biofilms and production of extracellular polymeric substances (EPS) occludes pore throats in the porous matrix, providing significant reductions in the effective hydraulic conductivity. Recently, MSE performed a field-relevant scale biofilm barrier demonstration at their test facility in Butte, MT. Bacteria and nutrients were injected using vertical wells spaced across a 180 foot long, 130 foot wide, and 20 foot deep test cell. The bacteria injected were selected based on their ability to produce copious amounts of EPS under simulated field conditions and were starved prior to injection to enhance their subsurface transport and survival. Resuscitation and growth of the bacteria resulted in a decrease in hydraulic conductivity in the test cell that reached a reduction greater than 99% within 85 days. This reduced hydraulic conductivity has been maintained for over one year. Overall, the results demonstrate that biofilm barriers are effective for manipulating hydraulic conductivity at a field-relevant scale.

INTRODUCTION

The first step in the remediation of contaminated groundwater is often containment of the plume to prevent further migration. The contained plume can then be treated using in-situ or ex-situ methods. Traditional barrier methods, such as sheet piling and slurry walls are difficult to install, are usually not cost effective at depths greater than fifty feet, and are difficult to remove when no longer needed. Biofilm barriers are a novel technology for the containment of contaminated groundwater. These barriers are formed by the injection of bacteria and nutrient solutions into the subsurface. The growth of the bacteria and production of extracellular polymeric substances occludes pore throats in the subsurface matrix, resulting in a decrease in effective hydraulic conductivity. Similar approaches have been used in the petroleum industry for enhancing secondary oil recovery (Premuzic and Woodhead, 1993). Biofilm barriers can be installed cost effectively with little surface disturbance. They also have no obvious depth limitation and can be removed when no longer necessary. These characteristics make biofilm barriers a promising alternative to traditional strategies for the containment of contaminated groundwater.

Laboratory evaluations of biofilm barrier technology have demonstrated drastic reductions in the effective hydraulic conductivity of porous media can be achieved with biofilm barriers (Cunningham et al., 1997, James et al., 2000). Furthermore, these studies demonstrated biofilm barriers could withstand heavy metals (strontium and cesium) and a chlorinated solvent (carbon tetrachloride). Although these studies demonstrated the feasibility of using biofilm barriers for the containment of contaminated groundwater, they were conducted using small-scale laboratory systems. This study was conducted to evaluate the use of biofilm barrier technology at a field-relevant scale. Overall, the results demonstrate that biofilm barrier technology is feasible for the containment of groundwater contaminants.

MATERIALS AND METHODS

Test Cell Construction. Construction of the Test Cell began in Fall 1999. Following a survey of the site, soil was excavated to a depth of 20 feet in the central 50-foot by 100-foot portion. The sides had a 2:1 slope, causing the excavation to be 180 feet long and 130 feet wide at the surface (see Figure 1). The excavation was lined with a polyvinyl chloride (PVC) liner. The excavated material was then screened into two fractions, particles greater than 20 mesh and those less than 20 mesh. Because it was a wet screening process, the fraction less than 600 mesh was removed with the wash water.

A two-foot lift of the finer material was placed on the bottom of the test cell followed by a series of four-inch diameter PVC perforated pipes used for initial filling of the cell with groundwater. A network of perforated four-inch diameter PVC pipes was laid on the western side slope to form the crossflow inlet. The crossflow outlet consisted of similar pipe on the eastern boundary of the cell. The coarser soil fraction was then placed back into the hole in approximately two-foot lifts and compacted after each lift. Due to a shortage of course material from the excavation, additional coarse fill material was imported from another site and used to fill the South end of the Test Cell. This imported material was larger grained than the coarse material excavated from the Test Cell Site. After thirteen feet of this soil was placed into the hole, a five foot layer of the fine soil was placed on the top and compacted.

The site was surveyed again and the well locations were marked (see Figure 1). Eleven four-inch diameter injection wells were drilled along the center north-south axis. These wells were screened entirely in the coarse soil fraction interval (about 12 feet). Midway between these wells, ten one-inch centerline monitoring wells (C-1 through C-10) were drilled and were screened at the same interval. Six two-inch diameter wells (P-1 through P-6) were installed to test cell to house pressure transducers that would monitor water level. These were completed with five-foot screens in the center of the coarse soil fraction interval. Finally, eight additional one-inch diameter wells (T-1 through T-8) were installed for tracer tests. These were screened at the same interval as the injection wells.

A gradient was established across the test cell by filling the cell with groundwater and pumping the water into the crossflow inlet at about three gallons per minute. Concurrently, water was pumped from the crossflow outlet at the same rate and discharged to an existing containment pond. This resulted in a groundwater velocity of about one foot per day. The injection wells were connected to the nutrient tanks using one-inch diameter hose. The hose was heat traced and insulated and enclosed in an insulated channel. Construction was completed in December, 1999.

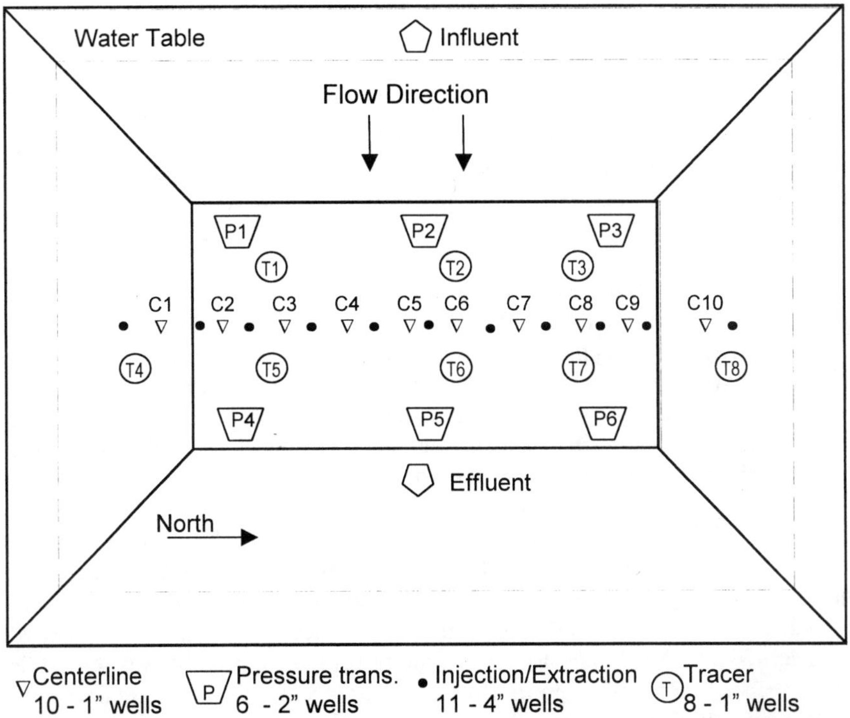
Water Table
Influent
Flow Direction
P1
P2
P3
T1
T2
T3
C1
C2
C3
C4
C5
C6
C7
C8
C9
C10
T4
T5
T6
T7
T8
P4
P5
P6
Effluent
North
Centerline 10 - 1" wells
Pressure trans. 6 - 2" wells
Injection/Extraction 11 - 4" wells
Tracer 8 - 1" wells

FIGURE 1. Diagram of test cell showing well locations. Groundwater flow was maintained by injecting water at the influent manifold while simultaneously extracting from the effluent manifold. The biofilm barrier was established by injecting at the injection/extraction wells and monitored at the centerline (C) wells. The tracer (T) wells were installed for conducting tracer tests, while the pressure (P) wells housed pressure transducers for monitoring water levels.

Biofilm Barrier Formation. Formation of the biofilm barrier was initiated by the introduction of bacteria and nutrient solutions to the injection wells located across the center of the test cell. The solution of bacteria consisted of a starved cell suspension of *Pseudomonas fluorescens* sp. CPC211A. This *Pseudomonas* strain was isolated by MSE from a petroleum-contaminated aquifer and was capable of petroleum hydrocarbon (BTEX) degradation. This strain was selected based on its growth and copious production of EPS under simulated field conditions. The starved cell suspension of CPC211A was produced using a proprietary protocol developed by MSE to enhance the transport and survival of bacteria introduced to the subsurface. The nutrient formulation for resuscitation of the starved bacteria included carbohydrate (molasses) as a source of carbon and energy, nitrate as a nitrogen source and alternative electron acceptor once oxygen had been depleted from the biofilm barrier, and other required sources of minor and trace nutrients.

Enumeration of Bacteria. The populations of bacteria in the vicinity of biofilm barrier were monitored by plate counts of groundwater samples collected from the centerline wells. The samples were collected using a peristalitic pump and sanitized tubing, after three well volumes had been purged. The samples were serially diluted using buffered water and plated on R2A Agar for enumeration of heterotrophic bacteria. Phenotypic profiles of bacteria from the enumeration plates were obtained using the bioMérieux API 20NE identification system.

Hydraulic Conductivity Determinations. The vacuum slug test method was used to measure values of hydraulic conductivity at numerous locations within the biofilm barrier. This method was selected due to the small diameter (1-inch) of the piezometers (centerline wells). The slug tests were conducted using a special wellhead apparatus, a high-speed data logger and pressure transducer, and a water level indicator. The wellhead apparatus was designed to seal the piezometer, while at the same time, allowing access for water level measuring instruments and providing a method to release the vacuum to begin a test.

To conduct a test, the well head apparatus was placed on a piezometer and secured. Depth to the static water level was measured and recorded to determine the available height the water level can be raised inside the piezometer. The transducer was then placed down the piezometer near the bottom, followed with the electric water level indicator. The water level indicator probe was set 5.0 feet above the static water level to indicate the desired water level was attained. The wellhead apparatus ports were then either closed or sealed and the vacuum line connected to a vacuum pump. The vacuum inside the well was increased until the water level meter indicated the water had been raised to the desired level. Once the water level was at the desired level, a small bleeder valve was opened to maintain a constant vacuum, holding the water level relatively steady and allowing the aquifer to equilibrate. The data logger was then activated and the vacuum released through a 2-inch ball valve on the wellhead apparatus, allowing the water level in the well to recover. The data was downloaded from the data logger to a personal computer and was analyzed using the SuperSlug™ software to determine values of hydraulic conductivity.

RESULTS AND DISCUSSION

Biofilm barrier formation in the test cell was initiated by introducing starved bacterial cells and nutrient mixture at injection wells located across the center of the cell (Figure 1). Solutions were applied to the test cells using a gravity-fed system with a hydraulic head of three to six feet. Addition of the nutrient and bacteria was conducted for two days and then a series of nutrient-only treatments were applied to the test cells, as indicated in Table 1.

TABLE 1. Injections of bacteria and or nutrients for establishment of biofilm barrier

Injection	Start Date	Duration (days)	Volume Injected (Gallons x 1000)
Bacteria/nutrient	12/10/99	2.0	65
Nutrient	12/16/99	4.5	46
Nutrient	12/27/99	3.2	39
Nutrient	1/13/00	8.1	33
Nutrient	1/31/00	2.1	20
Nutrient	2/14/00	2.1	20
Nutrient	2/23/00	1.4	20

Enumeration of bacteria in groundwater samples from the centerline wells during biofilm barrier indicated a substantial increase in the population of heterotrophic bacteria (Figure 2). The morphology of some of the predominant colonies on enumeration plates was consistent with those of the inoculated strain, *Pseudomonas fluorescens* CPC211A. In addition, several bacteria isolated from the enumeration plates had phenotypic profiles (API20NE) which matched that of *P. fluorescens* CPC211A. These results suggest that the introduced *Pseudomonas* strain became one of the predominant bacteria in the biofilm barrier.

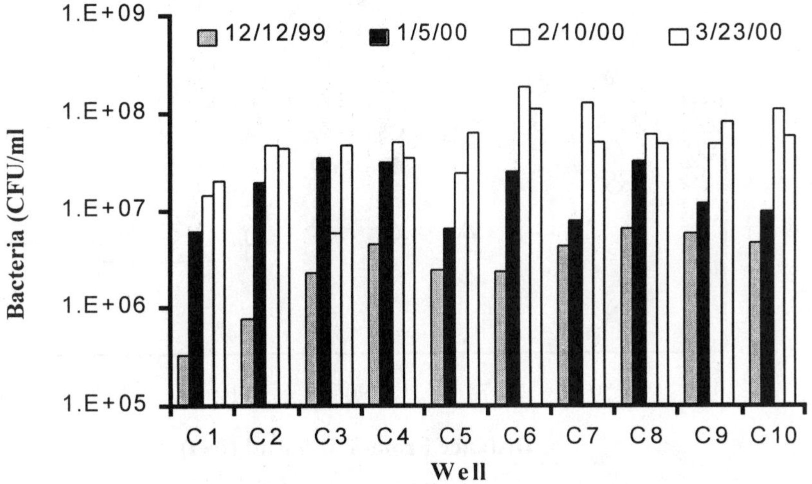

1.E+09
1.E+08
1.E+07
1.E+06
1.E+05
12/12/99
1/5/00
2/10/00
3/23/00
Bacteria (CFU/ml)
C1 C2 C3 C4 C5 C6 C7 C8 C9 C10
Well

FIGURE 2. Populations of heterotrophic bacteria in samples obtained from the centerline wells (C1-C10) during biofilm barrier formation. A substantial increase in the populations was observed. The colony morphologies and phenotypic characteristics of some of the predominant bacteria were consistent with those of the inoculated strain.

The average initial hydraulic conductivity, measured from the pressure wells (P1-P6), was 2.1 x 10^{-2} centimeters per second. By the end of nutrient injection sequence the average hydraulic conductivity at the centerline wells had been reduced by 99.4% and continued to decrease over the monitoring period. The hydraulic conductivities determined for each well tested are shown in Figure 3. The hydraulic conductivity reductions were greatest in the Central to Northern side of the test cell (wells C4-C10). These results indicate there may have been bypass flow on the South side of the test cell, where the initial hydraulic conductivity was higher due to the larger sized fill used in this area during test cell construction. Nonetheless, hydraulic conductivity was reduced by more than two orders of magnitude throughout the biofilm barrier. Overall, these results demonstrate an effective biofilm barrier can be established at a field-relevant scale.

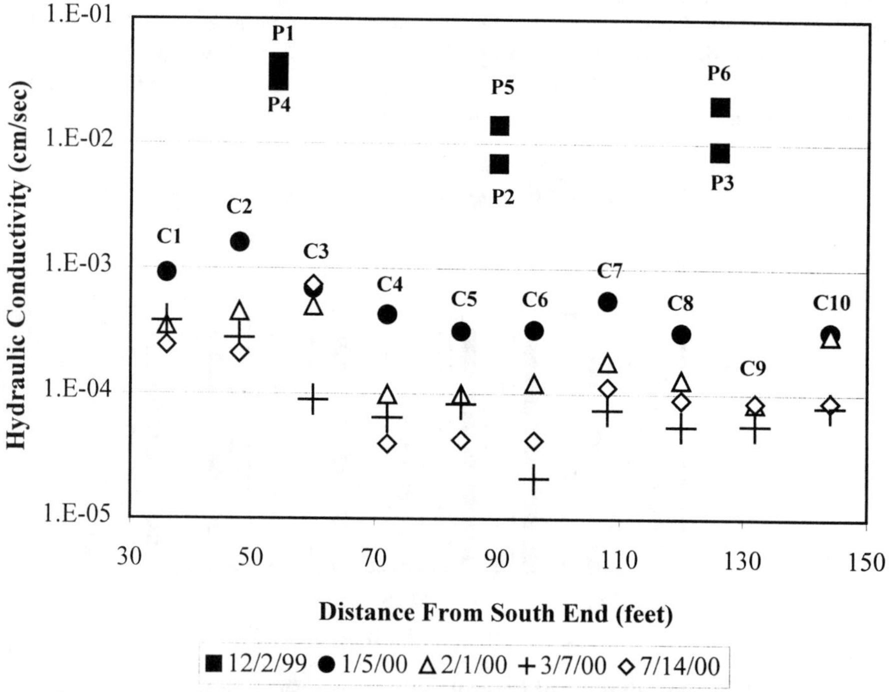
1.E-01
1.E-02
1.E-03
1.E-04
1.E-05
Hydraulic Conductivity (cm/sec)
P1
P4
P5
P2
P6
P3
C1
C2
C3
C4
C5
C6
C7
C8
C9
C10
30
50
70
90
110
130
150
Distance From South End (feet)
■ 12/2/99 ● 1/5/00 △ 2/1/00 + 3/7/00 ◇ 7/14/00

FIGURE 3. Hydraulic conductivites at the pressure (P) wells before biofilm barrier formation and centerline (C) wells after biofilm barrier formation. The average initial hydraulic conductivity in the test cell was 2.1 x 10^{-2} cm/sec. Installation of the biofilm barrier resulted in hydraulic conductivity reductions of up to two orders of magnitude.

CONCLUSIONS

This study provided a demonstration of biofilm barrier feasibility at a field-relevant scale. Hydraulic conductivity was reduced by two to three orders of magnitude and this reduction has been maintained with little maintenance. Bacteriological evaluation indicated that the inoculated *Pseudomonas* strain was one of the predominant bacteria in the biofilm barrier. The barrier is currently undergoing evaluation to determine long-term maintenance (feeding) requirements. The next step in the development of this technology will be a pilot-scale demonstration at a contaminated site.

ACKNOWLEDGMENTS

The authors would like to thank Al Cunningham, J.W. (Bill) Costerton, Paul Sturman, Bryan Warwood, and Laura Boegli for their contributions to the development of this technology. This work was conducted under U.S. Department of Energy Contract Number DE-AC22-96EW96405.

REFERENCES

Cunningham, A., B. Warwood, P. Sturman, K. Horrigan, G. James, J.W. Costerton, and R. Hiebert. 1997. "Biofilm Processes in Porous Media - Practical Applications. In: P.S. Amy, D.L. Haldeman (Eds.), *The Microbiology of the Terrestrial Deep Subsurface*, pp. 325-344. Lewis Publishers: New York, NY.

James, G. A., B. K. Warwood, and R. Hiebert. 2000. "In-Situ Biological Barriers to the Spread of Pollution." In: J. J. Valdez (Ed.) *Bioremediation*, pp. 1-13. Kluwer Academic Publishers: Boston, MA.

Premuzic E., and A. Woodhead (Eds). 1993. *Microbial Enhancement of Oil Recovery-Recent Advances*. Elsevier, New York, NY.

BIOWALL IN SITU GROUND-WATER TREATMENT

Paul W. Becker (ExxonMobil Refining & Supply, Linden, NJ)
Brent B. Archibald (ExxonMobil Refining & Supply, Linden, NJ)
James H. Higinbotham (ExxonMobil Refining & Supply, Fairfax, VA)
Patrick C. Madden (Engineering Consultant, Whippany, NJ)

ABSTRACT: The engineered biowall is a promising new technology for treating contaminated ground water. This *in situ* treatment process places a permeable reaction zone at the edge of a plume or property line and biologically treats contaminants as they slowly migrate through the zone. Biowall research at ExxonMobil has focused on treating light aromatic (BTEX) contaminants, particularly benzene. It has included laboratory operations as well as a field unit demonstration for several years at a refinery site. Economic screening results indicate that biowalls can provide 20-50% savings over conventional pump and treat systems. Important factors to consider in weighing the advantages of biowalls are ground-water velocity, the depth of contamination, and the availability of existing wastewater treatment facilities. Where applicable, biowalls provide some important advantages over other ground-water remediation technologies. Among these are low cost, design flexibility, control of offsite migration, and minimal operating and maintenance requirements. ExxonMobil's first commercial biowall facility is scheduled for start-up in 2001.

INTRODUCTION

During the past decade, permeable reactive barriers, or reaction walls, have emerged as an important *in situ* option for ground-water remediation. One of the simplest and least expensive types of reaction walls is the biowall, which is an aerobic biological system designed to remove dissolved hydrocarbons. This technology, shown schematically in Figure 1, places a biological treatment cell within or at the edge of a plume or property line and treats contaminants as they slowly migrate through the zone. The subsurface reactor is filled with crushed stone, which serves as a medium for biofilm growth. Air and possibly nutrients are introduced to effect the biological reactions.

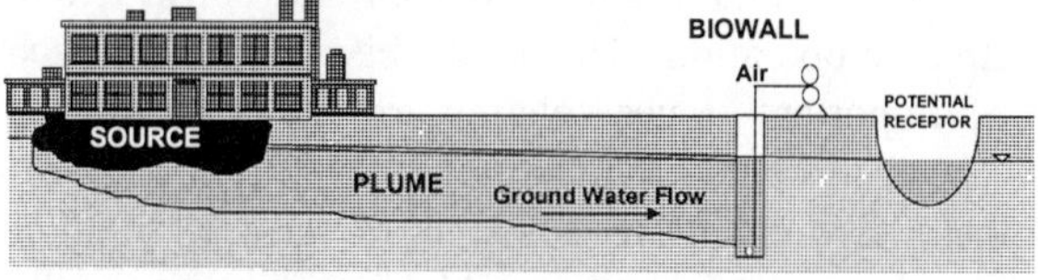
BIOWALL
Air
POTENTIAL
RECEPTOR
SOURCE
PLUME
Ground Water Flow

FIGURE 1. Biowall conceptual application.

Biowalls are also known by several other names, including: trench biosparge, microbial fence, funnel and gate, bubble curtain, bioscreens, and sparge curtain. Biowalls are typically not intended as a permanent source removal approach but rather as a plume management remediation concept. They are best implemented where source remediation is technically, operationally, or economically infeasible.

ExxonMobil interest in biowalls began in 1993 when they were identified as one of several innovative environmental technologies that could be cost-effectively utilized at petroleum sites (refineries, terminals and service stations). Active research began in 1994 through a cooperative effort with the New Jersey Hazardous Substance Management Research Center and Stevens Institute of Technology. These initial investigations quickly established "proof of concept" by confirming that an effective biofilm of hydrocarbon degraders could be deposited and maintained on inert media and achieve high contaminant removal in relatively short residence times. On the basis of this information, a field pilot test facility was constructed within an area of a northeastern USA refinery where dissolved benzene had been detected in ground water. This pilot test unit has been in operation since 1996, and has consistently reduced influent benzene concentrations to below the drinking water criterion of 1 μg/L.

BIOWALL APPLICABILITY AND CONCEPTUAL DESIGN

The general applicability of biowalls and the details of engineering design depend strongly on site conditions; principally the depth of contamination, the soil permeability and homogeneity, and the proximity of existing wastewater treatment facilities. At aquifer depths exceeding 25 feet (7.6 m), excavation costs increase substantially, reducing the attractiveness of biowall technology. At ground-water superficial velocities[1] above 1 foot/day (0.3 m/day), the reactor volume to achieve the required residence time becomes excessive. When treatment facilities are close at hand, it may be cheaper to send the ground water there than to build a biowall. Thus, it is important to conduct site-specific economic screening before applying biowall technology.

Biowall design configurations typically fall into one of three categories: an engineered wall or trench, a sparge line or zone, or a so called "funnel and gate", as developed at the University of Waterloo (Starr and Cherry, 1994), wherein ground water is channeled into a reactive zone for treatment.

The first category is essentially an excavated trench filled with a remediation media extending the full width and depth of the contaminated aquifer. This approach has many advantages, including simple design and construction, low cost, and relatively non-intrusive operation. However, this simple approach does not address a number of important issues relating to the contaminated aquifer. Among these are the need to contain and remove LNAPL (light non-aqueous phase liquid), the potentially extensive costs to effectively monitor performance, and the possibility that contaminants could bypass treatment by "short-circuiting" through "lenses" of highly permeable material. In addition, long trenches make it difficult to achieve uniform air and/or nutrient distribution. Obviously, the more extensive the plume area to be covered, the more costly and less appropriate the simple trench approach becomes.

---

[1] Also known as Darcy Flux, this is the volumetric flow rate per unit cross-sectional area of flow.

The second configuration, using a series of sparge wells, is essentially a simplification of the single trench approach that may be applicable for shallow aquifers in higher permeability soils. It is simpler, lower cost, and even less intrusive than the trench, but the concerns expressed above are even more significant for the sparge line. LNAPL would be difficult to deal with, performance monitoring could be more extensive, and the potential for short-circuiting would be much greater with a sparge line than with a continuous trench.

In practice, the most widely applicable configuration is the "funnel and gate," because this third approach allows for the most advantageous reactor design. Using a system of barriers or collection trenches as funnels to capture and channel ground water into a reactive zone for remediation provides many advantages relative to the simple trench or sparge wells. It also provides the best opportunity to tailor the reactor design to the specific conditions encountered at a given site. LNAPL collection can be done immediately upstream of the reactor, where it can be accomplished most efficiently. Channeling the ground water eliminates "short-circuiting" and allows for accurate and efficient performance monitoring at the outlet of the bioreactor. Finally, concentrating ground-water flow into a small reactor or treatment cell substantially improves the air and nutrient addition capability.

FIELD PILOT DEMONSTRATION

By 1995, research at Stevens Institute had successfully proven the biowall technology in the laboratory environment (Christodoulatos et al., 1996) and results were promising enough to justify a field pilot test. The principal goals of the field demonstration were to: 1) prove that an *in situ* aerobic biowall can treat ground water to target levels under actual site conditions; and 2) define broad applicability, geometry, and scale-up factors for the technology. The test site was selected based on previous hydrogeologic and contaminant evaluations conducted at a northeastern USA refinery. A detailed evaluation determined that the site had benzene concentrations in the range of 500 μg/L and no other significant

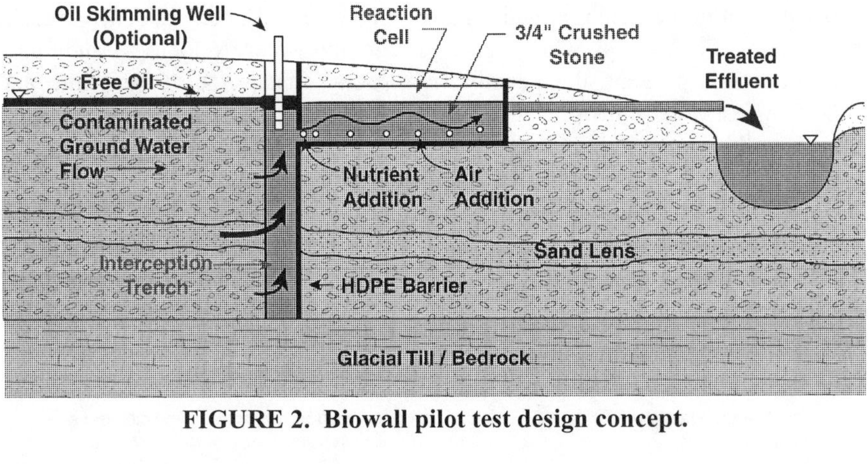
Oil Skimming Well (Optional)
Reaction Cell
3/4" Crushed Stone
Treated Effluent
Free Oil
Contaminated Ground Water Flow
Nutrient Addition
Air Addition
Sand Lens
Interception Trench
HDPE Barrier
Glacial Till / Bedrock

FIGURE 2. Biowall pilot test design concept.

contaminants. The geology of the test site consists of clay/silty clay, fill and till, with thin silty/clayey sand layers. The aquifer is less than 20 feet (6 m) deep and the hydraulic conductivity at the site is in the optimum range for testing the biowall concept.

The pilot facility, which is similar to the schematic on Figure 2, includes a gravel-filled reaction cell where air and (potentially) nutrients can be added through vertical or horizontal screens. A 20 foot (6 m) deep, 15 foot (4.6 m) wide, 3 foot (0.9 m) thick, gravel-filled collection trench brings ground water up and into the reactor. The reactor itself is fashioned from a used 8 foot x 22 foot (2.4 m x 6.7 m) steel dumpster buried to a depth of about eight feet (2.4 m). The downstream side of the collection trench and all sides of the reactor are lined with HDPE (high density polyethylene). For the purposes of the test program, effluent ground water is collected in a sump well and pumped to a storage tank for disposal. In actual application, the effluent from the reactor could be piped as a point discharge into surface water, or could be reintroduced to the aquifer via a subsurface infiltration or dispersion zone. An underflow baffle at the entrance to the reactor acts to prevent any free oil on the surface of the ground water from entering the reaction cell. To date, this capability has not been needed. The pilot facility was also designed to allow the potential addition of nutrients and/or solid oxidants, but current operations have been able to reduce ground-water contaminants to below drinking water criteria without using either of these options.

MAJOR FIELD PILOT LEARNINGS

The major purpose of the field program was to gather information on key biowall design issues: 1.) quantifying the rate of benzene degradation, 2.) achieving adequate oxygen transfer and distribution, and 3.) measuring the rate of vaporization of volatile organic compounds. In all areas, the results are encouraging.

Benzene degradation has been extremely rapid, reaching concentrations under 1 μg/L in less than half the reactor volume. Figure 3 is a plot of benzene concentration in the field pilot unit as a function of travel time (residence time) in the reactor. Over a wide range of ground-water velocities and inlet concentrations, benzene degradation is a strong function of travel time and follows a first-order kinetic relationship. The calculated first-order rate constant[2] has generally been in the range of 10-15 day^{-1}.

The air distribution system, after initial experimentation covering a range of injection points and air rates, has consistently maintained dissolved oxygen (DO) levels in the biowall reactor at or near saturation levels. Ground water entering the biowall is anaerobic, with DO levels less than 0.2 mg/L. To maintain sufficient DO levels throughout the reactor, a multi-point air injection system was

---

[2] Defined by the expression: $C/C_o = e^{-kt}$, where C = concentration, t = residence time, and k = kinetic constant.

needed, each carefully controlled to balance aeration requirements with the need to avoid stripping of volatiles.

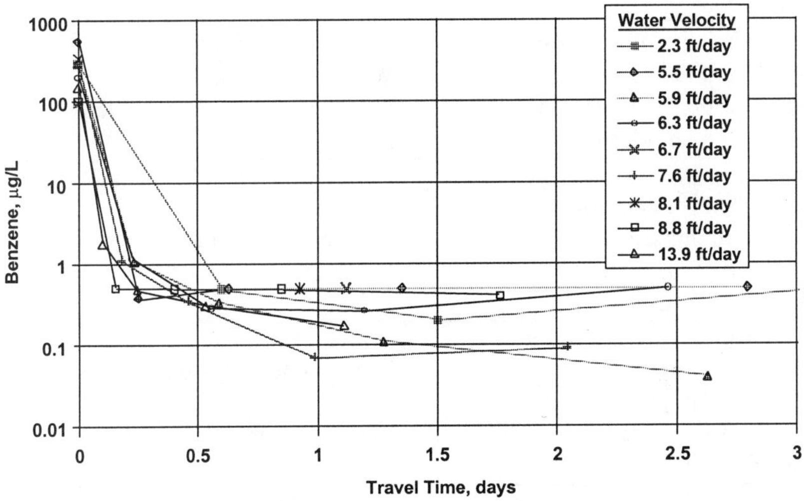
Water Velocity
2.3 ft/day
5.5 ft/day
5.9 ft/day
6.3 ft/day
6.7 ft/day
7.6 ft/day
8.1 ft/day
8.8 ft/day
13.9 ft/day
1000
100
10
1
0.1
0.01
Benzene, µg/L
0
0.5
1
1.5
2
2.5
3
Travel Time, days

FIGURE 3. Benzene degradation in field unit.

Because of the extremely rapid benzene degradation rate initially observed, concerns were raised about the possibility that benzene was merely being vaporized. This concern was addressed through direct vapor measurements, as well as theoretical stripping calculations. In both cases, vaporization of benzene and other volatiles was proven to be negligible. Benzene was never detected in vapor space samples taken above the reactor, and calculations assuming that the air leaving the reactor is at the analytical detection limit indicate that this would account for less than 1% of the benzene in the inlet water.

SCOPING OUT A POTENTIAL BIOWALL PROJECT

The first full-scale, commercial biowall facility is planned for start-up in 2001. The design issues identified for this "pioneer" plant are essentially the same as those that must be resolved for any application. These issues and the types of data required to resolve them are summarized in Table 1.

To ensure that adequate data are available to quantify the design parameters identified in Table 1, the following activities are essential to ensure its success.

Site Characterization. Site characteristics are needed to confirm general biowall applicability, measure hydraulic conductivity, overall heterogeneity, and determine the nature and extent of the contaminant plume. This information will also provide an estimate of ground-water flow rate and direction and assess its variability.

TABLE 1. Biowall design issues.

Type of issue	Data / information required
<u>Hydrogeological</u>	
Speed / direction of ground-water flow	Hydraulic conductivity,
Depth of aquifer	Water table delineation
Volume ground water to be treated	Plume geometry, flow modeling
Soil type and homogeneity	Soil cores
<u>Chemical</u>	
Contaminants of concern	Analytical results
Degradation kinetics	Lab treatability study
Oxygen demand	Chemical loading, treatability study
Need for nutrients	
<u>Regulatory</u>	
Clean-up criteria	Regulations / risk-based limits
Permitting requirements	
Location of sensitive receptors	Historical information, site assessment
Monitoring requirements	
<u>Operational</u>	
Need for LNAPL recovery?	Well data
Long-term site disposition	Business plans
Seasonal changes	Weather, regional ground-water
Availability of water treatment facilities	fluctuations
<u>Construction</u>	
Soil type	Soil borings
Depth of water table	Well data
Depth to confining layer	Soil borings
Location of underground obstructions	Historical information

Laboratory Treatability Studies. Studies are needed to quantify degradation kinetics for contaminants of concern and estimate the bioreactor residence time required. Care must be taken, (through refrigeration, etc.) to ensure that contaminants do not decrease by natural degradation or vaporization during shipment. It may become necessary to spike the ground water being tested with contaminants of concern at the laboratory site.

Ground-water Modeling. Modeling is generally required to estimate the size of the collection area required for plume capture. It can also determine the impact of the biowall system on the existing water table. Ultimately, modeling of extreme cases will set the design rate and bioreactor volume required.

Economic Screening. Economic screening is recommended to confirm that a biowall is the best approach for ground-water remediation and/or containment and to identify the most cost-effective design.

Communication with Appropriate Regulatory Jurisdictions. This is required to determine the clean-up end point and to ensure the acceptance of the biowall concept.

BIOWALL PIONEER PLANT

Because of the exceptional performance of the field pilot facility, the possibility of utilizing the reactor as part of a permanent facility was considered along with other options during economic evaluations. Ultimately, expanding the field pilot unit as shown in Figure 4 proved to be the least expensive option for ground-water treatment. A design feasibility study determined that the pilot unit reaction cell had the capacity to handle expected full-scale ground-water rates and the collection trench could be cost-effectively expanded. Further investigation at the site indicated that most of the plume was located east of the pilot unit, and slightly closer to the reservoir. By installing a 400-foot sheet pile barrier at the leading edge of the plume and a 200-foot collection trench through the center of the plume, the contamination could be contained and remedial goals met. Unlike the idealized layout shown in Figure 2, an infiltration gallery or dispersion zone was added to induce passive, gravity-driven hydraulic flow through the system.

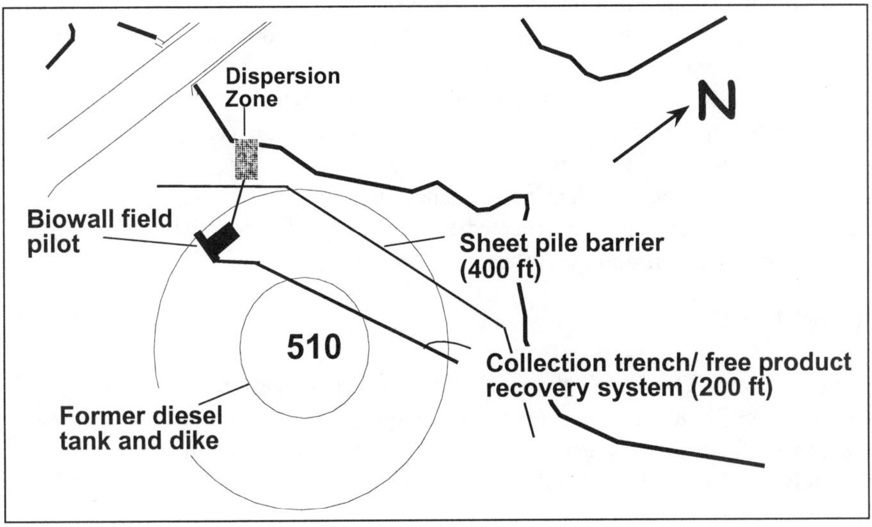
Dispersion Zone
N
Biowall field pilot
Sheet pile barrier (400 ft)
510
Collection trench/ free product recovery system (200 ft)
Former diesel tank and dike

FIGURE 4. Pioneer biowall area.

BIOWALL ECONOMICS

Economic screening results indicate that biowalls can provide 20-50% savings over conventional pump and treat systems. These savings were generated for perimeter treatment of a dissolved hydrocarbon plume several hundred feet wide at a northeast USA refinery location. The study compared three biowall options and four pump and treat options. The biowall cases explored the single trench option, a trench with bioreactor option, and a third case in which an effluent water infiltration trench was added to the bioreactor case. The pump and treat cases looked at two collection options and two treatment options, for a total of four cases. The collection options compared a trench similar to the one used for the biowall with a series of wells manifolded together. The two treatment options considered were sending the extracted water to the refinery waste water treatment plant (WWTP) or treating it onsite in an activated carbon system. For the conditions at this site, the biowall cases were all less costly than the pump and treat cases, with savings ranging from 20-50%. However, all of the cases were sensitive to site conditions and location (relative to the WWTP), making it necessary to reassess these economic comparisons for other sites.

SUMMARY

Overall, biowall technology offers many advantages. By protecting the perimeter of a given property, the biowall approach protects against the most significant risk – offsite migration of contaminants to a sensitive receptor. In addition, biowalls can minimize intrusive investigations and remediations at active facilities, where they might unrealistically hamper current operations.

Biowalls also offer cost incentives over pump and treat alternatives and have minimal long-term operating and maintenance requirements. In addition, with their ability to handle complex hydrogeology, biowalls provide design flexibility to cover a range of site specific conditions.

Biowalls are a direct ground-water treatment option that provides complete NAPL control. If necessary, nutrient addition can be accomplished readily, without significant cost or complexity.

Based on the excellent performance demonstrated over several years of field operation, biowalls are expected to play a significant roll in ground-water remediation at ExxonMobil facilities worldwide.

REFERENCES

Christodoulatos, C., G.P. Korfiatis, N. Pal, and A. Koutsospyros. "In Situ Groundwater Treatment in a Trench Bio-Sparge System", *Journal of Hazardous Waste and Hazardous Materials*, Special Issue on Innovative Soil and Groundwater Remedial Technologies, January 1996.

Starr, R. and J.A. Cherry. 1994. "In Situ Remediation of Contaminated Ground Water: The Funnel-and-Gate System." *Groundwater*. *32*(3): 465-476.

BIODEGRADATION OF A NAPHTHALENE PLUME IN A FUNNEL-AND-GATE™ SYSTEM

Philippe Lamarche (Royal Military College of Canada, Kingston, Ontario, Canada)
François Lauzon (Department of National Defence, Ottawa, Ontario, Canada)
Michel Tétreault (Royal Military College of Canada, Kingston, Ontario, Canada)
James F. Barker (University of Waterloo, Waterloo, Ontario, Canada)

ABSTRACT: A Funnel-and-Gate™ system was installed at Canadian Forces Base Borden to remediate a naphthalene plume in a shallow aquifer. The aim of the study was to investigate the effectiveness of an in-situ bioreactor at remediating the anoxic naphthalene plume under denitrifying conditions. The treatment system consisted of an impervious sheet pile wall intercepting the plume, with a gate in the middle. The gate held four pairs of removable cassettes. Three pairs of cassettes contained a silica sand/granular activated carbon mixture; the other cassettes were filled with nitrate-releasing concrete briquettes. To allow a thorough monitoring of the process, sample tubing was installed at different depths throughout the gate. A number of multilevel wells were also installed upstream and downstream of the gate. For the first phase of the study, nitrate briquettes cassettes were installed upstream of the other three pairs of cassettes. Monitoring results revealed a decline of naphthalene concentrations, that ranged from 1 to 5 mg/L upstream of the gate, to 180 µg/L or less (down to non-detectable levels, < 10 µg/L) downstream. The relative contribution of denitrifiers and aerobes to treatment was not clear, however.

INTRODUCTION

In May 1997, a Funnel-and-Gate™ system was installed in a shallow aquifer at Canadian Forces Base Borden. The system intercepted a naphthalene plume. The intent was to investigate the effectiveness of an in-situ bioreactor to remediate the anoxic naphthalene plume under denitrifying conditions.

The plume was created in 1991 by the University of Waterloo (UW) to investigate natural attenuation processes for complex biodegradable mixtures (King, 1997). A volume of sand containing coal tar creosote was buried below the water table, in a sandy aquifer. The coal tar source was located above a landfill leachate plume that contaminates this aquifer.

The plume has been monitored extensively since its creation (Kerr, 2001, personal communication; King, 1997). As of 1995, most components of the plume seemed to have stabilised or regressed. However, naphthalene continued to advance without any signs of regression. In the core of the plume, naphthalene concentrations above 5 mg/L were measured (King, 1997). The core of the plume had a low level of dissolved oxygen (DO; mean value 0.33 mg/L). The low DO level could be responsible, in part, for the low rates of natural attenuation of the naphthalene. Remediation of the plume was of interest to UW.

Laboratory-scale research on the hydraulic performance of in situ bioreactors in soil has been carried out at the Royal Military College of Canada (RMC) since 1992. Attempting to remediate an existing plume, such as the one at CFB Borden, with an in situ system was of interest to RMC. An in situ bioreactor with addition of electron acceptor to enhance biodegradation of the naphthalene was selected. Nitrate was chosen as electron acceptor, since oxygen addition to water is difficult, and large amounts of DO would have been required to degrade the naphthalene completely.

A laboratory-scale feasibility study was carried out in 1996-97 (Anello, 1997). It showed that naphthalene biodegradation under denitrifying conditions was possible. Measurable biodegradation rates were predicted for an in situ bioreactor 1.2 m long.

The bioreactor was installed in a Funnel-and-Gate ™ system. UW was interested in the hydraulic performance of Funnel-and-Gate™ systems. They provided a research gate for the project. The system design and some monitoring results are presented in this paper.

MATERIALS AND METHODS

The treatment system was installed in two steps. The infrastructure (Funnel-and-Gate™), designed by UW and C^3 Environmental, was installed by C^3 Environmental from May 5 to May 15, 1997. The bioreactor was installed by RMC from July 30 to August 5, 1997. Details of the installation are presented in Lauzon (1998a; 1998b). Modifications made to the system are presented in Lamarche et al. (1999).

Funnel-and-Gate™. The funnel is a low hydraulic conductivity cutoff wall made from Waterloo Barrier™ sheet piling. Waterloo Barrier™ is custom rolled sheet piling with a sealable cavity at the joints. The cavity is sealed with grout. The wall extended a little over 5 m on either side of a rectangular cell 2.3 m wide by 3.5 m long. Sheet piling was installed to a depth of 5.2 m for the wall and 7.9 m for the cell.

The cell was excavated to a depth of 5.5 m. The cassette casing (steel frame that holds the reactor cassettes; described below) was lowered in the excavation (Figure 1). A bentonite seal was created between the casing and cell walls on both sides. Four fully screened 50 mm wells were installed, two upstream and two downstream of the casing. The cell was backfilled with pea gravel (upstream and downstream) and native material (on either side of the casing). The sheet piling on the upstream and downstream face of the cell was then removed, allowing water to flow through the casing.

Reactor. The gate that contained the bioreactor is a research gate; it had features that would not be needed otherwise.

The casing (Figure 1) is a steel box, 0.80 m × 1.95 m × 5 m (width × length × depth), with screened upstream and downstream face. The casing held eight cassettes, each 0.67 m × 0.30 m × 2.43 m, with screened upstream and downstream face. Four cassettes were inserted at the bottom of the gate, the other four on top (Figures 2 and 3).

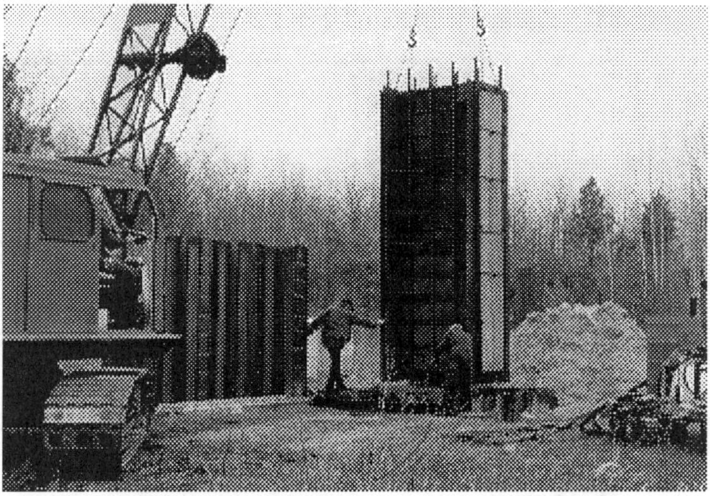

FIGURE 1. Cassette casing lowered into the excavated sheet piling cell.

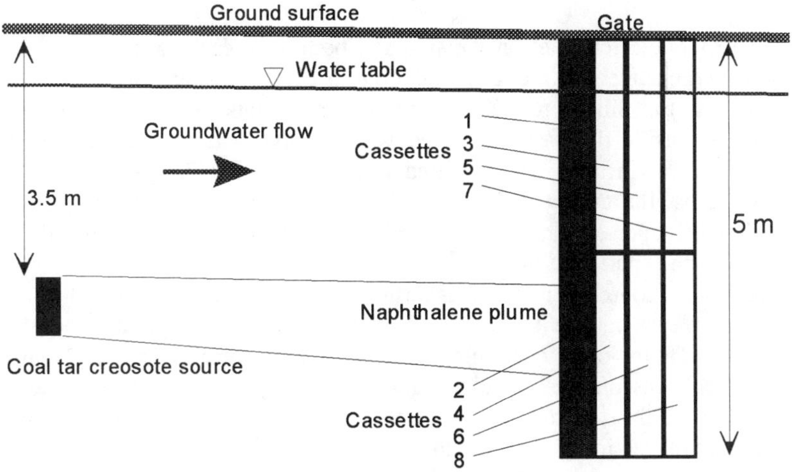
Ground surface
Gate
Water table
Groundwater flow
Cassettes
1
3
5
7
3.5 m
5 m
Naphthalene plume
Coal tar creosote source
Cassettes
2
4
6
8

FIGURE 2. Funnel-and-Gate™ elevation view, and contaminant plume.

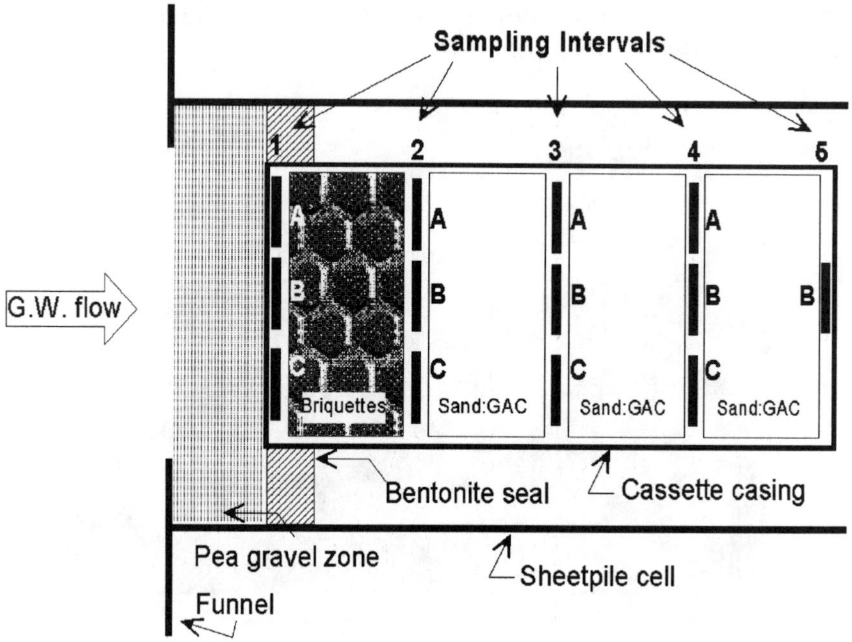
Sampling Intervals
1
2
3
4
5
G.W. flow
A
B
C
Briquettes
Sand:GAC
Sand:GAC
Sand:GAC
Bentonite seal
Cassette casing
Pea gravel zone
Sheetpile cell
Funnel

FIGURE 3. Funnel-and-Gate™ plan view, showing sampling points location.

A rectangular channel on the side and bottom of each cassette fits into a slightly wider channel on the side and bottom of the casing. This creates a rectangular cavity. Silicon tubing was inserted inside this cavity as the cassettes were lowered inside the casing. Fitted with a valve and pressure gauge, the silicon tubing was inflated with distilled water to 52 kPa (7.5 psi), sealing the cassettes against the casing wall. The seal around the two upstream pairs of cassettes failed upon installation; the other two seals held pressure for the remainder of the experiment in 1997.

A passive source of nitrate was desired. Kao and Borden (1997) had used concrete to release nitrate to water. Concrete briquettes (cylinders, 0.15 m diameter × 0.15 m long) containing sodium nitrate were prepared. Lauzon (1998a; 1998b) presents the ingredients and briquettes preparation procedure. On July 14, 1997, 200 briquettes were poured. On August 5, 160 briquettes were peeled off and placed in cassettes 1 and 2 (50% porosity).

The other three pairs of cassettes were filled with a medium-coarse silica sand containing 1% per volume of granular activated carbon (GAC). The GAC was shown to provide preferential attachment sites for the microorganisms, thereby decreasing the reduction in hydraulic conductivity resulting from microbial growth (Anello, 1997). The sand-GAC mix was poured in the cassettes dry (water was pumped out of the gate), to prevent segregation. The middle layer of the three bottom cassettes was bioaugmented with native biomass propagated as described by Anello (1997).

Initial Monitoring System. Detailed monitoring of the system was desired. The steel floor of the top cassettes prevented installation of monitoring points at the same location in the top and bottom cassettes. The reactor monitoring system was inserted in the gap between each pair of cassettes (Figure 3).

Frames were fabricated with perforated steel angle. The steel frame held Teflon PFE® tubing (3.2 mm external diameter) at desired sampling depths (Figure 4). The end of each length of tubing was curved upstream, and covered with nylon screen. Once the frame was inserted in the gate, the tubing ends rested against the effluent face of a cassette (the first sampling "interval" was just at the entrance of the gate; Figure 3). Three sets of 13 sampling points (A, B, and C) were installed at the two edges and the centre of each gap, except in the last gap downstream where gate configuration allowed only the central set of tubing to be installed (Figure 3).

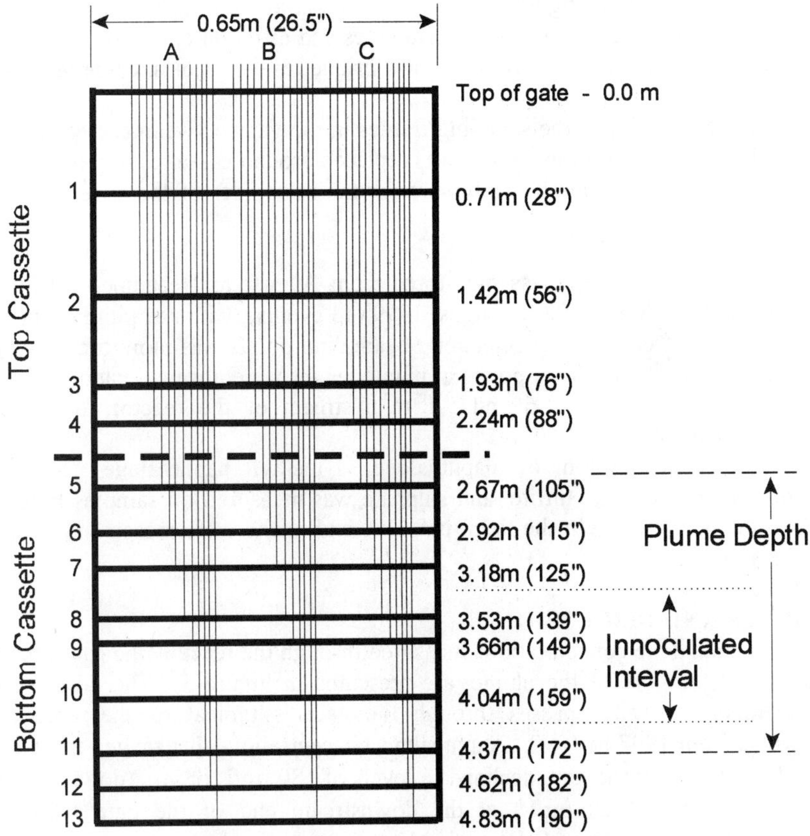
0.65m (26.5")
A
B
C
Top of gate - 0.0 m
Top Cassette
Bottom Cassette
1
0.71m (28")
2
1.42m (56")
3
1.93m (76")
4
2.24m (88")
5
2.67m (105")
6
2.92m (115")
7
3.18m (125")
8
3.53m (139")
9
3.66m (149")
10
4.04m (159")
11
4.37m (172")
12
4.62m (182")
13
4.83m (190")
Plume Depth
Innoculated
Interval

FIGURE 4. Gate multilevel sampling frame.

On December 4, 1997, a multilevel monitoring well was installed in the pea gravel 0.7 m upstream of the gate. Of the nine sampling depths, seven corresponded to sampling depths in the gate, the other two sampling points being deeper. The briquettes cassettes were removed from the ground.

Reactor and Monitoring System Modifications, 1998. In early June 1998, three pairs of multilevel monitoring wells (13 depths each) were installed in front of the gate, 0.5 m, 1.5 m, and 2.5 m upstream of the cutoff wall.

Because the data collected in 1997 suggested that biodegradation may be enhanced upstream of the nitrate briquettes, the gate was also modified (Lamarche et al., 1999). Between June 23 and June 25, 1998, cassettes in positions 1 to 5 were removed and emptied. Cast iron pipes were welded to the floor of cassettes 1 and 5. These pipes provided a conduit through which multilevel monitoring wells were installed into the bottom cassette (plume height). Cassettes in position 1 and 2 were filled with clean silica sand, while cassette 5 was filled with its original sand-GAC mix.

Concrete briquettes (660) had been poured between June 2 and 5. The new briquettes had a slightly different composition and dimensions (0.1 m diameter × 0.1 m long). The smaller briquettes and different composition yielded a higher nitrate release rate. Over 600 new briquettes were put in cassettes 3 and 4 on June 24.

The cassettes and the sampling frames in the gaps between cassettes were re-installed. The silicon tubing seal was replaced around cassettes in positions 1-2 and 5-6. The new seals held pressure for the remainder of the experiment in 1998-99.

Monitoring. The system was monitored from August to December 1997 (19 events), and from May 1998 to August 1999 (13 events), with sampling events in February and March 1999. A selection of sampling points was monitored. In the reactor, the central multilevel points at plume height were always included. Two multilevel wells, located 3.6 and 5.1 m upstream of the reactor, were also monitored regularly.

The concentration of naphthalene, 1-methyl naphthalene, 2-methyl naphthalene, and nitrate, nitrite, and sulphate was measured on samples brought back to the laboratory. In the field, pH, DO, and nitrite were measured at many locations.

RESULTS AND DISCUSSION

Naphthalene concentrations measured through the reactor (the gate), at the elevation of the core of the plume, are presented in Figure 5. The decrease in concentration as water passes through the gate is typical of measurements performed from 1997 to 1999. Naphthalene concentrations that ranged from 1 to 5 mg/L upstream of the gate declined to levels of 180 µg/L or less (down to non-detectable levels, < 10 µg/L) at the downstream end of the gate. Similar observations were recorded for the other organics measured too.

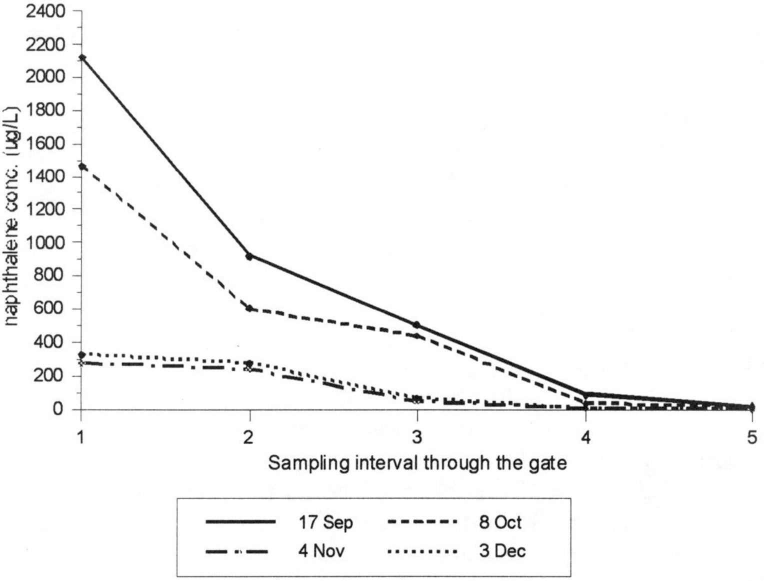
2400
2200
2000
1800
1600
1400
1200
1000
800
600
400
200
0
naphthalene conc. (µg/L)
1
2
3
4
5
Sampling interval through the gate
17 Sep
8 Oct
4 Nov
3 Dec

FIGURE 5. Naphthalene concentrations through the gate, at plume depth, 1997. The location of sampling intervals 1 to 5 is shown in Figure 3.

The occurrence of in situ biodegradation of any contaminant can be inferred from the following observations (Madsen, 1991; National Research Council, 1993; Seagren et al., 1996):

- Loss of contaminant in one or all phases (decrease of concentration in excess of hydrodynamic dilution);
- Associated decrease in nutrients and electron acceptors;
- Increase in total inorganic carbon (TIC; due to CO_2 production) or metabolites concentration
- Laboratory evidence that microbes present on site are able to degrade the contaminant;
- Field evidence of microbial distribution that correlates with contaminant distribution.

All these observations were made in the present study. The loss of contaminant (naphthalene, others) is shown by the data presented in this paper and elsewhere (Lauzon, 1998a; 1998b; Lamarche et al., 1999). Lauzon (1998a; 1998b) presents laboratory and field analyses results that show the loss of electron acceptor (nitrate) and an increase in nitrite (a denitrification intermediate) concentrations. The laboratory study presented by Anello (1997) established that microbes that can degrade the contaminant existed on site. Finally, microcosm studies carried out by the Biotechnology Research Institute (BRI) and Biological Activity Reaction Tests (BART™) presented evidence of a microbial distribution that correlates with the contaminant disappearance.

An active population of denitrifiers was found on site. Whether nitrate was used as nitrogen source or as electron acceptor was not clear from the data collected in 1997. Data collected since is being analysed to determine the role of the nitrate.

CONCLUSIONS

The data obtained present clear evidence of the fact that in situ biodegradation has been enhanced by the addition of nitrate, and that some denitrifying activity occurred in the reactor. It is not clear whether or not the nitrate was used mostly as electron acceptor or as a nitrogen source. However, nitrate-releasing concrete briquettes can be used to enhance in situ biodegradation.

ACKNOWLEDGMENTS

Students and staff at the University of Waterloo, C^3 Environmental, and the Royal Military College of Canada greatly contributed to the positive outcome of this research. Funding and encouragement was provided by numerous agencies and firms, including: the Department of National Defence (Canada), Shell Research (UK), National Science & Engineering Research Council (Canada), Beazer East, Geomatrix Consultants, Blasland, Bouck & Lee Inc., and Key Environmental.

REFERENCES

Anello, G. 1997. "The Study of Naphthalene Degradation under Aerobic and Denitrifying Conditions in a Bioreactor." M.Eng. Thesis, Royal Military College of Canada, Kingston, ON.

Kao, C., and Borden, R.C. 1997. "Enhanced TEX Biodegradation in Nutrient-peat Barrier System." *Journal of Environmental Engineering. 123*(1): 18-24.

King, M.W. 1997. "Migration and Natural Fate of a Coal Tar Creosote Plume." Ph.D. Thesis, University of Waterloo, Waterloo, ON.

Lamarche, P., J.A. Héroux, R. Caldwell, and D. Eagles. 1999. *Enhanced In-situ Biodegradation and Biocurtain Development*. Military Engineering Research Group report, MERG 13-99, Royal Military College of Canada, Kingston, ON.

Lauzon, F. 1998a. "In-situ Biodegradation of a Naphthalene Plume in a Funnel-and-gate™ System." M.Eng. Thesis, Royal Military College of Canada, Kingston, ON.

Lauzon, F. 1998b. "In-situ Biodegradation of a Naphthalene Plume in a Funnel-and-gate™ System." Department of Civil Engineering Research Report, Royal Military College of Canada, Kingston, ON.

Madsen, E.L. 1991. "Determining In Situ Biodegradation." *Environmental Science and Technology*. *25*: 1663-1673.

National Research Council. 1993. *In Situ Bioremediation: When Does it Work?* National Academies Press, Washington, DC.

Seagren, E.A., D.J. Holander, D.A. Stahl, and B.E. Rittman. 1996. "Innovative Evaluation Methods for Bioremediation." *Proceedings of the Conference on Non-aqueous Phase Liquids (NAPLs) in the Subsurface Environment: Assessment and Remediation.* ASCE, Washington, DC.

BIOLOGICAL ACTIVATED CARBON BARRIERS FOR THE REMOVAL OF CHLOROORGANICS/BTEX MIXTURES

A. Tiehm, M. Gozan, A. Müller, K. Böckle, H. Schell
(Water Technology Center, Karlsruhe, Germany)
H. Lorbeer, P. Werner (TU Dresden, Pirna, Germany)

ABSTRACT:. In our study, the combination of pollutant adsorption on granular activated carbon (GAC) and biodegradation was examined in order to develop a long lasting biobarrier in the subsurface. The groundwater plume, located in an abandoned East German chemical production area, is contaminated with a pollutant cocktail containing chloroethenes, chlorobenzene, and benzene-toluene-ethylbenzene-xylene (BTEX). Based on the pre-investigations, a sequential anaerobic/aerobic treatment is applied in order to achieve complete pollutant mineralization resulting in permanent bioregeneration of the GAC barriers. This paper addresses laboratory experiments on the sequential anaerobic/aerobic operation of the biological activated carbon barriers. Pollutant elimination by simultaneous adsorption and biodegradation, by adsorption only, and by biodegradation only was evaluated. Results confirmed the biobarrier concept. It was demonstrated that GAC loaded with trichloroethene (TCE) was completely regenerated by biodegradation. The elimination of TCE, monochlorobenzene (MCB), and benzene (BZ) was most pronounced in case of the simultaneous adsorption/biodegradation process.

INTRODUCTION

Chlorinated hydrocarbons and aromatic hydrocarbons like benzene, toluene, ethylbenzene, and xylene (BTEX) represent pollutants very often encountered in groundwater. These hazardous pollutants are normally eliminated by pump-and-treat systems which require high operating costs (Böckle and Werner, 1997). Passive techniques such as subsurface reactive walls, that utilise the natural groundwater flow, emerge as an alternative to the classical remediation procedures. Driven by the natural groundwater gradient, the groundwater transports the pollutants through reactive zones where elimination occurs.

The anaerobic cometabolic biodegradation of trichloroethene (TCE) via *cis*-dichloroethene (*cis*-DCE) and vinyl chloride (VC) to ethene is well known. The dechlorination rates of TCE to *cis*-DCE are faster than *cis*-DCE to VC and ethene (Middeldorp et al., 1999). The lower chlorinated metabolites, *cis*-DCE and VC, are biodegradable also under aerobic conditions. Both under anaerobic (Middeldorp et al., 1999) and aerobic (Freedman et al., 2000; Gao and Skeen, 1999; Schäfer and Bouwer, 2000) conditions, auxiliary substrates are required for the dechlorination of the chloroethenes. Monochlorobenzene (MCB) and benzene (BZ) are degraded without auxiliary substrates, but more rapidly in the presence of molecular oxygen. Therefore, a sequential anaerobic/aerobic treatment is applied in order to mineralise the pollutant mixture.

Objective. The research goal was to develop a long lasting Granular Activated Carbon (GAC) biobarrier. Such biobarrier systems, wherein pollutant adsorption and biodegradation occur simultaneously, exhibit the following advantages:

(i) The GAC is continuously regenerated resulting in a longer operation period as compared to elimination by adsorption only,

(ii) The pollutants and microorganisms accumulate in a small volume, where the processes are easier to control and to stimulate, as compared to the aquifer without the reactive wall.

Site Description. The site is located in Bitterfeld, Germany, in a former chemical production area. The groundwater is contaminated with a pollutant cocktail containing chloroethenes, chlorobenzenes, and BTEX.

MATERIALS AND METHODS

A laboratory scale experiment (Figure 1) consisting of 4 sets of sequential anaerobic and aerobic operation units was built to study adsorption and biodegradation of the model contaminants. A set of 2 vertical 0.164 L columns (diameter 3.5 cm and height 17 cm) was filled with Filtrasorb GAC TL 830 (Chemviron Carbon, Belgium). The GAC was washed in demineralized water to remove fines, followed by drying at 105°C. A similar set was filled with granular pumice stones (GPS) Type S, 4mm (Pumex, Italy). Two reference sets (GAC and GPS) were operated under sterile conditions.

After the pre-loading period, bacterial mixed cultures capable of utilising TCE, BZ, and MCB were injected into columns 1 and 3 (active columns). In the sterile columns (Set 2 and 4), NaN_3 was added continuously to hinder the growth of microorganisms. During the loading period, the flowrate to GAC columns varied from 50 to 90 L/d. For the bioregeneration experiment, the flowrate (medium and substrate) to each GAC and GPS column was adjusted to approximately 4 L per day. The pore volumes of the GAC and GPS columns were ± 57% and 64%, respectively.

Substrate and Medium. Mineral medium (158.5 mg KH_2PO_4, 22.5 mg $(NH_4)_2HPO_4$, 10 mg $MgHPO_4$ x $3H_2O$, 13.8 mg $NaHCO_3$, 14.8 mg Na_2SO_4 per litre, and trace elements of $FeSO_4$, $MnSO_4$, $CoCl_2$, $ZnSO_4$, $CaCl_2$, H_3BO_3, $NaMoO_4$, H_2SO_4 Na_2WO_4, $NaSeO_3$ and $NiCl_2$) flowed through a 30 m silicon tube (in Na_2SO_3 liquid) to eliminate dissolved oxygen (Arcangeli and Arvin, 1995) before entering a mixing bottle together with contaminant model.

The contaminant model consisted of TCE, MCB and BZ in volume ratio of 21.8:44.0:34.2 and was introduced to a mixing bottle at the rate of 0.03 mL/h. All the contaminants were products of Fluka Co. and had a purity ≥ 99.5%. Organic substrates were added to favour reductive dechlorination of the chloroethenes in the anaerobic zone. These auxiliary substrates were ethanol (pa ≥ 99.8%) and saccharose (microbiology purpose). H_2O_2 (30%) and $NaNO_3$ (≥ 98%), were added to stimulate biodegradation under aerobic-denitrifying conditions. The auxiliary substrates, H_2O_2, and $NaNO_3$ are products of Merck Co.

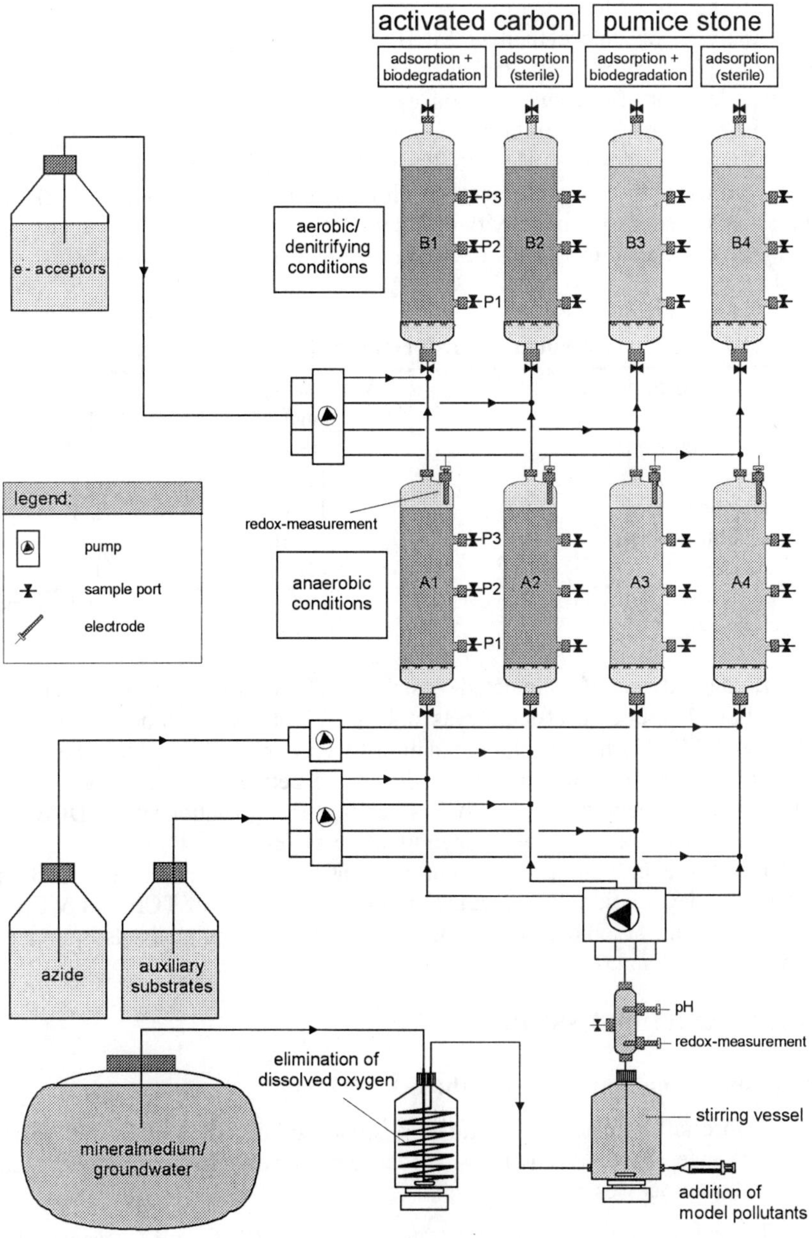
activated carbon
pumice stone
adsorption + biodegradation
adsorption (sterile)
adsorption + biodegradation
adsorption (sterile)
e - acceptors
aerobic/ denitrifying conditions
B1
B2
B3
B4
P3
P2
P1
legend:
pump
sample port
electrode
redox-measurement
anaerobic conditions
A1
A2
A3
A4
azide
auxiliary substrates
pH
redox-measurement
elimination of dissolved oxygen
stirring vessel
mineralmedium/ groundwater
addition of model pollutants

FIGURE 1. Experimental Set-up

Analyses. Measurement of redox potential and pH was done regularly with a multimeter (pH 91, WTW Co.). Determination of O_2 content was carried out with a dissolved O_2 meter (Oxi 330, WTW Co.). H_2O_2 level was monitored with peroxide test strips (Merck Co.).

Analysis of chloroethenes, MCB, and BZ was performed by a gas chromatograph from Hewlett Packard (Series II 5890) equipped with a flame ionisation detector (FID) and electron capture detector (ECD). Separation was accomplished in a 50-m capillary column (PONA, id 0.21 mm, methyl silicon film with 0.5 µm thickness, Hewlett Packard Co.). Ionic chloride concentration was determined with an ion chromatograph from Dionex Co. (model 2010I) with suppression system and conductivity detector.

Physicochemical properties of MCB, BZ, and the chloroethenes are shown in Table 1.

TABLE 1. Physico-chemical properties of the groundwater pollutants.

Compound	Water solubility [mg/L]	Log K_{ow}[a] [-]
Monochlorobenzene	484[b]	2.8[b]
Benzene	1780[b]	2.1[b]
Trichloroethene	1000[c]	2.3[c]
1,2 *cis*-Dichloroethene	800-3500[c]	1.5[c]
Vinyl chloride	1100[d]	1.4[d]

[a]K_{ow} = octanol-water partition coefficient at 25 °C, [b]McLeod & Mackay, 1999; [c]Ritter & Werner, 1991;[d]Scherb, 1978

Biodegradation is calculated based on Cl^- released. For the anaerobic column, the Cl^- found in effluent was only produced by the biodegradation of TCE to cis-DCE. No further biodegradation to vinyl chloride nor to ethene was found. For the aerobic column, the Cl^- difference between influent and effluent stand from the biodegradation of MCB since no biodegradation of *cis*-DCE was observed. Adsorption of the contaminant was calculated by taking into consideration the influent and effluent concentration of contaminants and the calculated biodegradation. The accumulated biodegradation of TCE and MCB in the anaerobic and aerobic columns, as well as the accumulated adsorption are shown in Figure 4 and 6.

RESULTS AND DISCUSSION

Adsorption of Contaminants onto the GAC column.

Before adding bacterial cultures, a stream of BZ and MCB was passed through the anaerobic GAC columns in order to pre-load the columns and to study the adsorption processes.

The curves in Figure 2 demonstrate that more MCB than BZ was adsorbed on GAC. Breakthrough of BZ was observed earlier after 190 L, followed by MCB after 400 L of throughflow. After 1,000 L of contaminant stream, effluent concentrations of both pollutants were constant. Most of the adsorption sites of the GAC were fully engaged. The amounts of MCB and BZ adsorbed were 324.8 mmol and 28.4 mmol, respectively.

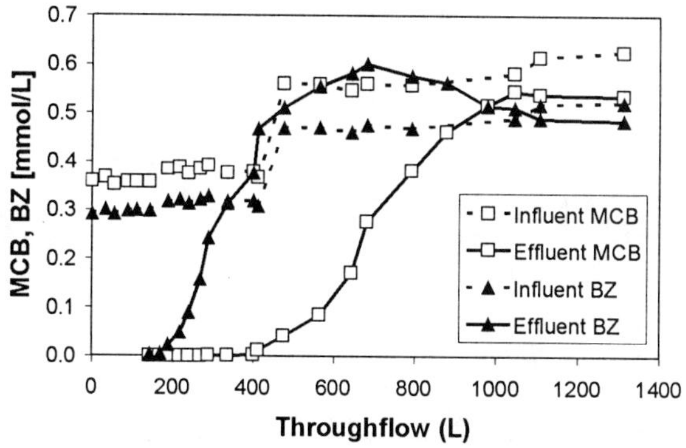

MCB, BZ [mmol/L]
Throughflow (L)
Influent MCB
Effluent MCB
Influent BZ
Effluent BZ

FIGURE 2. Breakthrough of MCB and BZ through GAC.

MCB has a lower water solubility and higher log K_{ow} than BZ (Table 1). Therefore, MCB displaced BZ during the loading period. This finding is consistent with adsorption isotherms (Tiehm et al., 2000) that also demonstrate the better adsorption of the more hydrophobic pollutants to GAC.

Biobarrier: Anaerobic Stage.

After pre-loading the anaerobic GAC columns with MCB and BZ, TCE was added into the contaminant model solution, which streamed through the 4 column sets. After an additional 37 days of loading, inoculation of the anaerobic columns (active GAC and GPS) was done with a mixed culture dechlorinating TCE.

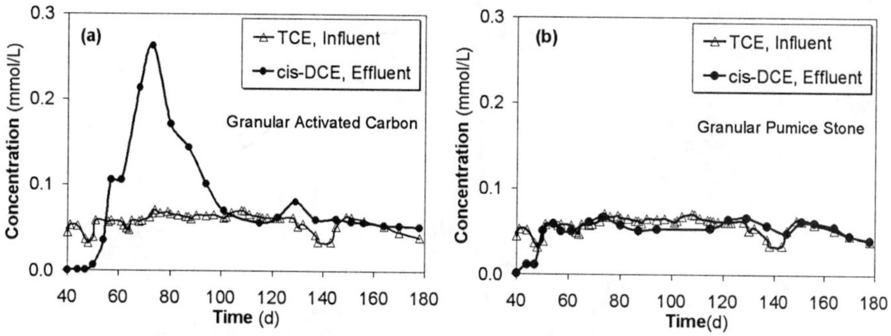

(a)
TCE, Influent
cis-DCE, Effluent
Granular Activated Carbon
Concentration (mmol/L)
Time (d)
(b)
TCE, Influent
cis-DCE, Effluent
Granular Pumice Stone
Concentration (mmol/L)
Time(d)

FIGURE 3. TCE and *cis*-DCE in the anaerobic zone of the (a) GAC and (b) GPS active columns.

Microbial dechlorination of TCE resulted in the formation of *cis*-DCE in both columns (Figure 3). In the GAC column, the effluent concentration of *cis*-DCE significantly exceeded the influent concentration for about 40 d (Figure 3a). This demonstrates a sufficient bioavailability of the TCE initially adsorbed by GAC. On the contrary, the influent TCE was stoichiometrically transformed to *cis*-DCE in the GPS column (Figure 3b).

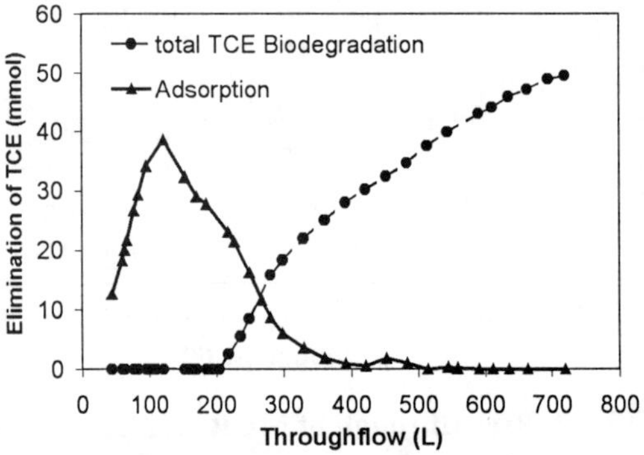
total TCE Biodegradation
Adsorption
Elimination of TCE (mmol)
0
10
20
30
40
50
60
0
100
200
300
400
500
600
700
800
Throughflow (L)

FIGURE 4. Bioregeneration of TCE in the GAC anaerobic column.

The GPS column was suitable for microbial colonization, but not for the temporary accumulation of TCE. Figure 4 illustrates the periods of TCE retardation and bioregeneration in the active GAC column.

Biobarrier: Aerobic Stage.

The aerobic stage is aimed to eliminate MCB and BZ as well as to further decompose the metabolites from TCE degradation. H_2O_2 and nitrate were added as electron acceptors. In order to avoid toxic effects, the concentration of H_2O_2 was increased gradually from 10 mg/L up to 100 mg/L.

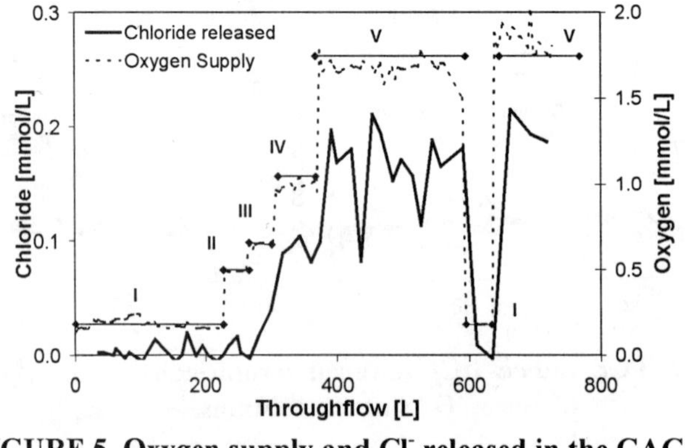
Chloride released
Oxygen Supply
Chloride [mmol/L]
Oxygen [mmol/L]
I
II
III
IV
V
V
I
0.0
0.1
0.2
0.3
0.0
0.5
1.0
1.5
2.0
0
200
400
600
800
Throughflow [L]

FIGURE 5. Oxygen supply and Cl^- released in the GAC aerobic column. The notations I, II, III, IV, and V indicate the O_2 level from 10, 30, 40, 60, and 100 mg H_2O_2/L.

At low concentrations of H_2O_2/oxygen, no biodegradation of MCB, *cis*-DCE, nor BZ was observed. Obviously, the remaining organic compounds, added in the anaerobic column as auxiliary substrates, consumed most of the available oxygen. Aerobic dechlorination started after increasing the H_2O_2 level to 40 mg/L. Further increments to 100 mg/L H_2O_2 clearly increased the formation of chloride due to MCB degradation. The short decrease of hydrogen peroxide after 600 L throughput resulted in an immediate drop of the MCB degradation, thus clearly demonstrating the important role of oxygen (Figure 5).

The mass balance with regard to MCB in the aerobic column is depicted in Figure 6. Most of the MCB was eliminated by adsorption on GAC. Bioregeneration of the loaded GAC started after 300 L throughput when H_2O_2 addition was increased. However, the aerobic dechlorination processes were limited by oxygen availability since the bioregeneration of MCB loaded GAC occurred only slowly. Furthermore, aerobic degradation of *cis*-DCE in the aerobic GAC column was not observed although a degrading mixed culture was enriched and inoculated (data not shown).

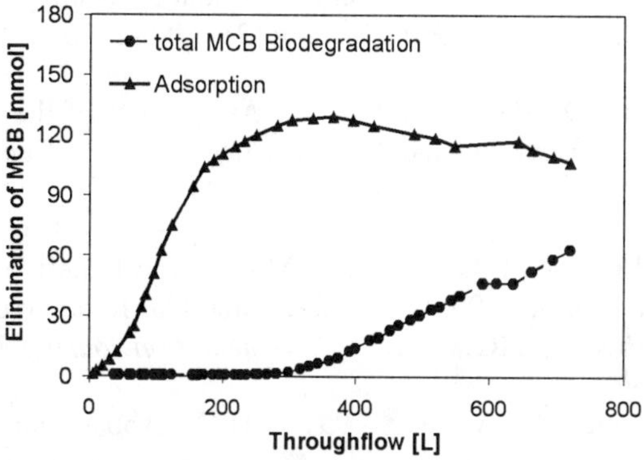
180
150
120
90
60
30
0
Elimination of MCB [mmol]
total MCB Biodegradation
Adsorption
0
200
400
600
800
Throughflow [L]

FIGURE 6. Bioregeneration of MCB in the GAC aerobic column.

A mass balance of BZ revealed that the overall elimination in the microbiologically active GAC columns (26%) was higher than in the sterile columns (10%). This might be due to biodegradation of BZ in the aerobic column, or to a better adsorption possible due to GAC bioregeneration by TCE and MCB degradation.

CONCLUSION

As a conclusion, results so far demonstrate the feasibility of the sequential anaerobic/aerobic biological activated carbon barrier concept. Present work focuses on the optimization and reduction of auxiliary substrates and on process implementation in the field.

ACKNOWLEDGEMENTS

The study is part of the SAFIRA research programme and is funded by the German Ministry of Education and Research (BMBF No. 02WT9937/7). Misri

Gozan gratefully acknowledges a scholarship of the German Academic Exchange Service (DAAD).

REFERENCES

Arcangeli, J.P. and E. Arvin. 1995. "A Membrane Deoxygenator for Study of Anoxic Process. A Rapid Communication.", *Wat. Res.* *29*(9): 2220-2222.

Böckle, K. and P. Werner. 1997. "Microbial Regeneration of Activated Car-bon Loaded with Volatile Chlorinated Hydrocarbons (VCH)." In *4th International In Situ and On-Site Bioremediation Symposium* *4*(5): 97-101. Louisiana.

Freedman, D.L, A.S. Danko and M.F. Verce. 2000. "Substrate Interactions During Aerobic Biodegradation of Methane, Ethene, Vinyl Chloride and 1,2-Dichloroethenes." In *Pre-print Book 2 of 1st World Water Congress of the International Water Association*, pp. 461-468. Paris.

Gao, J. and R.S. Skeen. 1999. "Glucose-induced Biodegradation of *cis*-DCE under Aerobic Conditions." *Wat. Res.*, *33*(12): 2789-2796.

McLeod, M. and D. Mackay. 1999. "An Assessment of the Environmental Fate and Exposure of Benzene and the Chlorobenzenes in Canada." *Chemosphere* *38*(8): 1777-1796.

Middeldorp, P.J.M., M.L.G.C. Luijten, M.H.A. van Eekert, S.W.M. Kengen, G. Schraa, A.J.M. Stams. 1999. "Anaerobic Microbial Reductive Dehalogenation of Chlorinated Ethenes (A REVIEW)." *Bioremediation Journal* *3*(3): 151-169.

Ritter, R.A. and P. Werner. 1991. "HKW-Abbau im Boden." *Ecomed, Landsber/Lech.*

Schäfer, A. and E.J. Bouwer. 2000."Toluene Induced Cometabolism of *cis*-DCE and VC under Conditions Expected Downgradient of a Permeable Fe(0) Barrier." *Wat. Res.* *34*(13): 3391-3399.

Scherb, K. 1978. "Untersuchungen zur Ausdampfung einiger niedermolekula-rer Chlorkohlenwasserstoffe aus einem Flußgerinne." In *Münchner Beiträge zur Abwasser-, Fisch- Flußbiologie (Schadstoffe, Oberflächenwasser, Abwasser)* 30: 235-248.

Tiehm, A., S. Schulze, K. Böckle, A. Müller, H. Lorbeer and P. Werner. 2000. "Elimination of Chloroorganics in a Reactive Wall System by Biodegradation on Activated Carbon." In *Proceedings of the 7th FZK/TNO Conference on Contaminated Soil* 2: 924-931. Leipzig, Germany.

BIOBARRIER COMPRISED OF SOIL AND BAC: SUPPRESSION OF GREENHOUSE GASES

Yutaka Sakakibara (Waseda University, Tokyo, Japan)
Daisuke Kamimura (Gunma University, Gunma, Japan)

ABSTRACT: The feasibility of a biobarrier for suppressing diffused emissions of CH_4 and N_2O (greenhouse gases) from landfill sites or other pollution sources was investigated using laboratory-scale columns packed with soil and biological activated carbon (BAC) with different ratios. A gas mixture (10 % CH_4, 1 % N_2O, and 89 % atmospheric air) was fed into the bottom of the columns; and measurements and comparisons were made for efflux gas compositions under long-term and short-term gas loading conditions. Experimental results showed that below gas loading rates of 100 mmol-CH_4/m^2d and 10 mmol-N_2O/m^2d, the greenhouse gases were effectively decomposed or removed in the columns; thereby, the net global warming potentials (GWPs) were reduced below 10 % of the feed gas. This result and comparisons under different BAC ratios showed that the BAC/soil mixture may serve as a promising biobarrier to suppress the emission of CH_4 and N_2O. Furthermore, experimental data taken under simulated temporal variations in gas fluxes demonstrated that CH_4 was first adsorbed onto BAC/soil mixture and then gradually oxidized to CO_2. This simultaneous adsorption and biodegradation process is considered very important in functioning as effective barrier under natural conditions. For large emission fluxes above 500 mmol-CH_4/m^2d, produced methane should be recovered as an energy source because of its significant amount of combustion energy.

INTRODUCTION

Methane and nitrous oxide are greenhouse gases, which have extremely large global warming potentials (GWPs) in comparison with carbon dioxide (see Table 1). In addition, these gases are produced from various diffused sources, such as landfill sites, farmlands, wetlands, etc. Field investigations show that the

TABLE 1. Global warming potential (η_i) compared with CO_2.

Gas	Warming Effect (℃/ppb)	η_i (mol of gas/mol of CO_2)
CO_2	0.00001	1.0
CH_4	0.0002	20[1] ~ 30[2]
N_2O	0.001	200[2] ~ 280[1]

[1]Projected value for the next 20 years by IPCC.
[2]Projected value for 1980 to 2030 by Ramanathan et al. (1985).

emission fluxes of these gases were dynamically changed with time in accordance with atmospheric variations such as pressure changes (Young, 1990).

This study investigates the feasibility of a biobarrier for suppressing the emissions of these greenhouse gases from landfills or other pollution sites. The biobarrier is comprised of soil and biological activated carbon (BAC) at different ratios. Laboratory-scale column experiments were conducted at different gas loading rates (or emission fluxes), and the effects of the loading and the ratio of BAC content in columns on the suppression performances were investigated. GWP was calculated to evaluate the net performance.

Objective. The primary objective of this study is to determine the feasibility of the proposed biobarrier, which was comprised of BAC and soils. A criterion of 90 % of GWP reduction was selected to demonstrate the effective barrier performance.

MATERIALS AND METHODS

Experimental Setup. The laboratory-scale experimental apparatus used in this study is illustrated in Figure 1. Soil and BAC were packed into stainless steel columns at different BAC volume ratios. Column volume and inside diameter were about 0.3 L and 4 cm, respectively. The ratios of BAC were in the range of 0 to 50 vol %. Two identical columns were prepared for each ratio, and one column was autoclaved to estimate adsorption amounts of greenhouse gases.

A gas mixture, composed of 10% CH_4, 1% N_2O, and 89% atmospheric air, was used in this study. This gas mixture was fed into the bottom of the columns in two different gas loading modes: long-term and short-term loadings. In long-term loading, a fixed amount of the gas mixture was fed every day in the range of 0.25 to 5.1 mol/m^{2}d; whereas in the short-term loading, the gas loading rate was instantaneously increased for 12 hours. In each experiment, measurements were made for volumes and compositions of efflux gases emitted from the columns. Experimental apparatuses were placed in a room at 20 ±5 ℃. Details of the experimental conditions and procedures were shown elsewhere (Sakakibara et al., 1999).

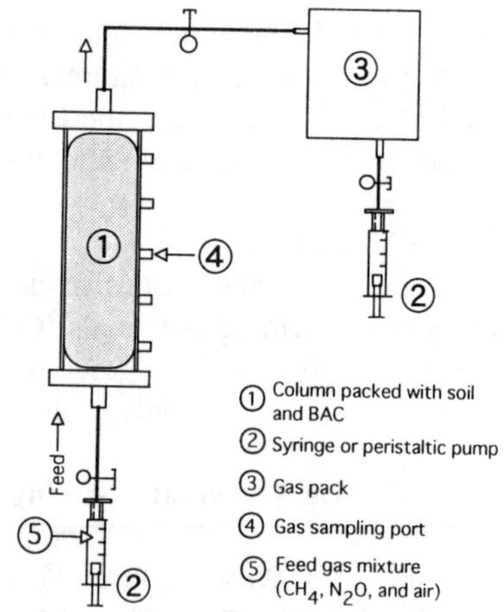
③
①
④
②
Feed
⑤
②
① Column packed with soil and BAC
② Syringe or peristaltic pump
③ Gas pack
④ Gas sampling port
⑤ Feed gas mixture (CH_4, N_2O, and air)

FIGURE 1. Illustration of laboratory-scale experimental apparatus.

BAC and Soil. Granular activated carbon (GAC) was submerged in an activated sludge reactor treating a synthetic organic wastewater. After one month of cultivation, GAC was taken out and used as BAC in this study. Adsorption capacities of BAC were about 1/2 to 1/3 of the original capacities for CH_4 and N_2O. Soil was obtained from a plot of forest in Gunma University.

Sampling and Analysis/Measurement. Gas composition was analyzed with TCD gas chromatograph (Shimadzu GC-8A) equipped with packed columns of activated carbon and molecular sieves. Helium gas was used as carrier gas.

RESULTS AND DISCUSSION

Composition of Efflux Gas. Figure 2 is a typical result of long-term loadings, where time course changes in efflux gas compositions are shown. Under the gas loading rates in this study, the compositions of CH_4, N_2O, and O_2 in the efflux gas were always lower than those of feed gas. Higher compositions of CO_2 mean it is produced in the columns. As loading rates were increased, the compositions of CH_4, N_2O, and O_2 in the efflux gas increased, whereas CO_2 decreased except in the initial phase of the experiment. These results demonstrated that greenhouse gases were removed for a long-term loading by the BAC/Soil mixtures. The CO_2 production as well as O_2 consumption suggests biological reaction in the column. In autoclaved columns, the extent of CO_2 production and O_2 utilization was very low.

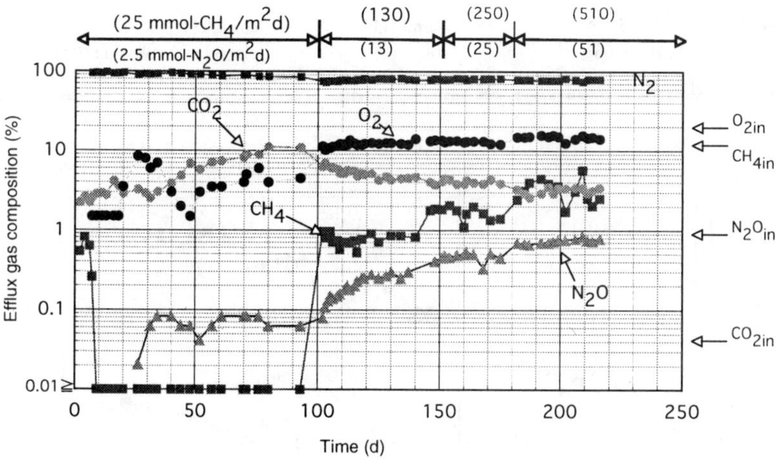
(25 mmol-CH4/m2d)
(2.5 mmol-N2O/m2d)
(130)
(13)
(250)
(25)
(510)
(51)
100
10
1
0.1
0.01≧
0
50
100
150
200
250
Time (d)
Efflux gas composition (%)
N2
CO2
O2
CH4
N2O
O2in
CH4in
N2Oin
CO2in

FIGURE 2. Time course changes in efflux gas compositions in a column packed with 50% BAC.

Reduction of GWP. The reduction level of GWP, ϕ, of efflux gases after passing the column is calculated by:

$$\phi = \frac{Q_{out}(\Sigma \eta_i C_{out}|_i + C_{out}|_{CO_2})}{\Sigma Q_{in} \eta_i C_{in}|_i} \cdot 100 \quad (\%) \qquad (1)$$

where η_i = global warming potential compared to CO_2 (Table 1).
$C_{in}|_i$ and $C_{out}|_i$= concentration of *i*-th component of gas mixture.
Q_{in} and Q_{out} = influx and efflux flow rate of gas.

Figure 3 shows the reduction of GWP, ϕ. This ϕ increased with increasing gas loadings, but more than 90 % of the GWP was reduced when the CH_4 and N_2O loadings were less than about 100 and 10 mmol/m^{2}d, respectively. Table 2 lists emission fluxes reported at various sites. Comparison of the actual emission fluxes of gases from various sources (Table 2) with the experimental results in Figure 3 indicates that the soil and BAC mixture may be useful as an effective biobarrier to suppress the emission of greenhouse gases at various sites.

TABLE 2. Emission flux of CH_4 and N_2O at various sites.

Gas	Flux (mmol/m^{2}d)	Source	Reference
CH_4	6～14,000	Landfill	Jones and Nedwell (1990)
〃	24～480	Landfill	Borjesson and Svensson (1997)
〃	2～100	Paddy soils	Kimura et al. (1991)
〃	1～16	Wetlands	Aselmann & Crutzen (1989)
〃	25～510	—	Loading in this study
N_2O	0.1～5	Farmland	Hosomi (1992)
〃	0.01～0.06	soil	Minami and Fukushi (1984)
〃	2.5～51	—	Loading in this study

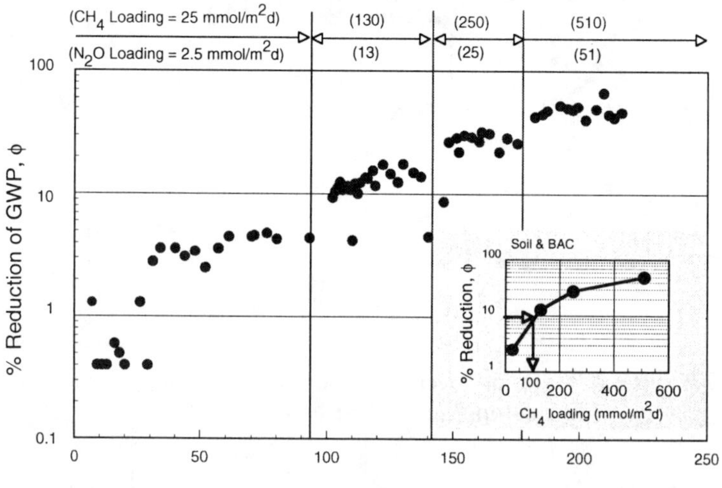
(CH4 Loading = 25 mmol/m2d)
(N2O Loading = 2.5 mmol/m2d)
(130)
(13)
(250)
(25)
(510)
(51)
% Reduction of GWP, φ
Time (day)
Soil & BAC
% Reduction, φ
CH4 loading (mmol/m2d)

FIGURE 3. Reduction of GWP levels at different emission fluxes.

When gas loadings or gas emission fluxes became very large, sufficient reductions in GWP would not be expected. For example, the reduction level was about 30 % or less at a loading of 500 mmol-CH_4/m^2d. However, this methane flux is equivalent to 100 m^3-CH_4/d for 1 ha, which is further equivalent to the energy consumption of 100 households in Japan (Tokyo Gas, 1998). It is noted that if methane fluxes are very large in comparison with the barrier performances, the produced methane should be recovered as energy sources.

Long-Term Loadings and BAC ratios. Figure 4(A) compares steady-state removal rates of CH_4, N_2O, and O_2 as well as the production rate of CO_2 at different gas loadings. These rates were obtained from mass balances for the influx and efflux gases at steady state conditions. In addition, CH_4 decomposition and O_2 utilization rates were obtained from differences between the corresponding rates in biotic and abiotic (autoclaved) columns. From Figure 4(A), it can be seen that for different loading conditions, the CO_2 production rates were nearly equal to CH_4 decomposition rates, and are half of the O_2 utilization rates. This result indicates that the following biological reaction takes place in columns.

$$CH_4 + 2O_2 \rightarrow CO_2 + 2H_2O \qquad (2)$$

Figure 4(B) shows the effect of BAC ratios on the O_2 utilization, CH_4 decomposition, and N_2O removal rates. These rates tended to increase with increasing BAC ratios, but they were nearly constant when BAC became larger than 20 to 30 % in the columns. N_2O was almost completely removed by the BAC/soil mixture. Biodegradation or denitrification of N_2O was not evaluated in this study because of the large amount of N_2 in feed gas.

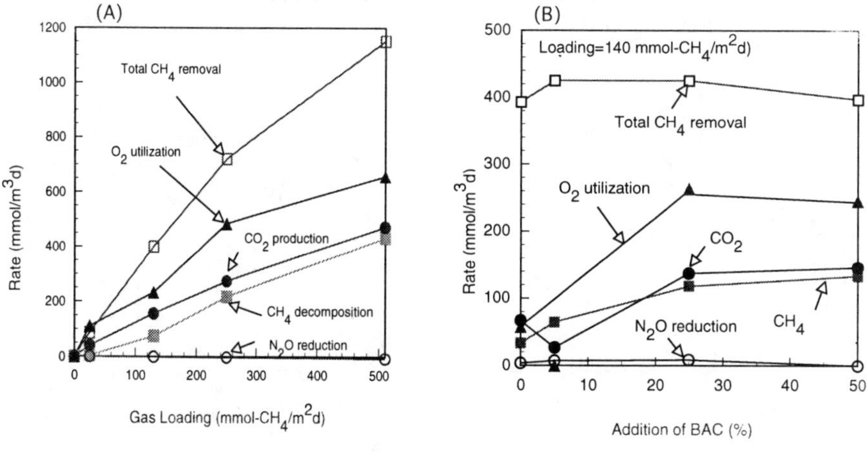
(A)
Total CH4 removal
O2 utilization
CO2 production
CH4 decomposition
N2O reduction
Rate (mmol/m3d)
Gas Loading (mmol-CH4/m2d)
(B)
Loading=140 mmol-CH4/m2d)
Total CH4 removal
O2 utilization
CO2
N2O reduction
CH4
Rate (mmol/m3d)
Addition of BAC (%)

FIGURE 4. Effects of (A) gas loading rates and (B) BAC addition on biological reaction rates.

Short-Term Loading. Young (1990) reported that emission fluxes of landfill gas changed with time dynamically in response to variations in atmospheric pressure. Figure 5 is a typical response to a simulated variation in gas loading rate (or gas emission flux), where the variation and the response of gas removals were shown in Figures 5(A) and 5(B), respectively. Experimental data showed that after the peaks of CH_4 and O_2 removals, CO_2 production became predominant. This means that CH_4 and O_2 were first adsorbed onto BAC/soil mixture and thereafter gradually oxidized to CO_2. This adsorption process is considered important in keeping a stable and high suppression ability, especially for short-term variations in emission flux. Furthermore, it is considered that the subsequent biological reaction is effective to regenerate adsorption capacities. We think the combination of adsorption and bio-regeneration processes in soil/BAC mixture is an important mechanism for the effective functioning of the biobarrier.

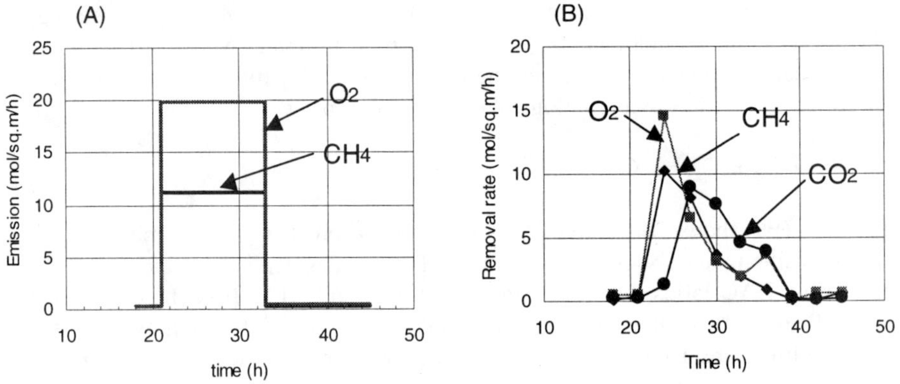
(A)
Emission (mol/sq.m/h)
time (h)
O2
CH4
25
20
15
10
5
0
10
20
30
40
50
(B)
Removal rate (mol/sq.m/h)
Time (h)
O2
CH4
CO2
20
15
10
5
0
10
20
30
40
50

FIGURE 5. Response to short-term variation in gas loading rate; (A) variation of gas loading and (B) responses of removal and production rates.

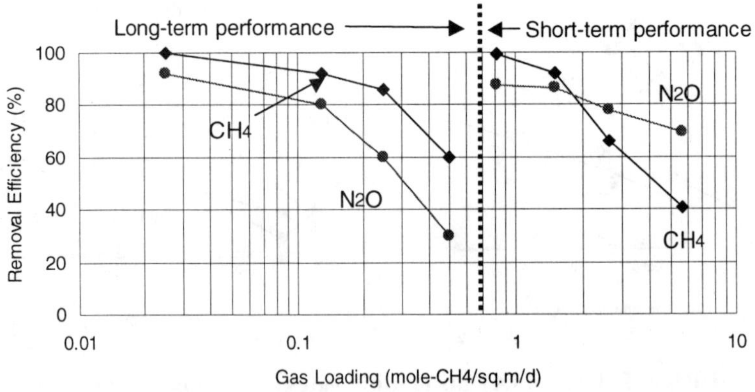
Long-term performance
Short-term performance
Removal Efficiency (%)
Gas Loading (mole-CH4/sq.m/d)
CH4
N2O
N2O
CH4
100
80
60
40
20
0
0.01
0.1
1
10

FIGURE 6. Comparison of gas removal rates in short- and long-term responses.

Figure 6 compares the removal efficiencies of CH_4 and N_2O in short-term and long-term loadings. This figure demonstrates that the suppressions for CH_4 and N_2O in short-term loadings were efficiently achieved even at very high loadings of 1 mol-CH_4/m^2d and 0.1 mol-N_2O/ m^2d, which are about one order of magnitude larger than those for long-term loadings. For short-term variations in flux, much larger suppression can be expected in the proposed BAC/soil biobarrier.

CONCLUSIONS

Experimental results in long-term loadings showed that, below the gas loading rates of 100 mmol-CH_4/m^2d and 10 mmol-N_2O/m^2d, the greenhouse gases were effectively degraded and the net GWP levels were reduced below 10 % of the feed gas. This result and the comparisons for different BAC/soil ratios demonstrated that the addition of GAC into cover soils may serve as a promising biobarrier.

Furthermore, experimental results in short-term loadings demonstrated that CH_4 was adsorbed onto GAC/soil mixture and was gradually oxidized to CO_2. This simultaneous adsorption and biodegradation process is considered an essential part in functioning as effective barrier under natural conditions. For large emission fluxes of CH_4 above several hundreds mmol/m^2d, the produced methane should be recovered as energy source because of its significant amount of combustion energy.

REFERENCES

Aselmann, I. and P. J. Crutzen. 1989. "Global Distribution of Natural Freshwater Wetlands and Rice Paddies, Their Net Primary Productivity, Seasonality and Possible Methane Emissions." *J. Atmos. Chem. 8*: 307-358.

Borjesson, G. and B. H. Svensson. 1997. "Seasonal and Diurnal Methane Emissions from a Landfill and their Regulation by Methane Oxidation." *Waste Management & Research. 15*: 33-54.

Hosomi, M. 1992. "Soils and Global Warming". *J. Jpn. Soc. Wat. Environ. 15*(11): 15-20 (in Japanese).

Jones, H. A. and Nedwell, D. B. 1990. "Soil Atmosphere Concentration Profiles and Methane Emission Rates in the Restoration Covers above Landfill Sites: Equipment and Preliminary Results". *Waste Management & Research 8*: 21-31.

Kimura, M., H. Ando, and H. Haraguchi. 1991. "Estimation of Potential CO_2 and CH_4 Production in Japanese Paddy Fields." *Environmental Science. 4* (1): 15-25.

Minami, K. and S. Fukushi. 1984. "Method for measuring N_2O flux from soil in the field." *Jpn. J. Soil Sci. Plant Nutr. 53*: 525-529 (in Japanese).

Ramanathan, V., R. J. Cicerone, H. B. Singh, and J. T. Kiehl. 1985. "Trace Gas Trends and Their Potential Role inch Climate Change." *J. Geophys. Res. 90*: 5547-5566.

Sakakibara, Y., Y. Tanaka, and R. Yoshida. 1999. "Emission Control of the Greenhouse Gases of CH_4 and N_2O by the Addition of Biological Activated Carbon into Soil." *J. Environ. Sys. Eng., JSCE. 629/VII*-12: 27-36 (in Japanese).

Tokyo Gas 1989. *http://www.tokyo-gas.co.jp/IR/corpoi.html.*

Young, A. 1990. "Volumetric Changes in Landfill Gas Flux in Response to Variations in Atmospheric Pressure." *Waste Management & Research. 8*: 379-385.

COMPARATIVE COST AND PERFORMANCE OF TWO NOVEL BIOLOGICAL PERMEABLE BARRIERS (BPB)

Fatemeh R. Shirazi, Ph.D.
(Stratum Engineering Inc., St. Louis, MO)

ABSTRACT: This investigation proposed and examined an innovative combination of two technical concepts that currently have strong market potential for groundwater remediation: in-situ bioremediation and permeable barrier for subsurface groundwater confinement/cleanup. Biological Permeable Barrier (BPB) is a new technology that creates an in-situ, passive, long-lasting, and cost effective bioremediation of organic-contaminated groundwater. In this investigation, a series of batch and column experiments were designed to simulate BPB and evaluate biodegradation of a target contaminant, 2,4,6 trichlorophenol (TCP) under aerobic conditions. The experiments were design to account for any significant changes in removal efficiency of TCP from groundwater due to hydraulic retention time (HRT), applied loading, availability of dissolved oxygen (DO), and toxic shock loads of TCP. The results of this investigation provided the basis for the conceptual design and cost evaluation for a contaminated site with polycyclic aromatic hydrocarbons (PAHs) in Louisiana.

Bio-beads were found to be an excellent permeable barrier media with 91%-100% TCP removal efficiency at loading up to 600-300 mg/L.d, respectively. The mixture of GAC-immobilized cells (3%) and silica sand (97%) offered 100 % removal for TCP loading up to 1200 mg/L.d (HRT= 14.8 minutes) by a combination of biological and physical adsorption mechanisms. Both media tolerated shock loads of TCP (>550 mg/L) and deficiency of DO of less than 2 mg/L and resumed their biodegradation activity in a matter of few days. BPB proved that it can overcome some of the problems associated with today's in situ bioremediation technologies by providing the best possible environment for microorganisms. Immobilization of organic-degrading bacteria shields the viable microorganisms from environmental stresses. The cost of BPB was determined to be substantially less than any other treatment technology for in situ bioremediation of contaminated groundwater.

INTRODUCTION

To date, the most common approach for large-scale in-situ bioremediation has been to inject nutrients to stimulate contaminant-degrading organisms (biostimulation). Biostimulation has not proven to be reliable due to biofouling, inhibition of the biodegradation reaction, and the difficulty of bringing the stimulated population and contaminants into contact. Another approach, called bioaugmentation, involves the addition of bacteria and nutrients to contaminated groundwater. In bioaugmentation, the microorganisms are exposed to the stress conditions in the environment where they are introduced. The losses of viable microorganisms as a result of stress conditions and migration of microorganisms are the major problems with this technology. ***Therefore, it is highly desirable to establish and maintain a biological system (in situ) with a high density of active microorganisms capable of biodegrading organic compounds under a control process that shields the viable cells and prevents their loss under environment stress conditions.***

This investigation aims to respond to this great need by providing a controlled and stable process (BPB). BPB entails immobilizing microbial organisms, which are acclimated to the target contaminants in unique polymeric beads (polyvinyl alcohol) called Bio-beads, or on granular activated carbon (GAC). The Bio-beads or mixture of GAC-immobilized cells and sand are then placed in an engineered trench across the flow path of the contaminated groundwater. Contaminated groundwater enters the BPB to which oxygen (under aerobic condition) and nutrients are supplied, and the remediated groundwater exits the BPB (Figure 1).

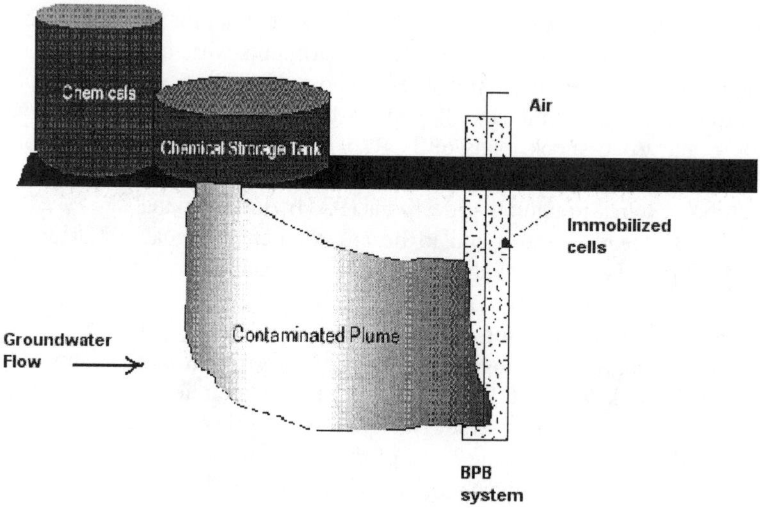
Chemicals
Chemical Storage Tank
Air
Immobilized cells
Groundwater Flow
Contaminated Plume
BPB system

FIGURE 1. BPB Concept

Ideally, a cost-effective BPB should be easy to construct and maintain, remain permeable, tolerate operational and environmental stresses, and capable to destroy a wide range of dissolved organic compounds in groundwater. In this investigation, Bio-beads and GAC-immobilized cells/sand were used to simulate a BPB system for a controlled and stable in-situ bioremediation process for over 300 days. The results of this investigation are presented.

Objectives. The main objective of this investigation is to explore a new concept for bioremediation of organic-contaminated groundwater using immobilized cells. Other objectives of this investigation are:

1. To evaluate and compare two candidates for a typical BPB system, (a) unique Bio-beads, (b) mixture of sand and cells immobilized on GAC, for

their ability to remove a target contaminant (TCP), under aerobic conditions from contaminated groundwater.

2. To investigate the performance of these two barriers under different operating conditions such as different concentrations (10 mg/L to 40 mg/L), and different flow rates (1 mL/min to 4 mL/min).

3. To evaluate the removal efficiency of these barriers under stressed conditions such as low dissolved oxygen and high TCP concentration (shock load of 550 mg/L).

4. To evaluate the ease of operation and cost of these two barriers under the same operating conditions.

METHOD AND MATERIALS

Groundwater was obtained from a water well located in the NE/4 NE/4 NE/4 Section of 9-T16N-R2E, Lincoln County, Oklahoma. All chemicals used in this study were reagent grade. The source of microorganisms for this study was activated sludge obtained from the Georgia-Pacific Leaf River Pulp Mill, New Augusta, Mississippi. The microorganisms were acclimated by feeding them TCP (10 mg/L) as their sole carbon source with continuous aeration and additional nutrients (phosphate buffer solution, magnesium sulfate solution, calcium chloride solution, ferric chloride solution. GAC-immobilized cells were prepared in accordance with the method used by Ehrhardt and Rehm (1985). The polyvinyl alcohol-boric acid method (Hashimoto and Furukawa 1987; Wu and Wisecarver 1992) was significantly modified to prepare long lasting polymer Bio-beads. Column studies were conducted to evaluate aerobic biodegradation of TCP (theoretical oxygen requirement of 0.89 mg of O_2 /mg TCP) under various operating conditions such as different flow rates (i.e. 1 to 4 mL/min.) and different TCP concentrations (i.e. 10 and 40 mg/L).

Feasibility Study of BPB Technology. A series of batch experiments were conducted to determine the physical characteristics of Bio-beads and GAC-immobilized cells/sand systems (Razavi-Shirazi 1997). The column experiments were designed to simulate and test BPB under a variety of operating conditions (Table 1). The experiments were carried out in total of four acrylic columns, and were set up as aerobic, continuous flow packed-bed reactors. Columns #1 and #2 consisted of 10 and 20 cm beds of Bio-beads (3-5 mm), respectively. Columns #3 and #4 consisted of 10 and 20 cm beds of 3% GAC immobilized cells and 97% clean silica sand. Phosphate buffer solution; magnesium sulfate solution; calcium chloride solution; and ferric chloride solution were added as nutrients to the influent feed bottle to maintain C:N:P of 100:10:3. A peristaltic pump and Tygon Tubing was used to deliver the groundwater into the base of the columns (upflow). Samples of the influent and effluent were taken for dissolved oxygen (DO), TCP, chloride (ICl), and pH determinations.

TABLE 1. Experimental conditions during column studies.

Column Experiments Parameters	Experiment No. 1 Period 1 Days 1-29	Experiment No. 2 Period 2 Days 33-58	Experiment No. 3 Period 3 Days 64-82	Experiment No. 4 Period 4 Days 83-96	Experiment No. 5 Period 5 Days 97-111	Experiment No. 6 Period 6 Days 113-127	Experiment No. 7 Period 7 Days 128-142	Experiment No. 8 Period 8 Days 150-164
Influent Concentration (mg/L)	10.0	20.0	20.0	20.0	30.0	20.0	20.0	40.0
Influent Flow Rate (mL/min.)	1.0	1.0	1.0	1.0	1.0	2.0	4.0	4.0
Residence Time(min)	Columns #1=49 #2=98 #3=59 #4=118	Columns #1=49 #2=98 #3=59 #4=118	Columns #1=49 #2=98 #3=59 #4=118	Columns #1=49 #2=98 #3=59 #4=118	Columns #1=49 #2=98 #3=59 #4=118	Columns #1=24.5 #2=49 #3=29.5 #4=58.9	Columns #1=12.3 #2=24.5 #3=14.8 #4=29.5	Columns #1=12.3 #2=24.5 #3=14.8 #4=29.5
Dissolved Oxygen (mg/L)	8.0-9.0	8.0-9.0	above 20.0	above 20.0	above 27.0	above 20.0	above 20.0	above 30.0
Loading Rate (g-TCP $L^{-1}.d^{-1}$)	Columns #1,#3=.074 #2, #4=.037	Columns #1,#3=0.15 #2, #4=.074	Columns #1,#3=.15 #2,#4=.074	Columns #1, #3=0.15 #2,#4=.074	Columns #1,#3=0.22 #2,#4=.11	Columns #1,#3=0.3 #2,#4=0.148	Columns #1,#3=0.6 #2,#4=0.3	Columns #1,#3=1.2 #2,#4=0.6

The 20-cm columns were twice subjected to a high influent TCP concentration (more than 550 mg/L), at a flow rate of 2.0 mL/min, and dissolved oxygen above 30.0 mg/L for 50.0 hours.

The short columns (10-cm) were subjected to a low influent dissolved oxygen concentration of less than 2 mg/L, at a flow rate of 2.0 mL/min, for 50.0 hours. During these 50.0 hours influent and effluent samples were taken to determine TCP, DO, pH, and ICl concentrations

Scale-Up Cost Estimate. The scale-up costs of this new BPB system using bio-beads or GAC immobilized cells/sand were investigated in details for a site contaminated with PAHs in Louisiana. For the purpose of developing and evaluating the costs of these barriers, it has been assumed that a funnel and gate design would be used at the site. The total cost included a gate containing Bio-beads or GAC immobilized cells. In the case of using GAC-immobilized cells, it is assumed that the gate would contain approximately 3 percent GAC and 97 percent silica sand. In the case of using Bio-beads, it is assumed that the gate would contain only 50 percent Bio-beads and 25 percent inert material. The nutrient and oxygen assumed be added to the gate to promote degradation of the contaminants to harmless by-products.

RESULTS AND DISCUSSION

Biodegradation of TCP for over 166 days of continuous operation. The results from the 166 days of continuous column experiments on both BPB materials proved that an elimination capacity of 100% TCP is feasible for Bio-beads for loads up to 300 mg L^{-1} d^{-1} (HRT= 24.5 minutes). The elimination capacities of both BPB materials are presented in Figure 3 and 4. At the loading rate of 600 mg L^{-1} d^{-1} (HRT=12.3 minute), the TCP removal efficiency of Bio-beads was reduced to 91%. The elimination capacity of GAC-immobilized cells of 100% TCP was feasible regardless of the organic load (up to 1200 mg L^{-1} d^{-1}) by combined adsorption and biological mechanisms. The results confirm that GAC, even with a substantial development of bacterial attachments and activities (biodegradation of TCP) maintains a substantial adsorption capacity.

Effects of High Shock Loads of TCP. During the first 50.0-hour shock loading of TCP, the degradation of TCP by the immobilized cells in the Bio-beads column was susceptible to the high shock load. The removal efficiency of the Bio-beads column decreased from 90% before shock load to 0% during shock load. The applied loading during the first 50.0-hour high shock load was 4,120 mg L^{-1} d^{-1}. The Bio-beads column recovered within 12 days as measured by decreasing TCP concentration in the effluent. The recovery time of the Bio-beads column from the second shock load was 5 days, which was much shorter than the first time.

The results also showed that the GAC-immobilized cells survived high shock loads (TCP > 550 mg/L). GAC protected immobilized cells from shock loading through rapid initial adsorption of TCP into pores and slow subsequent release of TCP by desorption. This desorption accompanied by biodegradation of the desorbed TCP (bioregeneration) after the first and second shock load (Razavi-Shirazi 1997).

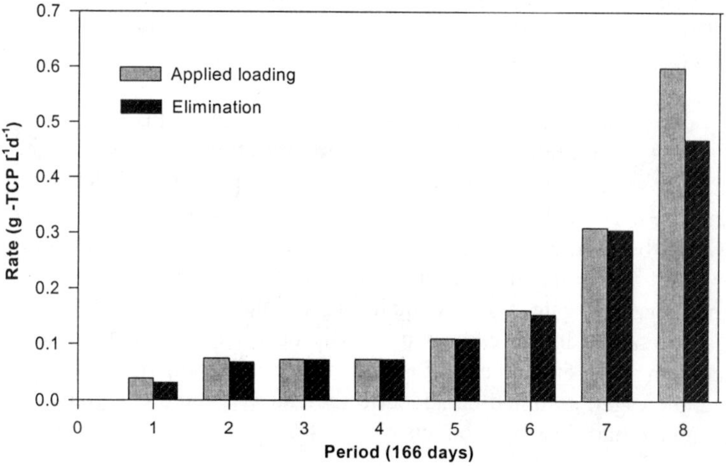
Applied loading
Elimination
Rate (g -TCP L-1d-1)
Period (166 days)

FIGURE 2. Removal of TCP by Bio-beads

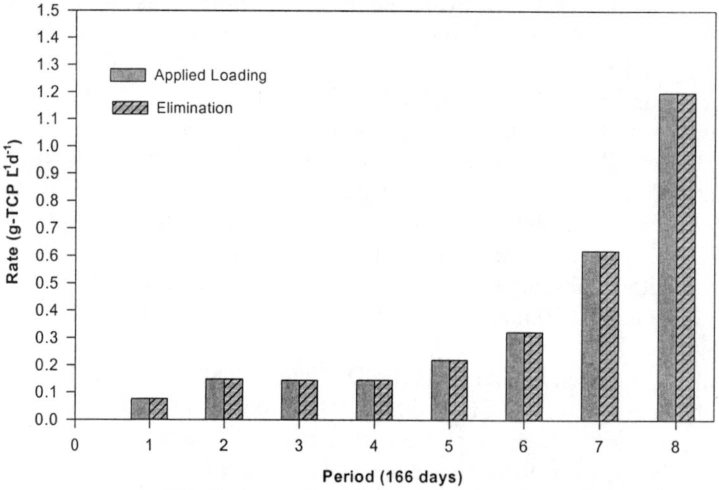
Applied Loading
Elimination
Rate (g-TCP L-1d-1)
Period (166 days)

FIGURE 3. Removal of TCP by GAC-immobilized cells/sand.

Effects of Low Dissolved Oxygen. To study the effects of low DO on TCP degradation, the oxygen supply to the 10-cm Bio-beads and GAC-immobilized cells/sand columns were discontinued twice during 74-day period (days 166-240) for 50 hours each time.

Increases and recoveries of Bio-beads column in terms of effluent TCP concentrations followed the same pattern. The recovery time of the Bio-beads column was shorter after the second interruption of DO. These results demonstrated the sensitivity of aerobic immobilized cells in the Bio-beads column and, at the same time, the tolerance of these cells toward the low DO influent.

During the interruptions of DO, the microorganisms immobilized on GAC were unable to biodegrade TCP as indicated by no change in chloride release or pH drops in the effluent. The removal mechanism for 100 % removal of TCP was mainly by absorption of TCP rather than its biodegradation. During steady state operations extra chloride was released in the effluent as the result of dehalogenation of TCP already adsorbed on GAC by attached microorganisms (bioregeneration). Bio-beads and GAC-immobilized cells were compared on the basis of removal efficiency, ease of operation and cost (Table 2.)

The structure and growth of microorganisms on GAC and Bio-beads were examined extensively by scanning electron microscopy (SEM) as shown in Figure 4 and 5.

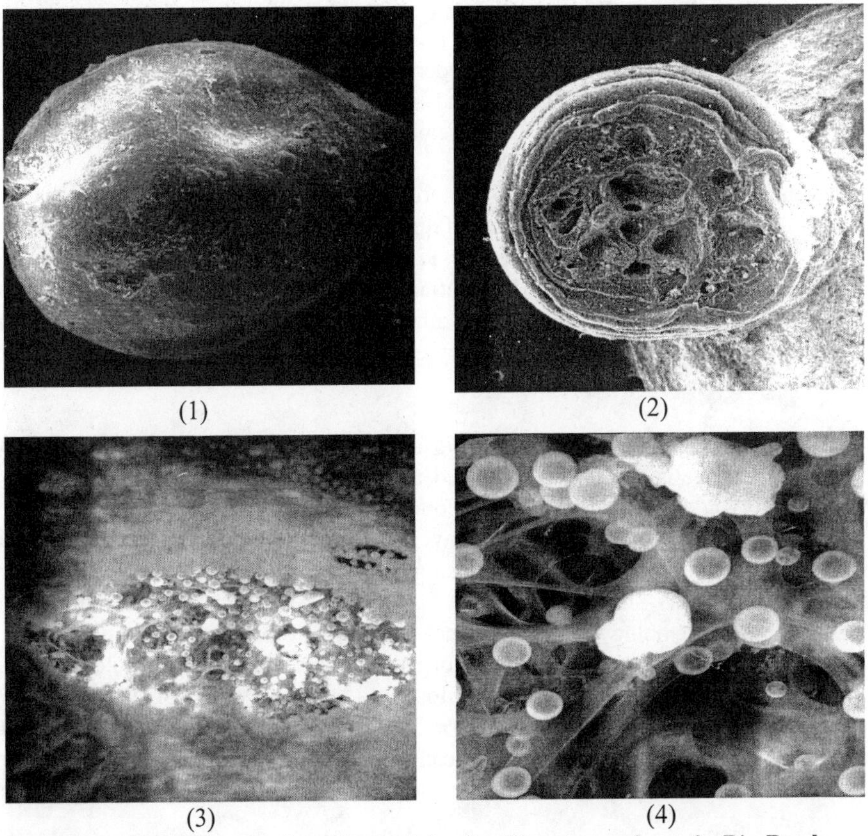
(1)
(2)
(3)
(4)

FIGURE 4. SEM of encapsulated bacterial cells. (1) outer surface of a Bio-Bead, 30X, (2) a cross-section of a Bio-bead at 26X, (3) one pocket of cells inside a Bio-bead, (4) population of cells inside a Bio-Bead, 3600X.

The outer surfaces of GAC before and after immobilization are shown in Figure 5. As shown in Figure 5, microcolonies and biofilm formation on the surface of the GAC can be seen by SEM.

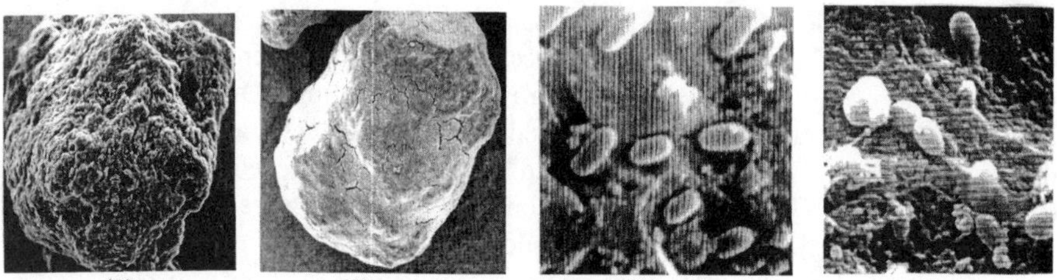

(1) (2) (3) (4

FIGURE 5. SEM of GAC before immobilization (1), GAC after immobilization of acclimated culture (2), microorganisms attached onto GAC surface after 45 days (3), microbial attachment and microcolonies formation (after 9 months, slime production) on GAC (4).

Scale-Up Cost. The cost of a permeable wall containing Bio-beads or GAC/Sand, cell cultures, nutrients and oxygen supply, and construction material for gate construction (sheet pilling) are presented in Table 2. The total volume of the gate is estimated to be 1500 ft^3. The cost of the Bio-beads for this gate (assuming 75% Bio-Beads and 25% inert material) is about $290,000. Additional costs associated with construction of the gate and proper nutrient and oxygen delivery systems are estimated to be $210,000. In the case of using 3%GAC-immobilized cell/sand as BPB material, the material cost is estimated to be $24,000 (including the culture). All other costs such as construction cost of the gate and nutrient and oxygen delivery systems will be $210,000 (same as Bio-Beads system).

REMARKS. For the first time, this unique Bio-beads system and 3% GAC immobilized cells/sand were evaluated as two new BPB materials in a side-by-side study under various operational conditions for over 240 days. The results of this research corroborated previous investigations as well as provided new basis for operation of BPB technology using Bio-beads and 3% GAC-immobilized cells/sand (Razavi-Shirazi 1997).

BPB technology by itself has no limitations for in-situ or ex-situ treatment (as a bioreactor). A wide range of organic compounds can be treated by Bio-Beads. The Bio-Beads can be tailored to co-treat other organics exist in contaminated groundwater. A shallow aquifer is more suitable for a funnel and gate configuration. Deeper aquifers can also take advantage of BPB technology by a unique delivery system currently under development by Stratum Engineering (Bio-Pak).

BPB technology is protected under U.S. Patent No.09/432,090 and PCT/US00/3006. Stratum Engineering Inc. has exclusive rights to this technology and is currently offering BPB technology to potential clients.

TABLE 2. Comparison Bio-beads and 3% GAC-immobilized cells/sand as two new BPB materials.

Basis of Comparison	Bio-beads immobilized cells system	3% GAC-immobilized cells/silica sand system
Removal efficiency	100 % load up to 300 mg $L^{-1}.d^{-1}$	100 % up to the loading of 1200mg $L^{-1}.d^{-1}$
Ease of operation	Easy to handle Remained firm and elastic. A few of the beads damaged during 240 days of continuous operation.	Easy to handle No physical damages were observed.
Tolerance	Survived high shock load and deficiency of DO, recovered from high shock load and low DO within 11-21days.	Survived high shock load and low DO Maintained 100% efficient.
Unit Cost	Mass production of the Bio-beads is $15 per Kg	GAC cost $ 10/ft^3 (3% GAC/ silica sand)
Estimated capital cost for a gate (20 width, 15 depth) with 5 feet thickness.	$500,000 including the construction cost of the gate. Expected life of this barrier is estimated to be 10-30 years.	$234,000 including the construction cost of the gate. Extra costs associated with excavation /replacement of barrier is anticipated every two years.
Overall assessment	Very favorable	Not applicable to a wide range of contaminants. Some additional costs are anticipated for replacement of the barrier.

REFERENCES

Ehrhardt, H.M., and H.J. Rehm. 1985. "Phenol Degradation by Microorganisms Adsorbed on Activated Carbon", *Applied Microbiology and Biotechnology. 21*: 32-36.

Hashimoto, S., and K. Furukawa. 1987. "Immobilization of Activated Sludge by PVA-Boric Acid Method", *Biotechnology and Bioengineering. 30*: 52-59.

Razavi-Shirazi, F., 1997. "Development of Biological Permeable Barriers for Removal of Chlorophenols (2,4,6-Trichlorophenol) in Contaminated Groundwater." *PhD Dissertation, Oklahoma State Univ., Stillwater, OK.*

Wu, K, and K. Wisecarver. 1992. "Cell Immobilization Using PVA Crosslinked with Boric Acid", *Biotechnology and Bioengineering. 39*: 447-449.

SORPTION AND MICROBIAL DEGRADATION OF TOLUENE ON A SURFACTANT-MODIFIED-ZEOLITE SUPPORT

Alana M. Fuierer (NMIMT, Socorro, NM)
Robert S. Bowman and Thomas L. Kieft (NMIMT, Socorro, NM)

ABSTRACT: We conducted laboratory studies to test the feasibility of using a reactive nutrient-amended microbial support system comprised of surfactant-modified zeolite (SMZ) to stimulate biodegradation of petroleum hydrocarbons. Toluene was used as a model hydrocarbon contaminant. Batch isotherm experiments showed that toluene sorption on SMZ and nutrient-amended SMZ (N-SMZ) was similar, resulting in a K_d of 13.0 L kg^{-1}, and toluene desorption reached equilibrium within 2 h. The first-order toluene biodegradation rate coefficient in solution-only aerobic microcosms was 0.24 h^{-1}, while in SMZ and N-SMZ slurry-phase cultures the rate coefficient was 0.13 h^{-1}. Transport simulations suggest that a 1-m wide N-SMZ permeable barrier could microbially degrade toluene to below the drinking water standard.

INTRODUCTION

BTEX compounds are major groundwater contaminants of concern, primarily as a consequence of accidental hydrocarbon spills and leaking underground storage tanks. Traditional *ex situ* remediation of BTEX compounds is slow, expensive and, often times, ineffective.

Several studies have focused on "*in situ* reaction systems" that combine a sorbent zone in the subsurface with contaminant degradation mechanisms in order to enhance remediation (Burris and Antworth, 1992). This study proposes an *in situ* biological treatment system, consisting of zeolite modified with a cationic surfactant (SMZ), that would allow establishment of an active microbial community, a continuous source of nutrients and electron acceptors, and protection against high concentrations of toxic chemicals.

Zeolites are aluminosilicate minerals characterized by cage-like molecular structures, high internal and external surface areas, and high cation exchange capacities (Breck, 1974). Bowman et al. (1995) have shown that natural zeolites become excellent sorbents for inorganic anions and organic species when the negatively charged surface is modified with a cationic surfactant. The sorption capacity allows SMZ to retain organic contaminants such as toluene, thereby reducing contaminant migration in the subsurface and protecting against high concentration fluxes. SMZ also retains inorganic anions such as nitrate and sulfate that can serve as electron acceptors (Li et al., 1998a). In addition, cationic nutrients such as ammonium and potassium can be preloaded onto SMZ. Li et al. (1998b) have shown that microorganisms can colonize SMZ but do not biodegrade the reactive surfactant coating.

Objective. The objective of our research was to examine the feasibility of using a nutrient-amended SMZ (N-SMZ) support to sorb and stimulate microbial degradation of dissolved petroleum products. We chose toluene as a model for a low molecular weight, aromatic petroleum component. We used laboratory experiments to investigate toluene 1) sorption on N-SMZ compared to SMZ; 2) desorption kinetics; and 3) biodegradation in the presence of SMZ and N-SMZ.

MATERIALS AND METHODS

Zeolite. The starting material for SMZ experiments was a natural clinoptilolite-rich zeolitic tuff with a particle size of 8-14 mesh (2.4 to 1.4 mm) from St. Cloud Mining Co. (Winston, NM). The starting material for N-SMZ experiments was a nutrient-amended zeolite called ZeoPro™ from ZeoponiX, Inc. (Louisville, Co.) with a particle size of 8-14 mesh. ZeoPro™ is manufactured using St. Cloud zeolite and contains by weight 0.1% N (as ammonium), 0.1% P (as hydroxyapatite), and 0.6% K.

SMZ and N-SMZ Preparation. The majority of SMZ was bulk-produced at the St. Cloud mine using hexadecyltrimethylammonium chloride (HDTMA-Cl). The details of the manufacturing process are described elsewhere (Bowman et al., 2001). All N-SMZ was prepared in the laboratory using HDTMA-Br. To achieve a target HDTMA loading of 130 mmol kg^{-1}, 100 g of ZeoPro™ and 250 mL of 52-mM aqueous HDTMA-Br solution were put into 500-mL centrifuge bottles and placed into a rotary shaker at 80 rpm and 25 °C for 24 h, a period shown to be sufficient to attain HDTMA sorption equilibrium (Bowman et al., 1995). The mixture was centrifuged and washed twice with purified water; then, the modified sample was air-dried. This procedure also was used to produce SMZ for the final biodegradation experiment.

Toluene Sorption/Desorption. We performed batch isotherm experiments to quantify sorption of toluene to SMZ and N-SMZ in comparison to untreated zeolite. Each sample contained 4.0 g of SMZ, N-SMZ, or untreated zeolite and 16 mL toluene solution (aq). In addition, a fourth treatment contained 4.0 g of SMZ and 16 mL toluene solution prepared with inorganic nutrient medium (Bushnell-Haas [B-H] broth (Difco, Detroit, MI)) to approximate the conditions of the biodegradation experiment more closely. Each liter of B-H broth contained 0.2 g $MgSO_4$, 0.02 g $CaCl_2$, 1.0 g KNO_3, 0.05 g $FeCl_3$, 1.0 g $(NH_4)_2HPO_4$, and 1.0 g KH_2PO_4. Five initial solution concentrations, ranging from 10 mg L^{-1} to 200 mg L^{-1} toluene, were prepared for each treatment. All samples were placed into a rotary shaker at 80 rpm and 25 °C for 24 h, and then analyzed with a gas chromatograph (GC).

We performed desorption kinetic studies on SMZ to determine if and how the desorption rate would affect biodegradation. Two desorption experiments were conducted with initial toluene concentrations of 60 and 110 mg L^{-1}. For both studies, the toluene solutions were prepared in B-H broth. Samples containing 4.0 g of SMZ and 16 mL toluene solution were shaken at 80 rpm at 25

°C. After a 24-h equilibration period, 8 mL of equilibrium solution was removed and 8 mL of toluene-free B-H broth was introduced into each vial. The samples were shaken again, then analyzed over time with a GC.

Culturing Toluene Degraders. We obtained a culture of aerobic, toluene-degrading microorganisms by a series of enrichments. The first enrichment culture was prepared in 70-mL glass serum vials using 50 mL B-H broth. Activated sludge (1 mL) from the wastewater treatment plant in Socorro, NM was added to the vial and the culture was enriched for toluene degraders by adding 100 mg L^{-1} toluene as the sole carbon source. The culture was incubated in a shaker at 80 rpm and 25 °C for 1 week. A 1-mL aliquot of culture medium was removed, transferred to vials containing B-H broth and toluene (100 mg L^{-1}), and incubated for 1 week. The latter enrichment step was repeated 3 times and the final stock culture was stored at 4 °C.

Toluene Biodegradation. We quantified toluene depletion in aerobic solution-only batch cultures to determine the biodegradation rate in the absence of SMZ and N-SMZ. Each microcosm contained 16 mL B-H broth, 100 mg L^{-1} toluene, and 0.2 mL inoculum. We also prepared uninoculated controls that contained 16 mL B-H broth and 100 mg L^{-1} toluene. All samples were incubated in a shaker at 80 rpm and 25 °C and sacrificed at various times over a 2-day period. A subsample of each was analyzed for toluene with a GC immediately.

We controlled the pH of all SMZ and N-SMZ slurry cultures by adjusting the medium to a starting pH of 11.5. For SMZ slurry cultures, the pH-adjusted nutrient medium contained the same nutrients as the B-H broth, but replaced 1.0 g $(NH_4)_2HPO_4$ and 1.0 g KH_2PO_4 with 2.0 g K_3PO_4. For N-SMZ slurry cultures, we controlled the pH by adding a nutrient-free solution of 2.0 g K_3PO_4/L H_20. Due to sorption of buffer components by the SMZ and N-SMZ, a starting pH of 11.5 yielded a pH of approximately 8.0 after a 24-h equilibration period.

Each sample contained 4.0 g of SMZ or N-SMZ pre-equilibrated with 16 mL of pH-controlled nutrient medium or nutrient-free solution. To each sample, 100 mg L^{-1} toluene and 0.2 mL inoculum were added. The microcosms were incubated in the shaker and sacrificed at various times over a three-day period. A subsample of each was analyzed for toluene with a GC immediately. We prepared and analyzed uninoculated controls in the same manner.

RESULTS AND DISCUSSION

Toluene Sorption/Desorption. The toluene sorption isotherms for both SMZ and N-SMZ were linear, reflecting a partitioning type mechanism (Figure 1). Therefore, the equilibrium distribution of toluene between the solid and aqueous phases is described as:

$$\frac{S_e}{C_e} = K_d \tag{1}$$

where C_e (mg L^{-1}) and S_e (mg kg^{-1}) are the equilibrium toluene concentrations in the aqueous and sorbed phases, respectively, and K_d (L kg^{-1}) is the solid-water distribution coefficient. Linear sorption isotherms on SMZ have been observed for other nonpolar organics (Bowman et al., 1995). A straight-line fit of the data resulted in the K_d values shown in Figure 1. A mean K_d of 13.0 L kg^{-1} was used to characterize toluene distribution between the solid and solution phases in the biodegradation studies.

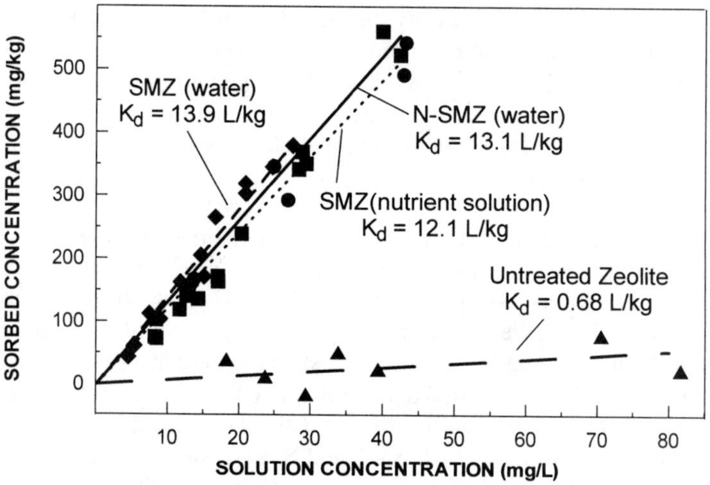
SORBED CONCENTRATION (mg/kg)
0
100
200
300
400
500
SMZ (water)
Kd = 13.9 L/kg
N-SMZ (water)
Kd = 13.1 L/kg
SMZ(nutrient solution)
Kd = 12.1 L/kg
Untreated Zeolite
Kd = 0.68 L/kg
10
20
30
40
50
60
70
80
SOLUTION CONCENTRATION (mg/L)

FIGURE 1. Sorption of toluene by SMZ, N-SMZ, and untreated zeolite. Fitted lines are based on linear regression through the origin.

Toluene desorption from SMZ was rapid and complete in less than 2 h at both concentrations (data not shown). Consistent with this rapid desorption rate, further analysis (not presented herein) indicated that biodegradation within slurry-phase microcosms was not desorption-rate limited. Furthermore, the final desorption equilibrium points fall on the sorption isotherm for SMZ (data not shown). This indicates that sorption of toluene by SMZ is a fully reversible process and suggests that a microbial support system comprised of SMZ could protect against high fluxes of contaminants without limiting access to the toluene for biodegradation. Since the sorption isotherms for SMZ and N-SMZ were very similar (Figure 1), desorption kinetics for SMZ and N-SMZ are also likely similar and the above observations should hold true for N-SMZ.

Toluene Biodegradation. Toluene was rapidly depleted in aerobic solution-only batch cultures (Figure 2). Figure 2 clearly indicates a lag phase of about 8 h followed by a period of biodegradation until t = 21 h. Uninoculated controls showed no decrease in toluene concentration over 45 h (data not shown). The samples incubated for more than 21 h maintained a mean toluene concentration in solution of 6 (± 2) mg L^{-1}. Therefore, the microorganisms were able to degrade 92 (±2) % of the toluene. Although it is uncertain why toluene was not completely degraded, it is possible that the microorganisms were lacking an essential nutrient.

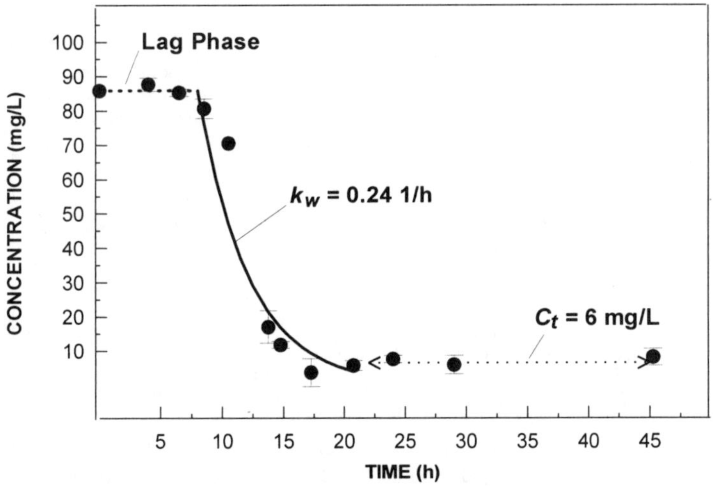
Lag Phase
k_W = 0.24 1/h
C_t = 6 mg/L
CONCENTRATION (mg/L)
100
90
80
70
60
50
40
30
20
10
5
10
15
20
25
30
35
40
45
TIME (h)

FIGURE 2. First-order approximation of toluene biodegradation kinetics in solution-only batch cultures. Points are means of triplicates; error bars represent standard deviation. Solid line based on linear regression of data according to $\boldsymbol{ln\ (C_t / C_0) = -\ k_w\ (t - t_{lag})}$.

Biodegradation of organic contaminants is often described by first-order kinetics. The following equation was used to calculate the first-order biodegradation rate constant in solution, k_w, over the period t = 8 to 21 h:

$$C_t = C_0 \exp^{\left[-k_w\left(t - t_{lag}\right)\right]} \tag{2}$$

where C_0 = initial concentration in solution (mg L^{-1}), C_t = concentration in solution at time t (mg L^{-1}), and t_{lag} = lag time (h) (Larson, 1984). A plot of $ln\ (C_t / C_0)$ vs. $(t - t_{lag})$ resulted in a k_w of 0.24 (± 0.05) h^{-1} (parentheses represent standard deviation at the 90% confidence level).

Figure 3 shows the depletion of toluene in slurry-phase batch cultures. The uninoculated controls maintained a mean concentration of 22 (± 1) mg L^{-1} throughout the experiment. For SMZ-inoculated samples, we observed a lag phase of approximately 8 h followed by a biodegradation period. The lag phase for N-SMZ samples was longer, about 20 h (Figure 3). Consistent with the solution-only cultures, we did not observe 100% biodegradation of toluene within either of the slurry-phase microcosms. After a 25-h incubation period, toluene degradation ceased within the SMZ samples and the toluene concentration in solution increased, reaching a plateau of 4 (± 1) mg L^{-1}. The final concentration was equivalent to an 80 (± 7) % degradation of toluene originally present. The toluene concentration in the N-SMZ samples incubated for more than 32 h also increased and then maintained a mean concentration of 10 (± 1) mg L^{-1}, resulting in a 54 (± 5) % degradation of toluene. No downward trend in concentration was observed for either treatment after the initial biodegradation period.

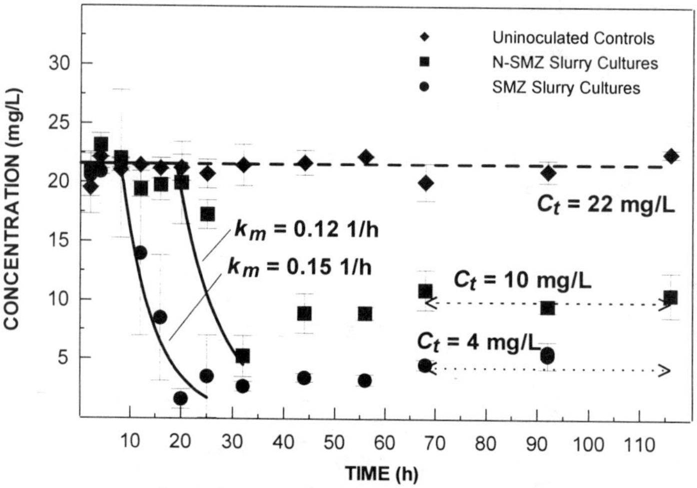

Uninoculated Controls
N-SMZ Slurry Cultures
SMZ Slurry Cultures
CONCENTRATION (mg/L)
TIME (h)
k_m = 0.12 1/h
k_m = 0.15 1/h
C_t = 22 mg/L
C_t = 10 mg/L
C_t = 4 mg/L

FIGURE 3. First-order biodegradation kinetics of toluene in slurry-phase batch cultures. Points are means of triplicates; error bars represent standard deviation. Solid lines are based on linear regression of data according to $ln(C_t/C_o) = -k_m(t-t_{lag})$.

Figure 3 shows that, although the N-SMZ samples had a longer lag time, biodegradation occurred at a similar rate in the SMZ and N-SMZ samples. Applying a first-order biodegradation model to the data, the mixed biodegradation-desorption rate constant (k_m) for the SMZ cultures was determined to be 0.15 (± 0.07) h^{-1} for t = 8 to 25 h. For the N-SMZ cultures, k_m was determined to be 0.12 (± 0.05) h^{-1} for t = 20 to 32 h. Statistical analysis confirmed that the k_m values for the SMZ and N-SMZ samples are not statistically different from each other, but are lower than the rate constant for solution-only cultures (k_w= 0.24 h^{-1}) at the 90% confidence level. The slower biodegradation rates in the presence of a sorbing solid phase are consistent with studies that have found biodegradation rates of organic contaminants to be slower in the presence of soil (Robinson et al, 1990; Zhang and Bouwer, 1997).

These experiments show that organic hydrocarbons such as toluene can be degraded in the presence of SMZ and N-SMZ. In addition, the use of N-SMZ eliminated the need to add nutrients to the system. Although we do not know what caused the incomplete degradation within the microcosms, one possibility is the lack of an unidentified nutrient or slow diffusion of a nutrient.

Our results suggest that N-SMZ could be used as a reactive permeable barrier placed in the path of a groundwater contaminant plume. To test the implementation of an *in situ* permeable reactive barrier, the 1-D modeling code CXTFIT2 (Toride et al., 1995) was used to predict the biodegradation of a groundwater toluene plume within a 1-m thick N-SMZ barrier. The simulation assumed the biodegradation rate constant we observed in the slurry-phase experiments (k_m = 0.13 h^{-1} for SMZ and N-SMZ). We used additional input data consistent with a permeable barrier pilot test by Bowman et al. (2001) (a dispersion coefficient of 0.23 m^2 d^{-1}; a pore-water velocity of 0.23 m d^{-1}; a volumetric water content of 0.6; and a bulk density of 1.0 kg L^{-1}).

The CXTFIT2 simulations are shown in Figure 4. The code predicted that a groundwater contaminant plume with an input concentration of 100 mg L^{-1} toluene would emerge from a 1-m N-SMZ barrier with a concentration < 1 mg L^{-1} once steady-state was attained. Federal drinking water standards allow a maximum concentration of 1 mg L^{-1} for toluene (National, 1995). Therefore, a N-SMZ barrier would remediate such a toluene plume below the federal requirement.

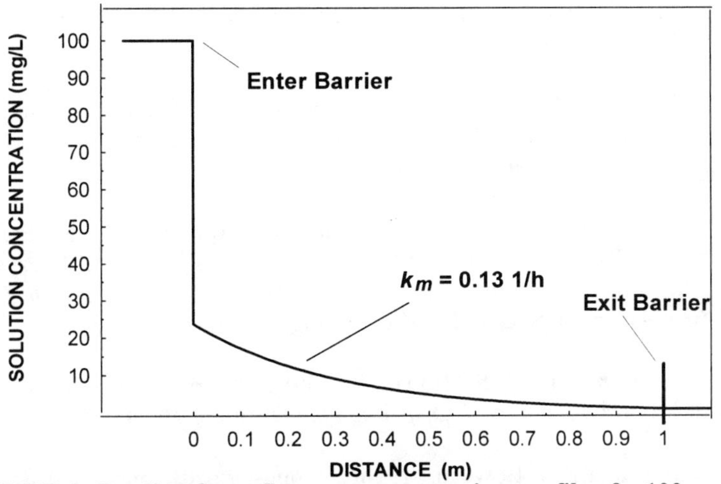
SOLUTION CONCENTRATION (mg/L)
100
90
80
70
60
50
40
30
20
10
Enter Barrier
k_m = 0.13 1/h
Exit Barrier
0 0.1 0.2 0.3 0.4 0.5 0.6 0.7 0.8 0.9 1
DISTANCE (m)

FIGURE 4. Predicted steady-state concentration profile of a 100 mg L^{-1} toluene plume as it passes through a 1-m N-SMZ reactive barrier.

N-SMZ has the potential to be used as the material for a permeable reactive barrier placed in the path of a contaminant plume. Because N-SMZ has a strong affinity for toluene, the microbial community will be protected from high contaminant concentrations. In addition, the reversible nature of toluene sorption should allow sustained degradation of toluene as it desorbs from N-SMZ. The nutrient portion of N-SMZ can also be customized and replenished to optimize biodegradation of various contaminants for site-specific conditions.

ACKNOWLEDGMENTS

This research was supported by a grant from the New Mexico Waste-management Education and Research Consortium (WERC).

REFERENCES

Bowman, R.S., G.M. Haggerty, R.G. Huddleston, D. Neel, and M. Flynn. 1995. "Sorption of Nonpolar Organics, Inorganic Cations, and Inorganic Anions by Surfactant-Modified Zeolites". In D.A. Sabatini, R.C. Knox, and J.H. Harwell (Eds.), *Surfactant-Enhanced Remediation of Subsurface Contamination,* pp. 54-64. American Chemical Society: Washington, DC.

Bowman, R.S., Z. Li, S.J. Roy, T. Burt, T.L. Johnson, and R.L. Johnson. 2001. "Pilot Test of a Surfactant-Modified Zeolite Permeable Barrier for Groundwater Remediation". In J.A. Smith and S.E. Burns (Eds.), *Physical and Chemical Remediation of Contaminated Aquifers*, pp. 161-185. Kluwer Academic Publishers: New York, NY.

Breck, D.W. 1974. *Zeolite molecular sieves: structure, chemistry, and use*. John Wiley and Sons: New York, NY.

Burris, D.R. and C.P. Antworth. 1992. "Insitu Modification of an Aquifer Material by a Cationic Surfactant to Enhance Retardation of Organic Contaminants". *J. Contaminant. Hydrol. 10*:325-337.

Larson, R.J. 1984. "Kinetic and Ecological Approaches for Predicting Biodegradation Rates of Xenobiotic Organic Chemicals in Natural Ecosystems". In M.J. Klug and C.A. Reddy (Eds.), *Current Perspectives in Microbial Ecology*, pp. 677-686. American Society for Microbiology: Washington, D.C.

Li, Z., I. Anghel, and R.S. Bowman. 1998a. "Sorption of Oxyanions by Surfactant-Modified Zeolite". *J. Dispersion Sci. Technol. 19*:843-857.

Li, Z., S. Roy, and R.S. Bowman. 1998b. "Long-term Chemical and Biological Stability of Surfactant-Modified Zeolite". *Environ. Sci. Technol. 32*:2628-2632.

National Primary Drinking Water Regulations, 40 CFR Parts 141 and 142 (1995).

Robinson, K.G., W.S. Farmer, and J.T. Novak. 1990. "Availability of Sorbed Toluene in Soils for Biodegradation by Acclimated Bacteria". *Water Res. 24*:345-350.

Toride, N., F.J. Leij, and M. van Genuchten. 1995. *The CXTFIT Code for Estimating Transport Parameters from Laboratory or Field Tracer Experiments, Version 2.0.* U.S. Salinity Laboratory, USDA, ARS: Riverside, California.

Zhang, W. and E.J. Bouwer. 1997. "Biodegradation of Benzene, Toluene and Naphthalene in Soil-Water Slurry Microcosms". *Biodegradation. 8*(3):167-175.

ABIOTIC AND BIOTIC Cr(VI) REDUCTION IN A LABORATORY-SCALE PERMEABLE REACTIVE BARRIER

C. Henny (University of Maine, Orono, Maine)
L.J. Weathers (Tennessee Technological University, Cookeville, Tennessee)
L.E. Katz (University of Texas, Austin, Texas)
J.D. MacRae (University of Maine, Orono, Maine)

ABSTRACT: The removal of Cr(VI) from the following column reactors was compared: (1) a column containing 5 g steel wool only (column Fe(0)), (2) a column containing 5 g steel wool which was seeded with a mixed culture of sulfate-reducing bacteria (SRB) and which was continuously supplied with 20 mM lactate (column Fe(0)+SRB+Lactate), (3) a column containing 5 g steel wool which was seeded with SRB but not supplied with lactate (column Fe(0)+SRB), and (4) a sterile control column packed with 5 mm-diameter glass beads. The influent sulfate concentration to all columns was 12.4 mM. Influent Cr(VI) concentrations of 380 and 960 μM, and 9.6 and 19.2 mM were investigated. By the end of the study on day 58, the instantaneous Cr(VI) removal efficiency had decreased to 20% in the Fe(0) only column, to 25% in the Fe(0)+SRB column, and to 90% in the Fe(0)+SRB+Lactate column. Based on the total mass of Cr(VI) removed during the study, the ratio of the overall performance of the column pairs was: Fe(0)+SRB compared to Fe(0): 143%; and Fe(0)+SRB+Lactate compared to Fe(0): 178%. Thus, seeding the columns with the SRB culture greatly improved their Cr(VI) removal performance.

INTRODUCTION

Chromium exists primarily in two oxidation states in natural waters and soil: hexavalent chromium [Cr(VI)] and trivalent chromium [Cr(III)]. Cr(VI), which predominates under oxidizing conditions, is typically present as an anion, either chromate (CrO_4^{2-}) at pH > 6.5 or bichromate ($HCrO_4^-$) at pH < 6.5. Cr(VI) is toxic and mutagenic to organisms, and is very soluble over a wide pH range in natural waters. In ground water, Cr(VI) movement has been found to be only slightly retarded by adsorption to aquifer material. Therefore the presence of hexavalent chromium in the environment, especially in groundwater, represents a serious health threat to humans. Cr(III), in contrast, predominates under more reducing conditions, is less soluble, and precipitates as oxides and hydroxides at pH values > 5. Cr(III) is also less toxic than Cr(VI). The predominant species of Cr(III) in the pH range 6.5 - 10.5 is $Cr(OH)_3$ (Rai et al., 1987). Cr(III) has a strong tendency to adsorb to surfaces (Sass and Rai, 1987).

Techniques for the treatment of Cr(VI)-contaminated waters involve reduction of Cr(VI) to Cr (III), followed by adjustment of the solution pH to near-neutral conditions to precipitate Cr(III) ions. Benefits of this process are twofold: the toxicity of chromium is reduced and the transport of this metal in ground water is lessened.

To date, no study has been conducted to evaluate chromium reduction by a system combining zero valent iron and sulfate reducing bacteria (SRB) although previous research has indicated that Cr(VI) is reduced by Fe(0) (Blowes et al., 1997; Gould, 1982; Powell et al., 1995) and SRB (Lovley and Phillips, 1994). Similar research by Weathers et al. (1997) has established the beneficial effects of combining anaerobic, methanogenic microbial communities and Fe(0) for the degradation of chlorinated organic compounds. The hypothesis of this work was that a system combining a mixed culture of sulfate reducing bacteria with zero valent iron would have a greater Cr(VI) removal efficiency and a greater total Cr(VI) removal capacity than a zero valent iron system without the microorganisms. Hence, the overall goal of this research was to compare the performance of these types of systems with regard to their Cr(VI) removal efficiency and total Cr(VI) removal capacity.

MATERIALS AND METHODS

A 58-day column experiment was conducted which compared the ability of the following column reactors to remove Cr(VI): (1) a column containing steel wool (SW) only, (2) a column containing SW which was inoculated with a lactate-enriched SRB stock culture and continuously supplied with lactate, (3) a column containing SW which was inoculated with the SRB stock culture but which was not supplied with lactate, and (4) a column packed with autoclaved, 5 mm-diameter glass beads which served as a control for the loss of Cr(VI) by sorption. The operation of the two columns containing SW and SRB promoted different modes of growth: the lactate-fed column promoted the growth of heterotrophic, autotrophic and, possibly, mixotrophic SRB. In contrast, the organic carbon-free environment of the lactate-starved column required the SRB to immediately switch to an autotrophic mode of growth in order to remain viable.

Experiments were conducted at 30°C using adjustable glass columns with an inside diameter of 2.25 cm and an adjusted length of 9.2 cm (ACE Glass). Three of the columns were packed with 5 g of superfine steel wool (Rhodes, Chicago, IL). Two of these columns were then seeded with sulfate reducing bacteria by injecting 40 mL of stock reactor cell suspension (200 mg/L volatile suspended solids) into each. The columns were maintained in a no-flow mode for 7 days to permit the bacteria to colonize the steel wool. Following this, influent was pumped into the columns through Teflon tubing from 50-mL glass, gas-tight syringes (Hamilton) with a syringe pump (Harvard Apparatus). The columns were fed in an upflow mode at a volumetric flow rate of 10 mL/d. The porosity of the steel-wool-packed columns was measured to be 0.90, resulting in a hydraulic residence time of 4 days. After six days, Cr(VI) was added to the feed solution at a concentration of 380 μM. The Cr(VI) concentration was increased to 960 μM after 14 days total, to 9.6 mM after 28 days total, and to 19.2 mM after 46 days total. Table 1 summarizes the Cr(VI) influent concentrations to the columns. Effluent samples were collected periodically for 58 days and analyzed for Cr(VI), sulfate and lactate. Cr(VI) was measured using the colorimetric diphenylcarbazide method (APHA, 1992). Sulfate was measured using an ion chromatograph. Lactate was measured by high performance liquid chromatography.

TABLE 1. Cr(VI) influent concentrations to the columns.

Phase	Time period (days)	Cr(VI) influent conc. (mM)
I	0-6	0.0
	6-14	0.38
II	14-28	0.96
III	28-46	9.6
IV	46-58	19.2

RESULTS AND DISCUSSION

During phases I and II of the study, corresponding to influent Cr(VI) concentrations of 380 and 960 μM, respectively, Cr(VI) was not detected in the effluent from any of the treatment columns (Figure 1). Thus the instantaneous Cr(VI) removal efficiency was 100% for all columns.

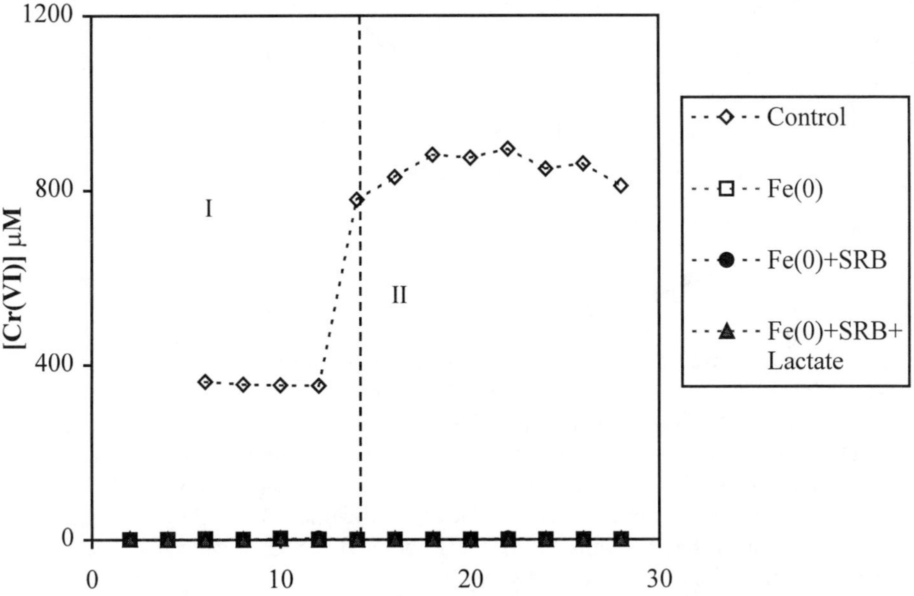
1200
800
400
0
[Cr(VI)] μM
I
II
0
10
20
30
Time (days)
Control
Fe(0)
Fe(0)+SRB
Fe(0)+SRB+ Lactate

FIGURE 1. Effluent chromium concentrations during Phases I and II.

During Phase I and II, the effluent sulfate concentration from columns Fe(0) and Fe(0)+SRB was about equal to that of the control, indicating the lack of sulfate reduction in these columns (Figure 2). In contrast, the sulfate concentration from the lactate-fed column during Phase II averaged about 38% of the control value, indicating that sulfate reduction was occurring in this column. Also, the lactate concentration in the effluent from this column was nondetectable on all but one sampling event during Phase II, even though the influent lactate concentration was about 20 mM (data not shown). Taken together, this data indicated that SRB were active in the lactate-fed column, but not the Fe(0)+SRB column, during Phase II.

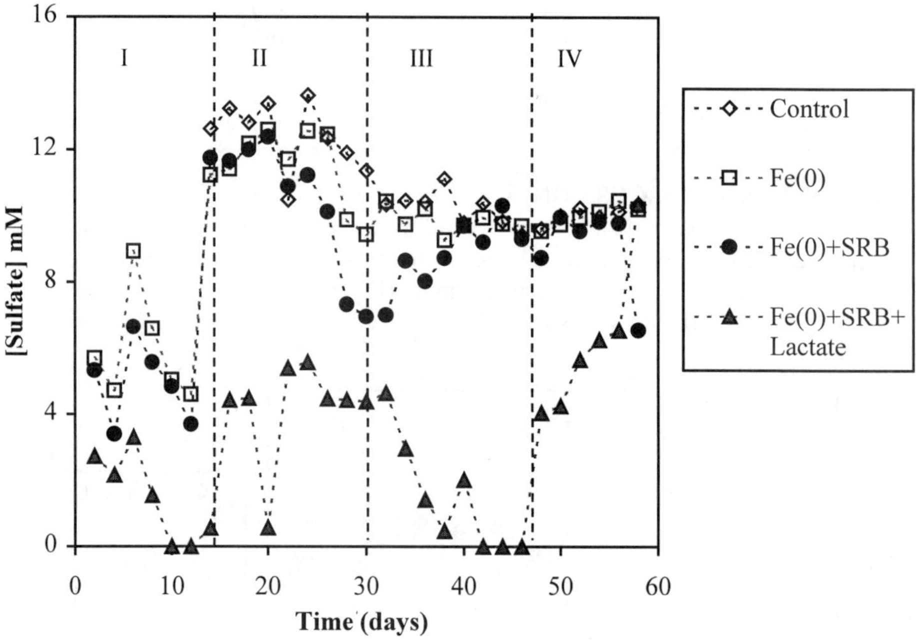
16
12
8
4
0
[Sulfate] mM
0
10
20
30
40
50
60
Time (days)
I
II
III
IV
Control
Fe(0)
Fe(0)+SRB
Fe(0)+SRB+
Lactate

FIGURE 2. Effluent sulfate concentration data.

During Phase III, corresponding to an influent Cr(VI) concentration of 9.6 mM, Cr(VI) levels in the effluent from the Fe(0) column increased starting on the sixth day at this influent concentration (Figure 3). As a result, the instantaneous Cr(VI) removal efficiency fell to about 45% and the cumulative Cr(VI) removal efficiency fell to 80% for the Fe(0) column during this period. In contrast, Cr(VI) was not detected in the effluent from the other treatment columns (Figure 3). Figure 4 indicates that columns Fe(0)+SRB and Fe(0)+SRB+Lactate performed 125% better than the Fe(0) column by the end of Phase III.

Sulfate levels decreased in the Fe(0)+SRB+Lactate column, reaching non-detectable levels on day 42 of the study (Figure 2), while the lactate concentration in the effluent from this column remained less than 1 mM throughout this period and was non-detect during half of this time (data not shown). Hence, as in Phase II, the data indicated that the SRB were biologically active in the Fe(0)+SRB+Lactate column, but not in the Fe(0)+SRB column.

The better performance of the Fe(0)+SRB column compared to the Fe(0) only column might have been due to the added sulfide in this column. The column was seeded with cell suspension from the stock reactor that contained sulfide and FeS. The lack of measurable sulfate reduction in the Fe(0)+SRB column appeared to discount the involvement of SRB in the reduction of Cr(VI) in this column. Because lactate was not provided to column Fe(0)+SRB, the microorganisms apparently did not have enough time to switch from heterotrophic to autotrophic growth using H_2 as electron donor and HCO_3^- as carbon source.

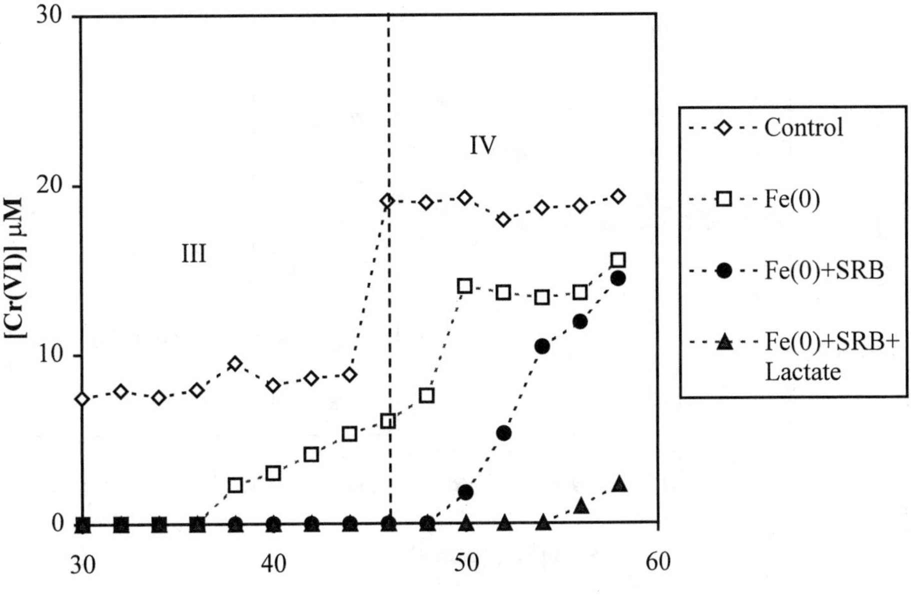
30
20
10
0
[Cr(VI)] μM
III
IV
30
40
50
60
Time (days)
Control
Fe(0)
Fe(0)+SRB
Fe(0)+SRB+
Lactate

FIGURE 3. Effluent Cr(VI) concentrations during Phases III and IV.

During Phase IV, corresponding to an influent Cr(VI) concentration of 19.2 mM, Cr(VI) levels in the effluent from the Fe(0)+SRB column increased starting on the second day at this influent concentration and on the eighth day in the effluent from the Fe(0)+SRB+Lactate column (Figure 3). Also, the Cr(VI) level in the Fe(0) only column reached a steady-state value of 14 mM during this time. As a result, by the end of the study period on day 58, the instantaneous Cr(VI) removal efficiency had decreased to 20% in the Fe(0) only column, to 25% in the Fe(0)+SRB column, and to 90% in the Fe(0)+SRB+Lactate column. The cumulative Cr(VI) removal efficiency values at the end of the study were 57% for the Fe(0) only column, 80% for the Fe(0)+SRB column, and 97% for the Fe(0)+SRB+Lactate column. The ratio of the overall performance of the column pairs, based on total mass of Cr(VI) removed, was: Fe(0)+SRB compared to the Fe(0) column: 143%; Fe(0)+SRB+Lactate compared to Fe(0): 178%; and Fe(0)+SRB+Lactate compared to Fe(0)+SRB: 122% (Figure 4), which indicated that seeding the columns with the SRB culture greatly improved their Cr(VI) removal performance.

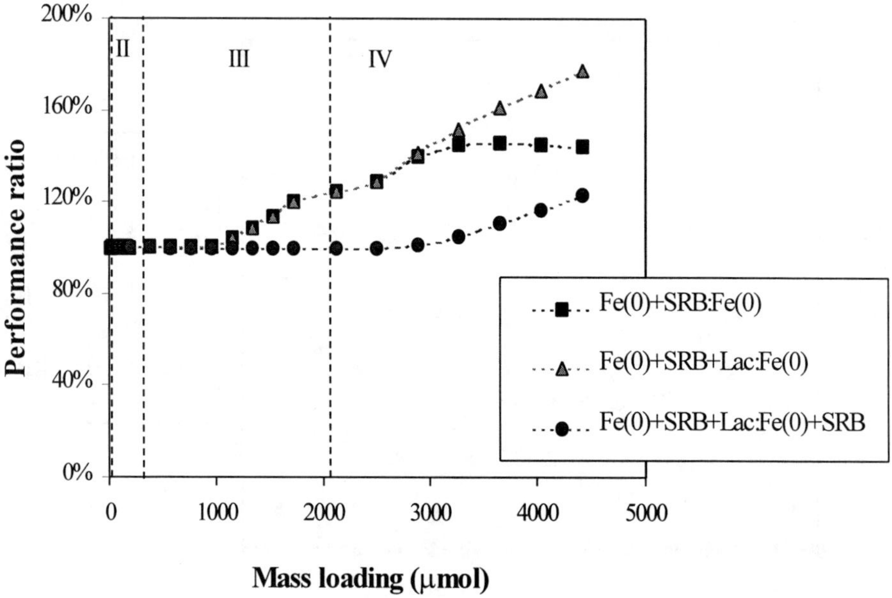
200%
160%
120%
80%
40%
0%
Performance ratio
II
III
IV
0
1000
2000
3000
4000
5000
Mass loading (μmol)
Fe(0)+SRB:Fe(0)
Fe(0)+SRB+Lac:Fe(0)
Fe(0)+SRB+Lac:Fe(0)+SRB

FIGURE 4. Performance ratio comparison for different column pairs, based on the total mass of Cr(VI) removed by each column.

ACKNOWLEDGEMENTS

This project has been funded by the United States Department of Energy (DOE) as part of the Environmental Management Science Program. The contents do not necessarily reflect the views and policies of the USDOE.

REFERENCES

APHA (American Public Health Association). 1992. *Standard Methods for the Examination of Water and Wastewater*, 16th Ed.

Blowes, D.W., C.J. Ptacek, and J.L. Jambor. 1997. "In-Situ Remediation of Cr(VI)-Contaminated Groundwater Using Permeable Reactive Walls: Laboratory Studies." *Environ. Sci. Technol., 31*, 3348-3357.

Gould, J.P. 1982. "The Kinetics of Hexavalent Chromium Reduction by Metallic Iron." *Water Research, 16*, 871-877.

Lovley, D.R. and E.J. Phillips. 1994. "Reduction of Chromate by *Desulfovibrio vulgaris* and its c3 Cytochrome." *Appl. Environ. Micro., 60*, 726-728.

Powell, R., R.W. Puls, S.K. Hightower and D.A. Sabatini. 1995. "Coupled Iron Corrosion and Chromate Reduction: Mechanisms for Subsurface Remediation." *Environ. Sci. Technol.*, *29*, 1913-1922.

Rai, D., B.M. Sass, and D.A. Moore. 1987. "Chromium(III) Hydrolysis Constant and Solubility of Chromium(III) Hydroxide." *Inorganic Chemistry*. *26*, 345-349.

Sass, B.M. and D. Rai. 1987. "Solubility of Amorphous Chromium(III)-Iron(III) Hydroxide Solid Solutions." *Inorganic Chemistry*. *26*, 2232-2234.

Weathers, L.J., G.F. Parkin and P.J. Alvarez. 1997. "Utilization of Cathodic Hydrogen as Electron Donor for Chloroform Cometabolism by a Mixed, Methanogenic Culture." *Environ. Sci. Technol.*, *31*, 880-885.

DEGRADATION OF TCE, Cr(VI), NO_3^-, AND SO_4^{2-} MIXTURES IN COLUMNS MIMICKING BIOAUGMENTED Fe^0 BARRIERS

Sumeet Gandhi, Byung-Taek Oh, Jerald L. Schnoor and Pedro J.J. Alvarez
(The University of Iowa, Civil & Environmental Engineering, Iowa City, Iowa)

ABSTRACT: Flow-through aquifer columns packed with a middle layer of granular iron (Fe^0) were used to study the applicability and limitations of bio-enhanced Fe^0 barriers for the treatment of contaminant mixtures in groundwater. Concentration profiles along the columns showed extensive degradation of most pollutants, mainly in the Fe^0 layer. One column was bioaugmented with *Shewanella algae* BRY, an iron reducing bacterium that could enhance the reactivity of Fe^0 by reductive dissolution of the passivating iron oxides. This strain did not enhance Cr(VI) removal, which was rapidly removed by Fe^0, but it did enhance nitrate removal (from 15% to 80%), partly because BRY has a wide range of electron acceptors, including nitrate. Sulfate was removed (55%) only in the column where Fe^0 was bioaugmented with sulfate reducers. Apparently, these bacteria used H_2 produced by Fe^0corrosion as electron donor to respire sulfate. Most of the trichloroethene (TCE) was degraded in the zone containing Fe^0 (50-70%), and bioaugmentation with *Shewanella algae* slightly increased the removal efficiency to about 80%. SEM pictures confirmed the microbial colonization of Fe^0 samples.

INTRODUCTION

The *in situ* application of granular iron (Fe^0) has become popular for the removal of redox-sensitive groundwater pollutants. Applications of permeable reactive Fe^0 barriers include the destruction of halogenated organic compounds and the immobilization of specific metals such as Cr(VI), U(VI), and Tc(VII) (Cantrell et al., 1995, Gillham and O'Hannesin, 1994; Johnson et al., 1996; Liang et al., 1996; Powell et al., 1995). Iron metal is a powerful reductant with a reduction potential of -0.44 V (eq 1). Thus, the transformation process is a redox reaction where the iron metal is oxidized and the target pollutant is reduced.

$$Fe^0 \rightarrow Fe^{2+} + 2e^- \qquad (1)$$

Early studies of Fe^0 systems focused primarily on abiotic processes. Nevertheless, the potential for microorganisms to enhance reductive treatment with Fe^0 has been conclusively demonstrated in recent laboratory experiments (Till et al., 1998; Weathers et al., 1997). H_2 is produced when Fe^0 corrodes anaerobically (i.e., $Fe^0 + 2H_2O \rightarrow H_2 + 2OH^-$), and microbial utilization of cathodic hydrogen has been shown to be a critical link between biogeochemical interactions and enhanced contaminant removal. Specifically, H_2 can serve as electron donor for the biotransformation of reducible contaminants.

The objective of this study was to further delineate the applicability and limitations of exploiting biogeochemical interactions in Fe^0 barriers for the treatment of common contaminant mixtures.

MATERIALS AND METHODS

Flow-through columns (30-cm in length and 2.5-cm in diameter) were used to simulate the removal of contaminant mixtures in Fe^0 barriers. Six columns were packed with a 5-cm layer of soil (where natural attenuation can occur) followed by an 18-cm layer of Fe^0 filings (representing a reactive barrier). One of the columns was sterilized with a biocide (Kathon CG/ICP, 6 mg/L) to discern degradation by Fe^0 alone. Another (non-sterile) column was run to determine if soil bacteria colonize the Fe^0 layer, presumably to feed on cathodic H_2 produced by anaerobic Fe^0 corrosion. This column also served as a baseline to evaluate the benefits of bioaugmentation. The third column was inoculated with *Shewanella algae* BRY (ATCC 51181, 10 mL of stock (20.4 mg protein/L) at each port) to determine if iron reducers enhance barrier reactivity by reductive dissolution of Fe(III) oxides (i.e., depassivation). The fourth column was inoculated with a sulfate-reducing enrichment (municipal anaerobic sludge [6.6 g VSS/L], 10 mL at each port) to enhance the removal of this pollutant, which does not react readily with Fe^0. The fifth column was prepared with soil and inert glass beads instead of Fe^0 to control for the effect of indigenous microorganisms on the degradation of the contaminants. The sixth column was packed with only glass beads to control for soil processes. All columns were fed synthetic groundwater (von Gunten and Zobrist, 1993) buffered with 5 g/L $CaCO_3$, purged with N_2/CO_2 (80/20 v/v), and amended with trichloroethene (TCE) (50 mg/L), Cr(VI) (10 mg/L), nitrate (25 mg/L as N), and sulfate (100 mg/L). A peristaltic pump was used to maintain flow at 3 mL/hr. The flow was increased to 12 mL/hr to prevent the columns from drying during sampling. Concentration profiles were taken through the different layers using side-sampling ports. Iron samples from the columns were analyzed by Scanning Electron Microscope (SEM) to observe microbial colonization and oxide formation.

TCE was analyzed using a Hewlett Packard 5890 gas chromatograph equipped with an electron capture detector, a DB-5 capillary column (J&W Scientific), and a HP 19395A headspace autosampler. Cr(VI) was measured using an ion chromatograph with a Dionex post column Reagent Delivery Module, a Spectra 100 UV-Vis detector, an Alltech 570 autosampler, and Dionex CG5a guard and CS5A analytical columns. Sulfate and nitrate were analyzed with a Dionex BioLC ion chromatograph. Separation was achieved with an AS4A column. The pH was measured in influent and effluent samples with a Fisher Scientific AB15 pH meter

RESULTS AND DISCUSSION

The pH increased slightly in all columns with Fe^0 from 8.0 (influent) to 8.5 (effluent). Contaminant concentrations were determined at different times along the lengths of all columns. Figures 1 to 4 show such profiles after 274 days, when the columns were operated at a hydraulic retention time of 12 hours. These profiles show extensive degradation of some pollutants, primarily in the Fe^0 layer.

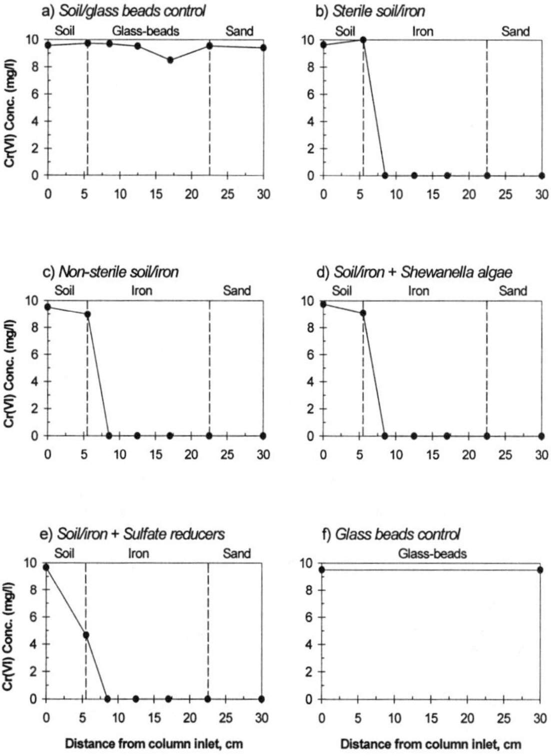
a) Soil/glass beads control
Soil
Glass-beads
Sand
b) Sterile soil/iron
Soil
Iron
Sand
c) Non-sterile soil/iron
Soil
Iron
Sand
d) Soil/iron + Shewanella algae
Soil
Iron
Sand
e) Soil/iron + Sulfate reducers
Soil
Iron
Sand
f) Glass beads control
Glass-beads
Cr(VI) Conc. (mg/l)
Distance from column inlet, cm

FIGURE 1. Cr(VI) Concentration profiles in various flow-through columns.

Figure 1 depicts the concentration profiles for Cr(VI) removal. Hexavalent chromium is reduced to trivalent chromium ion and precipitated as a relatively insoluble Cr(III)-hydroxide or a Cr(III)-bearing Fe(III)-hydroxide (Blowes et al., 1995). Powell et al. (1995) provided substantial evidence about the formation of Cr(III)-Fe(III)-hydroxide solid solution. In this experiment, Cr(VI) was removed only in the zone containing zero-valent iron. Cr(VI) was degraded within 3 cm of the influent into the simulated Fe^0 barrier. Bioaugmentation had no visible effect on Cr(VI) removal. Apparently, the abiotic process was so efficient that left little room for improvement by microbial participation.

Figure 2 shows nitrate removal along the length of the columns. Bioaugmentation enhanced nitrate removal, possibly due to the participation of denitrifiers that use cathodic H_2 as electron donor to respire NO_3^- (Till et al., 1998). Such organisms are ubiquitous, and were likely present in the soil (Figure 2c) and in the mixed sulfate-reducing culture that was added (Figure 2e). The beneficial effect of adding *S. algae* BRY (Figure 2d) may be explained, in part, by the fact that *Shewanella* species can utilize a broad range of electron acceptors, including nitrate. The enhanced removal of nitrate in this latter case could also be due to enhanced reactivity of the Fe^0 layer by reductive dissolution of iron oxides (observed by increased rust in the effluent) due to bacterial respiration.

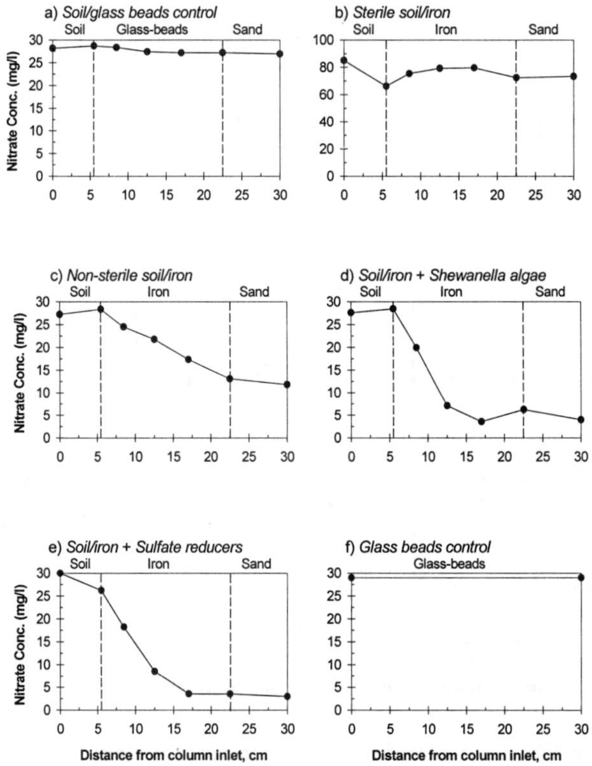
a) Soil/glass beads control
Soil
Glass-beads
Sand
b) Sterile soil/iron
Soil
Iron
Sand
c) Non-sterile soil/iron
Soil
Iron
Sand
d) Soil/iron + Shewanella algae
Soil
Iron
Sand
e) Soil/iron + Sulfate reducers
Soil
Iron
Sand
f) Glass beads control
Glass-beads
Nitrate Conc. (mg/l)
Distance from column inlet, cm

FIGURE 2. Nitrate concentration profiles in various flow-through columns.

Sulfate did not react readily with Fe^0 (Figure 3b) and was removed only in the column bioaugmented with sulfate reducers (Figure 3e). Indigenous sulfate reducers did not exert measurable activity in the viable (i.e., not poisoned) column (Figure 3c) during one year of operation. This illustrates the potential benefits of bioaugmentation, which can decrease the lag time associated with building up a critical biomass concentration needed for fast contaminant removal. Nevertheless, the benefits of bioaugmentation should be evaluated against the potential formation of cell aggregates and biofilms that reduce the available pore space and clog the barrier (Taylor and Jaffe, 1990). We did not observe any obvious evidence of such biofouling effects in our experiments. However, some clogging occurred due primarily to the precipitation of minerals that looked like the carbonate-containing species, siderite ($FeCO_3$) and aragonite ($CaCO_3$). Such species have been observed in Fe^0 barriers (Phillips et al., 2000), which suggests that high alkalinity in groundwater could hinder barrier longevity.

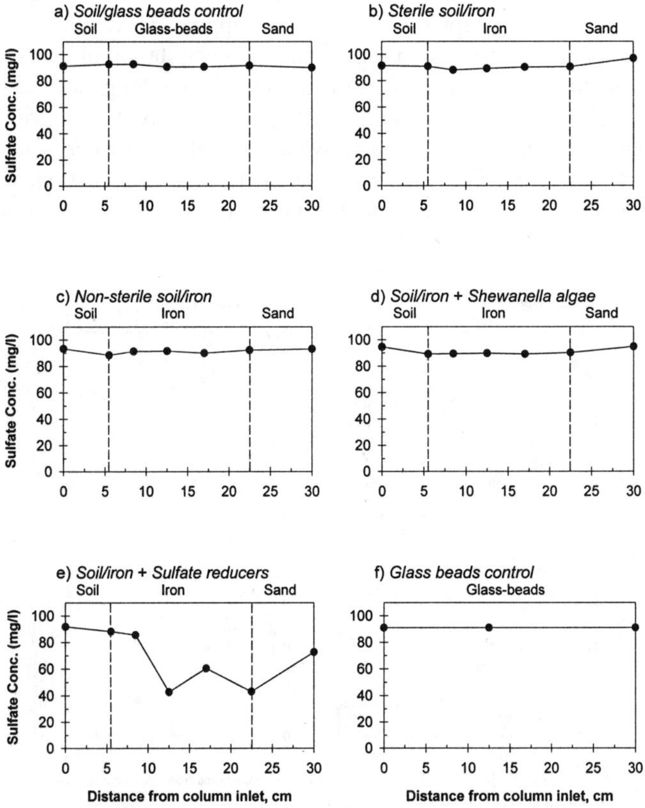
a) Soil/glass beads control
Soil
Glass-beads
Sand
b) Sterile soil/iron
Soil
Iron
Sand
c) Non-sterile soil/iron
d) Soil/iron + Shewanella algae
e) Soil/iron + Sulfate reducers
f) Glass beads control
Sulfate Conc. (mg/l)
Distance from column inlet, cm

FIGURE 3. Sulfate concentration profiles in various flow-through columns.

Figure 4e shows that there were no significant volatilization losses of TCE, demonstrating the integrity of the system. Most of the TCE was degraded in the zone containing iron. No chlorinated byproducts were detected in this work (limit of detection = 1 μg/L). This is consistent with previous studies who have reported that TCE dechlorination by Fe^0 yields primarily nonchlorinated aliphatics (mainly ethene and ethane) and chloride (Orth and Gillham, 1996). Bioaugmentation with iron reducers slightly enhanced TCE degradation, though this effect was not statistically significant. In addition, *cis*-dichloroethene (*c*-DCE), which is a signature of microbial dechlorination, was not detected. No TCE was detected in the effluent of columns containing Fe^0 when operated at a longer retention time of 2 days (data not shown). This retention time corresponds to a Darcy velocity of 1 cm/h which is within the range of typical groundwater flow velocities (Domenico and Schwartz, 1997). It should be pointed out that the removal efficiency of TCE decreased considerably with time, from 99% (in the first 4 months) to 55% (after 7 months). Schlicker et al. (2000) also observed a loss of iron reactivity towards TCE, and attributed it to passivation of the iron due to chromium precipitation and occlusion of reactive Fe^0 sites. This explanation for decreased reactivity is supported by point mapping SEM analysis of our Fe^0 samples (data not shown), which showed some precipitation of chromium oxides, and suggests that the presence of Cr(VI) may hinder the long term performance of Fe^0 barriers. However, we believe that the loss of reactivity was primarily due to precipitation of other minerals that were present in greater quantities (e.g., siderite, aragonite, and FeS in the column with sulfate reducers), as shown by semi-quantitative analysis by X-ray microscopy. Such precipitates could physically block reactive Fe^0 sites.

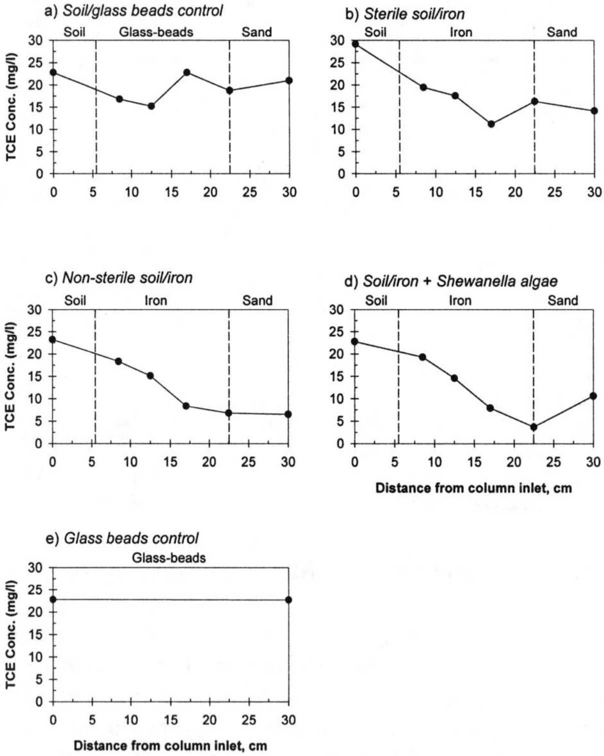
a) Soil/glass beads control
Soil
Glass-beads
Sand
b) Sterile soil/iron
Soil
Iron
Sand
c) Non-sterile soil/iron
Soil
Iron
Sand
d) Soil/iron + Shewanella algae
Soil
Iron
Sand
e) Glass beads control
Glass-beads
TCE Conc. (mg/l)
Distance from column inlet, cm

FIGURE 4. TCE concentration profiles in various flow-through columns.

SEM analysis showed abundant formation of amorphous and crystalline precipitates in Fe^0 samples from all columns. SEM analysis also confirmed microbial colonization on the iron surface. Samples from the viable (not-bioaugmented) column showed rod-shaped bacteria on the surface of iron oxides (Figure 5b), suggesting the possibility that these may be iron reducers, which use cathodic hydrogen as an electron donor to respire iron oxides. The column inoculated with the iron-reducing bacterium *Shewanella algae* BRY, which is a rod shaped bacterium, also showed colonization the Fe^0 surface (see arrows, Figure 5c). Cocci-shaped microorganisms where observed in the column inoculated with sulfate reducers (Figure 5d). This morphology is exhibited by several sulfate-reducer species, such as *Desulfosarcina* (which form similar clusters as shown in Figure 5d) and *Desulfococcus*. However, a rigorous characterization of any of the observed microorganisms was not conducted, which precludes their identification.

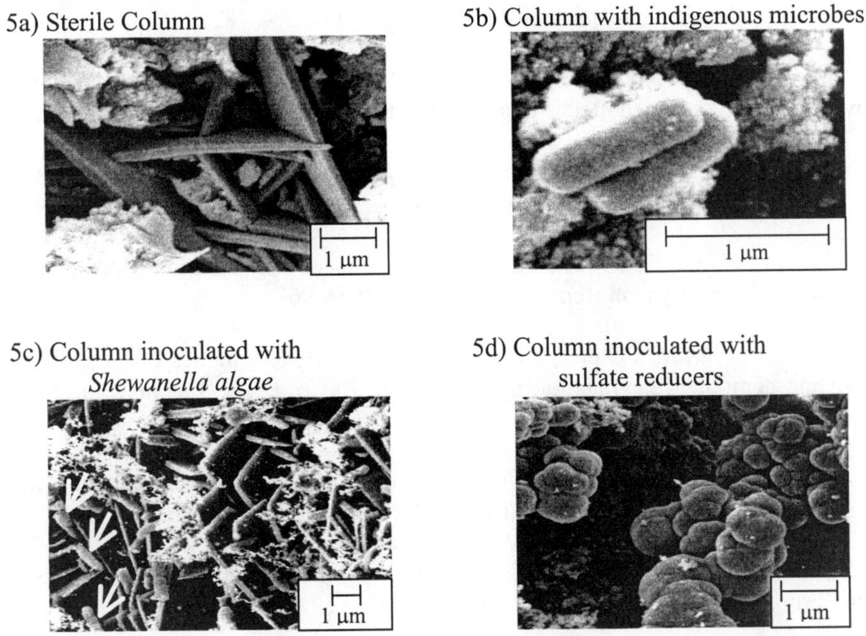
5a) Sterile Column
1 µm
5b) Column with indigenous microbes
1 µm
5c) Column inoculated with *Shewanella algae*
1 µm
5d) Column inoculated with sulfate reducers
1 µm

FIGURE 5. SEM pictures of Fe^0 samples from various columns.

CONCLUSIONS

Column experiments suggest that reactive iron barriers can effectively intercept and degrade mixtures of priority pollutants, and that microorganisms might enhance the removal efficiency in some instances. Bioaugmentation with specialized strains can also increase the range of pollutants that can be treated in iron barriers. Nevertheless, the effect of microorganisms on the long-term permeability and reactivity of the barrier is not well understood. Microorganisms may be able to mediate mineral dissolution or precipitation reactions that affect the availability of reactive sites, reduce barrier permeability by biofilm formation, or alleviate pore volume reduction caused by the formation and entrapment of H_2 gas. Clearly, a better understanding of these biogeochemical processes will form a stronger basis for barrier design.

ACNOWDLEGMENTS

We wish to thank Craig Just and all our Colleagues at ERF for technical assistance and valuable discussions. This work was funded by DOE and by the EPA HSRC for Regions 7 & 8.

REFERENCES

Blowes D.W., C. Ptacek, C.J. Hanton-Fong, and J.L. Jambor. 1995. *In situ* remediation of chromium contaminated groundwater using zero-valent iron. 209th *American Chemical Society National Meeting*. 35(1): 780-783.

Cantrell K.J., D.I. Kaplan, T.W. Wietsma. 1995. Zero-valent iron for the *in situ* remediation of selected metals in groundwater. *J. Hazard. Mater.* 42 (2): 201-212.

Domenico P.A. and F.W. Schwartz. 1997. *Physical and Chemical Hydrogeology*. 2nd Ed. John Wiley and Sons. New York, NY.

Gillham R.W. and S.F. O'Hannesin. 1994. Enhanced degradation of halogenated aliphatics by zero-valent iron. *Ground Wat.* 32: 958-967.

Johnson T.L., M.M. Scherer, and P.G. Tratnyek. 1996. Kinetics of halogenated organic compound degradation by iron metal. *Environ. Sci. Technol.* 30: 2634-2640.

Liang L., B. Gu, and X. Yin. 1996. Removal of technium-99 from contaminated groundwater with sorbents and reductive materials. *Sep. Technol.* 6: 111-122.

Orth W.S. and R.W. Gillham. 1996. Dechlorination of trichloroethene in aqueous solution using Fe^0. *Environ. Sci. Technol.* 30: 66-71.

Phillips D.H., B. Gu, D. Watson, Y. Roh, L. Lianf, and S.Y. Lee. 2000. Performance evaluation of a zerovalent iron reactive barrier: mineralogical characteristics. *Environ. Sci. Technol.* 34: 4169-4176.

Powell R.M., R.W. Puls, S.K. Hightower, and D.A. Sabatini. 1995. Coupled iron corrosion and chromate reduction: Mechanisms for subsurface remediation. *Environ. Sci. Technol.* 29: 1913-1922.

Schlicker O., M. Ebert, M. Fruth, M. Weidner, W. Wst, and A. Dahmke. 2000. Degradation of TCE with iron: the role of competing chromate and nitrate reduction. *Ground Wat.* 38: 403-409.

Taylor S.W. and P.R. Jaffe. 1990. Biofilm growth and the related change in the physical properties of a porous medium I. Experimental investigation. *Wat. Res.* 26: 2153-2159.

Till B.A., L.J. Weathers, and P.J.J. Alvarez. 1998. Fe(0)-supported autotrophic denitrification. *Environ. Sci. Technol.* 32(5): 634-639.

von Gunten U. and J. Zobrist. 1993. Bigeochemical changes in groundwater-infiltration systems: Column studies. *Geochim. Cosmochim. Acta.* 57: 3895-3906.

Weathers L.J., G.F. Parkin, and P.J.J. Alvarez. 1997. Utilization of cathodic hydrogen as electron donor for chloroform cometabolism by a mixed methanogenic culture. *Environ. Sci. Technol.* 31: 880-885.

DISSOLVED PLUME PCE REMEDIATION USING A COMBINATION OF ZERO VALENT IRON AND A HYDROGEN RELEASE COMPOUND

Nick M. Fischer, (Aquifer Technology, Englewood, CO)
Tom Reed, (URS, Wheat Ridge, CO)
Clint Madsen, Terracon, (Wheat Ridge, CO)
Tom Mascarenas, (Environmental Chemistry Services, Englewood, CO)

ABSTRACT: A Phase II Environmental Site Assessment revealed perchloroethylene (PCE) contamination in the groundwater underlying a shopping center in Aurora, CO. From site characterization completed, it was determined that the source of contamination was due to the dry cleaning chemical (PCE) discharged to a leaking sanitary sewer system. Contamination at this site appears to be limited primarily to PCE in groundwater. Under the Voluntary Cleanup and Redevelopment Act Application developed by the Colorado Department of Public Health and Environment, a corrective action plan was submitted and approved. After conducting an economic and technical feasibility study, it was determined that the most cost-effective and technically feasible remedial option, which could be readily applied and presented little long-term site disturbance was to use a Permeable Reactive Bed (PRB) combination of zero-valent iron (ZVI) and Hydrogen Release Compound (HRC®). Soil vapor extraction and in-situ air sparging was conducted, for a period of five months, to remediate the free-phase PCE source area. The off-site impact and dissolved plume were remediated using a combination of ZVI and HRC injections to form a dual-phase PRB. The placement of ZVI creates a permeable reactive zone to enhance the dechlorination of PCE via a series of abiotic reactions. HRC is used to accelerate in-situ bioremediation of PCE and daughter products that escape the influence of the zero-valent iron. The aquifer was composed mostly of silty clay with interbedded sand lenses. Sand was used to backfill utility corridors running throughout the site. Groundwater velocity is to the southwest estimated at approximately 7 to 65 ft/yr. The treatment area is 1,500 ft^2. Iron-filing remedial implementation activities were performed approximately 11 months following the SVE/air sparging application. 735 lbs of iron were injected via nine injection points, with PCE concentrations at this time averaging 680 μg/L. After seven months, PCE concentrations were reduced to an average of approximately 340 ug/L, a decrease of 50%. At this point, 300 lb HRC was injected via seven application points. PCE concentrations 1 year following the HRC applications decreased to an average of 8 μg/L, representing over a 95% reduction of the residual contaminant.

INTRODUCTION

Subsurface Stratigraphy, Sieve Analyses and Hydraulic Conductivity Tests. Based on observations made during completion of the soil borings, subsurface conditions can be generalized as follows, with reference to Figure 1: for wells TW-1 through TW-5 installed on site, road base and silty sand were generally observed beneath the paved surface to depths ranging from approximately four (4) to seven (7) feet below site grade. This material was underlain by clay with varying amounts of silt and sand that extended to depths ranging from approximately 14 to 24 feet below site grades. The clay material was in turn underlain by weathered shale material to the bottoms of the borings at approximately 24 to 26 feet below site grades.

For wells TW-6 through TW-9, installed in Smoky Hill Road, sands were encountered below the pavement to depths ranging from 11 to 14 feet below site grades. The sand was underlain by silt and clay with varying amounts of sand and silt to the bottoms of the borings at approximately 26 feet below site grades. The groundwater surface in the wells was observed in both the sand and clay zones.

For wells TW-10 through TW-12, which were installed south of Smoky Hill Road, clay with varying amounts of silt and sand was observed from beneath the surface pavement to depths ranging from approximately 22 to 23 feet below site grades. The clay material was underlain by weathered shale material to the bottoms of the borings at approximately 24 to 25 feet below site grade.

A sieve analyses was performed on soil samples collected from the on-site soil boring cuttings during the initial phase drilling. The sieve analyses results indicated that soils are approximately one (1) percent gravel, 28 percent sands, 28 percent silts and 43 percent clays. The results were indicative of extremely fine-grained soil and suggest relatively low hydraulic conductivities.

Slug tests were performed on two (2) on-site monitoring wells after the initial drilling. Hydraulic conductivities were estimated to range from approximately 6.3×10^{-5} cm/sec to 1.9×10^{-3} cm/sec. These relatively fast hydraulic conductivities, when considering the sieve analyses results, may suggest fracturing in the soils, allowing for contaminant migration along fracture pathways.

Based on the range of hydraulic conductivities, groundwater gradients and effective porosities, seepage velocities in the vicinity of the site may range from approximately 6.5 to 655 feet per year. It is estimated that the lower velocity seepage velocity range is likely at the subject site based on the distance from the apparent source area to the outlying monitoring wells (approximately 250 to 300 feet).

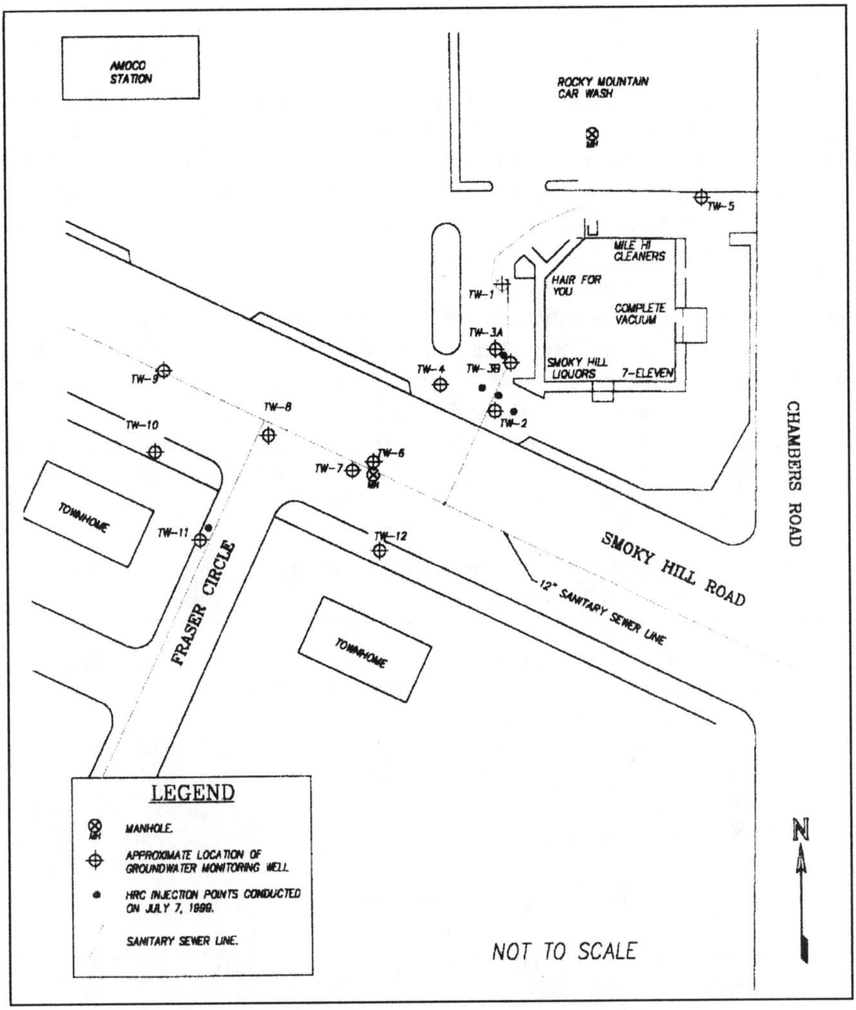
AMOCO STATION
ROCKY MOUNTAIN CAR WASH
TW-5
MILE HI CLEANERS
HAIR FOR YOU
COMPLETE VACUUM
SMOKY HILL LIQUORS
7-ELEVEN
TW-1
TW-3A
TW-3B
TW-4
TW-2
TW-9
TW-10
TW-8
TW-6
TW-7
TW-11
TW-12
TOWNHOME
TOWNHOME
FRASER CIRCLE
SMOKY HILL ROAD
CHAMBERS ROAD
12" SANITARY SEWER LINE
LEGEND
MANHOLE
APPROXIMATE LOCATION OF GROUNDWATER MONITORING WELL
HRC INJECTION POINTS CONDUCTED ON JULY 7, 1999.
SANITARY SEWER LINE.
NOT TO SCALE
N

FIGURE 1 - SITE AND STUDY AREA DIAGRAM

Remaining Onsite and Offsite PCE Remediation. After the onsite source area was remediated, the remaining onsite and offsite perchloroethylene (PCE) groundwater contamination required remediation. After conducting an economic and technical feasibility study, it was determined that the most cost effective and technical remedial option to use was to conduct a ZVI iron guar mixture to form a permeable reactive bed (PRB). An injection technology was used for the PRB. The first zero valent iron (ZVI) guar mixture PRB injection did not appear to be sufficient to remediate the remainder of the onsite and offisite PCE contamination. To complete the remaining onsite and offisite PCE contamination, the injection of a hydrogen release compound (HRC) into the same PRB was conducted, as well as additional ZVI-guar injections.

Economic and Technical Feasibility Study to Control Offsite Migration. Groundwater remedial alternatives were evaluated during this process. Both in-situ and ex-situ conceptual designs and cost estimates were prepared for various systems and clean up methodologies, based on parameters estimated from field-testing and assessment activities. The remedial alternatives matrix which was initially considered is presented in Table 1.

Based on a review of the alternative matrix and the study process, the selected remediation alternative for groundwater remediation was utilizing a PRB combination of ZVI and HRC. This in-situ combination was considered desirable based on time to cleanup, cost and because problems associated with groundwater recovery, groundwater treatment and groundwater disposal are eliminated. As well, this in-situ alternative creates little long-term site disturbance and did not require capital investment and operation/maintenance costs for remediation equipment and buildings.

TABLE 1 – COST CAPARISONS FOR REMEDIAL TECHNOLOGIES

Remedial Technology	Cost
• In-situ SVE with horizontal wells, and in-situ air-sparging with vertical wells	$116,000 to $126,000
• In-situ SVE with horizontal wells, in-situ air-sparging with horizontal wells	$134,000 to $144,000
• In-situ SVE with horizontal wells, groundwater recovery with vertical wells, low-profile aeration, disposal to sanitary sewer	$149,000 to $159,000
• In-situ dechlorination of groundwater using iron funnel and gate system	$128,000 to $138,000
• In-situ dechlorination of groundwater using ZVI-guar injection & HRC	$24,000 to $30,000

ZVI Dechlorination Theory. The placement of zero-valent iron creates a permeable reactive zone to enhance the declhorintation of PCE. The mechanism of the PRB is via re-dox reactions which dechlorinates the PCE and PCE byproducts. Reductive treatment with ZVI is driven by the oxidation of ZVI which releases electrons, and then used to reduce chlorinated solvents.

HRC Remediation Theory and Assessement. HRC is used to enhance in-situ biodegradation rates for chlorinated aliphatic hydrocarbons (CAHs) such as PCE and the degradation byproducts. HRC consists of a proprietary food grade polylactate ester that, upon being deposited in the subsurface, slowly releases lactate and its derivatives. Lactate is then metabolized to hydrogen, which in turn drives reductive dechlorination of CAHs. The use of HRC for groundwater remediation should result in a comparatively simple and cost-effective remediation alternative for the site that would otherwise require unacceptably long periods of time for natural attenuation, or high levels of capital investment and operating expense for more active remediation technologies.

To assess the HRC injection, groundwater samples were collected from both background and impacted monitoring wells. Groundwater samples have been analyzed for volatile organic compounds (VOCs), heterotrophic aerobic and anaerobic plate counts, total iron, and metabolic acids. In addition, groundwater samples were also collected for anions / cations, and alkalinity. Dissolved oxygen (DO), reduction-oxidation (redox), pH, methane, and carbon dioxide measurements were conducted in the field. In order for the PCE degrading anaerobic bacteria to thrive; an environment with low DO concentrations, reduced conditions (negative redox values), and nutrients in sufficient quantities, are all required to support bacteria growth.

METHODS

Remedial Implementation. A PRB injection model (PRBIM) was developed to assess site conditions for remediation, and has developed preliminary values for a pilot test. The PRBIM estimates travel times and reactive bed thicknesses required for the dechlorination of chlorinated solvents. Using this model and consulting URS, an injection scheme was developed.

ZVI remedial implementation activities were performed at the site on December 15, 16, 1998, and again February 2 and 4, 2000. The approximate amount of 92 pounds of guar, 915 gallons of water, and 915 pounds of iron-guar mixture was injected via direct-push technology at an injection pressure of approximately 80 psi, at twelve (12) injection points. Three (3) to four (4) injections were completed at various depths for each injection hole. URS provided personnel and developed field equipment to inject the slurry. The iron was mixed with guar powder and water forming a slurry in a mixing bin, and utilizing a positive displacement pump injected the iron-guar slurry via direct-push technology (DPT). An enzyme was added in the injected slurry to break down the guar as to not affect groundwater flow (the enzyme breaks down the guar in a couple of hours time). From conducting DPT soil sampling it appeared that the radius of injection ranged from 1 to 5 feet depending on placement of injection and subsurface soil classification.

HRC remedial implementation activities were performed at the site on July 7, 1999, and again November 12, 1999. The approximate amount of 300 pounds of HRC was injected via direct-push technology at an injection pressure of approximately 80 psi, at seven (7) injection points. Three (3) to four (4) injections were completed at various depths for each injection hole. The HRC injection was again conducted with a positive displacement pump via DPT.

RESULTS

VOCs. The HRC and ZVI PRB has reduced PCE concentrations at interior wells TW-2 from 680 to 4 micro grams per liter (µg/L), TW-6 from 450 to 150 µg/L, and at exterior well TW-11 from 180 to 11 µg/L. Degradation byproducts trichloroethene and 1,2-dichloroethene have been observed. These daughter products may degrade at a slower rate than the parent products depending on aquifer conditions and contaminant desorbtion that may be taking place. The event on December 17, 1998 is probably an artifact of iron injection, as rebound occurred, later to be addressed by the HRC.

TABLE 2 - SUMMARY OF PCE AND DEGRADATION BYPRODUCTS

Well	Date	PCE (µg/L)	TCE (µg/L)	1,1 DCE (µg/L)	1,2 DCE (µg/L)	VC (µg/L)
TW-2	7/18/96	286	ND	ND	ND	ND
TW-2	12/5/97	490	ND	ND	ND	ND
TW-2	2/26/98	680	ND	ND	ND	ND
TW-2	12/17/98	9.3	ND	ND	ND	ND
TW-2	12/31/98	200	ND	ND	ND	ND
TW-2	1/14/99	310	5.6	ND	ND	ND
TW-2	1/24/99	370	ND	ND	ND	ND
TW-2	3/16/99	65	ND	ND	ND	ND
TW-2	7/28/99	160	ND	ND	ND	ND
TW-2	8/15/99	90	ND	ND	ND	ND
TW-2	9/6/99	46	6.0	ND	ND	ND
TW-2	12/23/99	5.5	8.3	ND	ND	ND
TW-2	12/3/00	4.0	3.0	ND	41	ND
TW-2	3/2/01	4.0	34	ND	27	ND
TW-6	10/15/96	200	ND	ND	ND	ND
TW-6	2/26/98	310	ND	ND	ND	ND
TW-6	12/17/98	370	ND	ND	ND	ND
TW-6	12/31/98	830	ND	ND	ND	ND
TW-6	1/14/99	590	ND	ND	ND	ND
TW-6	12/30/99	400	ND	ND	ND	ND
TW-6	6/8/00	320	ND	ND	ND	ND
TW-6	12/29/00	150	ND	ND	ND	ND

note: 1st iron wall application = 12/15/98. 1st HRC application = 7/7/99.

TABLE 2 - SUMMARY OF PCE AND DEGRADATION BYPRODUCTS

Well	Date	PCE (μg/L)	TCE (μg/L)	1,1 DCE (μg/L)	1,2 DCE (μg/L)	VC (μg/L)
TW-11	11/12/96	28	ND	ND	ND	ND
TW-11	9/19/97	18	ND	ND	ND	ND
TW-11	10/14/97	34	ND	ND	ND	ND
TW-11	1/19/98	32	ND	ND	ND	ND
TW-11	2/26/98	34	ND	ND	ND	ND
TW-11	8/18/98	39	ND	ND	ND	ND
TW-11	11/3/98	55	ND	ND	ND	ND
TW-11	12/31/98	50	ND	ND	ND	ND
TW-11	1/14/99	65	ND	ND	ND	ND
TW-11	3/19/00	68	ND	ND	ND	ND
TW-11	8/3/99	31	ND	ND	ND	ND
TW-11	12/28/99	180	ND	ND	ND	ND
TW-11	4/11/00	48	ND	ND	ND	ND
TW-11	6/7/00	11	ND	ND	ND	ND
TW-11	9/17/00	11	ND	ND	ND	ND
TW-11	11/30/00	ND	ND	ND	ND	ND

note: 1st iron wall application = 12/15/98. 1st HRC application = 7/7/99.

Heterotrophic Aerobic and Anaerobic Plate Counts. Groundwater samples were collected in both impacted and background areas. The addition of HRC increased anaerobic heterotrophic plate counts in the area in which HRC injection was conducted by a factor of 3,000. Anaerobic bacteria counts in background monitoring wells averaged 10,000 cfu/mL from four samples collected, whereas in areas which HRC was injected monitoring wells TW-2 and TW-11 averaged 30,000,000 from seven samples collected.

TABLE 3 - SUMMARY OF HETERTROPHIC BACTERIA COLONIES

Monitoring Well	Date	Aerobic (cfu/mL)	Anaerobic (cfu/mL)
TW-2	2/7/00	62,000,000	33,000,000
TW-2	3/27/00	15,000,000	14,000,000
TW-2	8/28/00	49,000,000	51,000,000
TW-11	9/14/00	6,800,000	1,600,000
TW-11	11/11/00	8,000,000	8,000,000
TW-11	8/28/00	87,000,000	77,000,000
TW-11	10/15/00	29,000,000	26,000,000
Background Wells			
TW-4	9/14/00	310,000	35,000
TW-4	11/11/00	21,000	7,300
TW-10	3/27/00	180,000	<100
TW-12	10/15/00	16,000	6,000

Metabolic Acids. Groundwater samples were collected from groundwater monitoring well TW-11 on January 19, 2000 and from TW-2 on February 7, 2000 and analyzed for metabolic acids (acetic, butyric, lactic, propionic, and pyruvic) as an additional assessment tool for biological degradation. It appears that the injection of HRC has released metabolic acids as expected.

TABLE 4 - SUMMARY OF METABOLIC ACIDS

Metabolic Acid	TW-2 Concentration (mg/L)	TW-11 Concentration (mg/L)
Acetic	817	84
Butyric	1090	72
Lactic	1700	36
Proprionic	1070	101
Pyruvic	<0.1	0.2

DO, Redox, and pH. Since the HRC injection began, the DO and redox measurements clearly show a significant change of conditions of DO and redox at monitoring wells TW-2 and TW-11, and moderate changes of DO and redox at interior well TW-3A. The HRC injection changed the environment from an oxidized state to a reduced state; thus potentially creating an anaerobic environment for PCE degradation. HRC addition decreased pH values in the area of injection at monitoring wells TW-2 and TW-11 from an average of 7.1 to 6.1. This was due to HRC and the release of metabolic acids. The system is still in the generally recognized optimal zone of 6-8.

TABLE 5 - SUMMARY OF FIELD MEASUREMENTS

Well	Date	DO (mg/L)	Re-dox (mv)	pH
TW-2	6/5/98	3.7	212	7.2
TW-2	7/28/99	1.0	-187	7.0
TW-2	8/13/99	0.8	-213	6.5
TW-2	9/21/99	0.6	-182	6.1
TW-2	1/20/00	0.5	-155	6.7
TW-2	4/30/01	0.5	-196	7.0
TW-3A	7/28/99	1.6	-74	7.0
TW-3A	8/13/99	1.4	-88	7.0
TW-11	6/5/98	3.8	193	7.2
TW-11	7/28/99	2.4	177	7.0
TW-11	8/13/99	2.2	156	6.5
TW-11	1/20/00	0.6	-165	5.9
TW-11	12/3/00	0.8	-87	6.9

Methane and Carbon Dioxide. As another parameter to assess abiotic (nonbiological) and biotic (biological anaerobic and aerobic degradation); methane and carbon dioxide were also measured in the field with a portable GeoTech GA90 Infra-red detector. Methane and ethane are generally end-products of anaerobic degradation. If hydrogen is present in excess this can lead to the excessive production of methane which can be inhibitory to reductive dechlorination. Slow release hydrogen compounds can limit this effect to the advantage of the desired reactions. Carbon dioxide is generally an end product of both aerobic and anaerobic degradation. Methane was not observed in any of the monitoring wells, however, carbon dioxide was measured at approximately 6% at TW-2 (background appears to be approximately 0.3%).

TABLE 6 - SUMMARY OF METHANE AND C02 MEASUREMENTS

Well	Date	Methane (%)	CO2 (%)
TW-1	8/2/99	ND	0.3
TW-2	8/2/99	ND	6.0
TW-3	8/2/99	ND	0.4
TW-4	8/2/99	ND	0.4
TW-5	8/2/99	ND	0.3
TW-6	8/2/99	ND	0.2
TW-7	8/2/99	ND	0.3
TW-8	8/2/99	ND	0.3
TW-9	8/2/99	ND	0.2
TW-10	8/2/99	ND	0.3
TW-11	8/2/99	ND	0.3
TW-12	8/2/99	ND	0.3

ND = not detected

CONCLUSION

After seven months following ZVI injection, PCE concentrations at impacted areas were reduced to an average of approximately 340 μg/L, a decrease of 69%. At this point, 300 lbs HRC was injected via seven application points. The addition of HRC increased both aerobic and anaerobic heterotrophic plate counts in the area in which HRC injection was conducted by a factor of 3,000. Anaerobic bacteria counts in background monitoring wells averaged 10,000 cfu/ml, whereas in areas which HRC was injected monitoring wells TW-2 and TW-11 averaged 30,000,000. Since the HRC injection began, the DO and redox measurements clearly show a significant change of conditions at monitoring well TW-2 and TW-11, and moderate changes at TW-3A. One year following the HRC applications, and some additional ZVI applications, PCE concentrations decreased to an average of 8 μg/L, representing over a 95% reduction of the residual contaminant. It appears that these technologies were cost effective and technically feasible, with little long-term site disturbance, and were readily implementable and effective.

COMBINATION OF IRON AND MIXED ANAEROBIC CULTURE FOR PERCHLOROETHENE DEGRADATION

Xiaohong Luo (NRC Research Associate at U.S. EPA, Ada, Oklahoma)
Guy W. Sewell (U.S.EPA, Ada, Oklahoma, USA)

ABSTRACT: The potential for enhancing reductive dechlorination of carbon tetrachloride and chloroform by incubating methanogenic enrichment together with iron has been demonstrated conclusively in early studies. However, the results with perchloroethene (PCE) are less clear. Laboratory batch experiments were performed to investigate PCE degradation in a system which combined zero-valent iron and a mixed anaerobic culture. The results demonstrated that in an organic-carbon-poor subsurface environment, anaerobic bacteria could utilize cathodic hydrogen as an electron donor for methane generation and PCE biotransformation. Compared to iron-only or culture-only systems, PCE degradation was enhanced in the system coupling iron and the anaerobic culture. The amount and type of zero-valent iron affected both the abiotic PCE reduction and the anaerobic biotransformation of PCE. In the system with 50 g/L iron B and enrichment, PCE was completely transformed to environmentally benign ethylene (70%) and ethane (30%), while in the system with 50 g/L iron A and enrichment, 67% of PCE was transformed to ethylene (44%), ethane (10%) and acetylene (13%). With lesser amounts of iron A (5 g/L) in the system, only 16% of the PCE was transformed to harmless products (ethylene, ethane and acetylene), while with the iron B (5 g/L) system, this proportion was 55%. The results of this study could be helpful in developing advanced remediation strategies for sites contaminated with PCE.

INTRODUCTION

Chlorinated aliphatic hydrocarbons (CAHs) are frequent groundwater contaminants. Their presence in the environment is of great concern because of their potential carcinogenicity. Permeable reactive barriers (PRB) are one of the most promising technologies for in-situ clean up of CAH-contaminated groundwater, with zero valent iron as the most frequently used medium (Sacre, 1997). Anaerobic bacteria also have the potential ability to completely dechlorinate CAHs to environmentally benign end products when suitable environmental conditions, including available electron donors, are present. Due to the advantages and disadvantages of these two technologies, the combination of iron and anaerobic reductive biodechlorination is receiving more attention. It is hypothesized that using anaerobic bacteria together with iron PRB could increase the rate and extent of transformation of CAHs. Early studies have demonstrated that coupling iron with either mixed anaerobic culture or pure cultures of methanogens (*Methanosarcina barkeri*, *Methanosarcina thermophila,* and *Methanosaeta concillii*) could enhance the rate and extent of carbon tetrachloride (CT) and chloroform (CF) degradation (Weathers et al., 1997; Novak et al., 1998). However, results with perchloroethene (PCE) are less clear (Gregory et al., 2000).

Since PCE and trichloroethene (TCE) are the chlorinated organic contaminants most frequently treated with permeable reactive barriers, we performed a study to evaluate PCE degradation by a combination of iron and mixed anaerobic culture. The objectives of the research were: (1) to determine if cathodic hydrogen could be utilized by anaerobic bacteria as an electron donor to support PCE biotransformation in an organic-carbon-poor subsurface environment, and (2) to evaluate the effects of the type and amount of iron on PCE degradation in the combination systems.

MATERIALS AND METHODS

Batch tests were conducted in 160 mL glass serum bottles containing a 60 mL headspace and a 100 mL liquid volume. Basal medium, enrichment culture, PCE and iron filings were anaerobically delivered to each bottle. All the bottles were sealed with Teflon-faced, gray butyl rubber stoppers and aluminum crimp caps to maintain anaerobic conditions. Low concentrations of PCE (2.7 µmol/bottle) were used in the first 49 days of the experiment to allow the anaerobic enrichment adapt to PCE. On the 50th day of incubation, the PCE concentration was increased to approximately 30.0 µmol/bottle.

Chemicals. PCE (99.9%, HPLC Grade, SIGMA Chemical Co.) was used as the original contaminant. PCE, trichloroethene (TCE), dichloroethenes (DCEs) (99.9%, HPLC Grade, SIGMA Chemical Co.), ethylene, methane, and hydrogen gases (99%, Scott Specialty Gases) were used as analytical standards. Peerless iron filings (iron A, BET surface area: 0.882 m^2/g) and electrolytic iron filings (iron B, BET surface area: 0.089 m^2/g, 100 mesh, Fisher Scientific) were used as reactive media and potential electron donors, which were not “cleaned” before use to remove the oxides that passivate the surface.

Enrichment and Growth Medium. Enrichment of anaerobic bacteria was developed by mixing 500 g Pinellas soil, 900 mL reduced Byrds Mill medium (RBMM) and 200 mg/L NaCl (Sewell and Hightower, 1998). Ten microliter PCE and toluene were added at the beginning of the incubation and periodically whenever solution concentrations became undetectable. The enrichments were stored in an anaerobic glove box. RBMM was made of Byrds Mill spring water and distilled water (1/1, v/v), supplemented with $(NH_4)_2HPO_4$. Resazurin was added as a redox indicator. After adjusting the pH to 7, the prepared medium was autoclaved for 45 min at 121°C and placed in an anaerobic glove box overnight. Sodium sulfide (Na_2S) was then added to give a final concentration of 1 mM.

Analytical Methods. PCE, TCE, DCEs and vinyl chloride (VC) determination was performed with a model 5890 series gas chromatography (Hewlett-Packard, Palo Alto, CA) equipped with an electron capture detector using liquid subsamples. Measurement of methane, ethylene, acetylene, and H_2 were conducted with KAPPA-5 Gas Analyzer equipped with a reduction gas detector and flame ionization detector (Trace Analytical Inc, Menlo Park, CA), using headspace gas samples. Equilibrium H_2 concentration in the liquid phase was then calculated by the use of Henry's law.

RESULTS AND DISCUSSIONS

Cathodic Hydrogen Utilization by Anaerobic Bacteria as Electron Donor for Methane Production and PCE Biotransformation. Figure 1a-d visualizes the hydrogen, methane, ethylene, and ethane profiles in the following bottle systems: control, iron B only, enrichment only, enrichment plus iron B. No organic electron donor was added to these systems. Cathodic hydrogen from the anaerobic corrosion of iron B was the only detected electron donor available to sustain anaerobic bacterial growth. In the bottle system with only anaerobic enrichment, little methane was formed (Figure 1b) and only low levels of the end products (ethylene and ethane) of PCE biotransformation were detected (Figure 1c and 1d), presumably due to the limited availability of electron donor. In the bottle system with only iron B, anaerobic corrosion of iron metal resulted in higher measurable H_2 concentrations (Figure 1a). In this system, abiotic reductive dechlorination occurred, no H_2-utilizing methanogenic activity was observed and H_2 concentrations remained higher than 100μM throughout the incubation (Figure 1a) . As a comparison, in the system incubated with both iron B (same amount as iron B only system) and enrichment, anaerobic bacteria were very active when the electron donor (cathodic hydrogen) was present. Large amounts of methane were formed (Figure 1b) and more ethylene and ethane were produced from PCE biotransformation (Figure 1c and Figure 1d), both processes should consume cathodic hydrogen as an electron donor, and thus explaining the much lower H_2 concentration in the system as compared to the iron only system (Figure 1a).

Effects of Type and Amount of Iron on Anaerobic PCE Degradation. Two types of iron filings were used in an effort to observe their effects on PCE degradation. Figure 2a-d show the end products concentrations from PCE degradation and the concurrent methane formation. At the 5 g/L concentration , iron A demonstrated less abiotic reductive capacity for PCE. The production of ethylene, ethane, and acetylene was only about 40% of that observed with the iron B system (Figure 2a-c). As for the enhancement of microbial activity, iron A showed limited effects in the systems incubated with both enrichment and iron A. Only a small amount of methane was formed (Figure 2d) and ethylene, ethane and acetylene production showed little difference from the system with iron A only (Figure 2a-c). However, iron B significantly enhanced microbial activity and PCE degradation in the system amended with enrichment and iron B. Methane production was 10 times higher than that observed in the other systems (Figure 2d), and ethylene, ethane and acetylene production was greatly enhanced compared to the systems with either iron B only or enrichment only (Figure 2a-c).

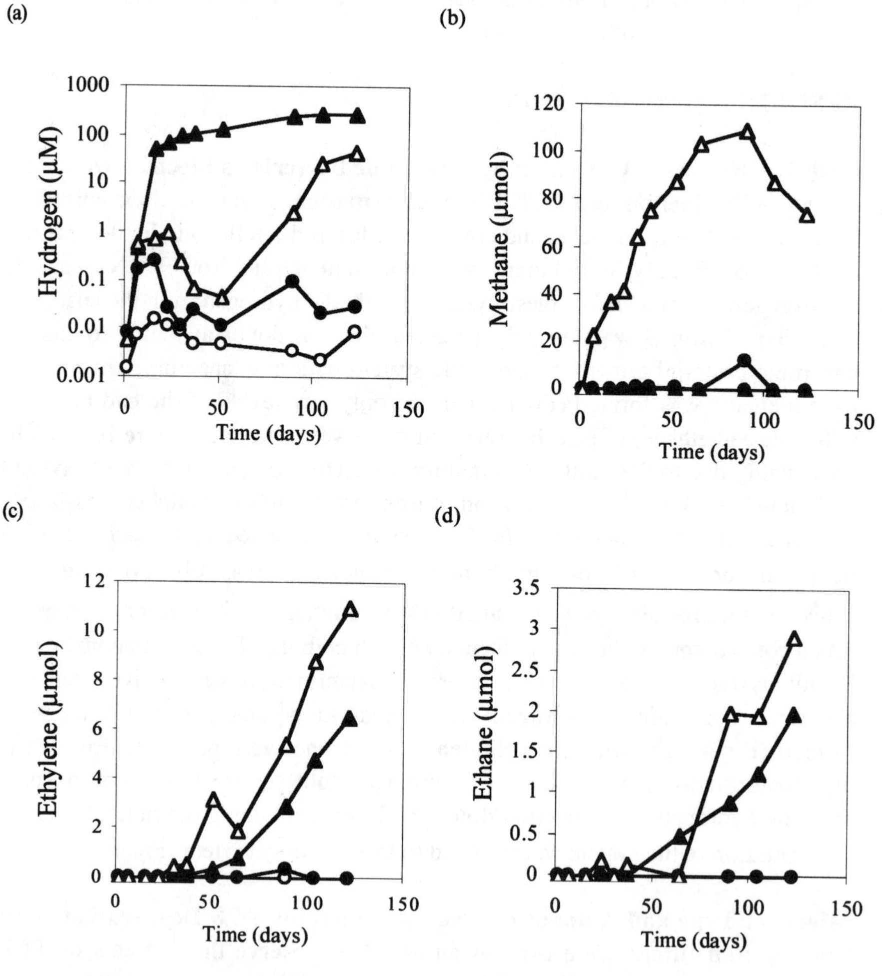
(a)
Hydrogen (µM)
1000
100
10
1
0.1
0.01
0.001
0
50
100
150
Time (days)
(b)
Methane (µmol)
120
100
80
60
40
20
0
0
50
100
150
Time (days)
(c)
Ethylene (µmol)
12
10
8
6
4
2
0
0
50
100
150
Time (days)
(d)
Ethane (µmol)
3.5
3
2.5
2
1.5
1
0.5
0
0
50
100
150
Time (days)

FIGURE 1. Environmentally benign products production from PCE degradation and concurrent methane formation with cathodic hydrogen as electron donor (O control; ● enrichment only; ▲ iron B (5 g/L) only; Δ enrichment + iron B (5 g/L))

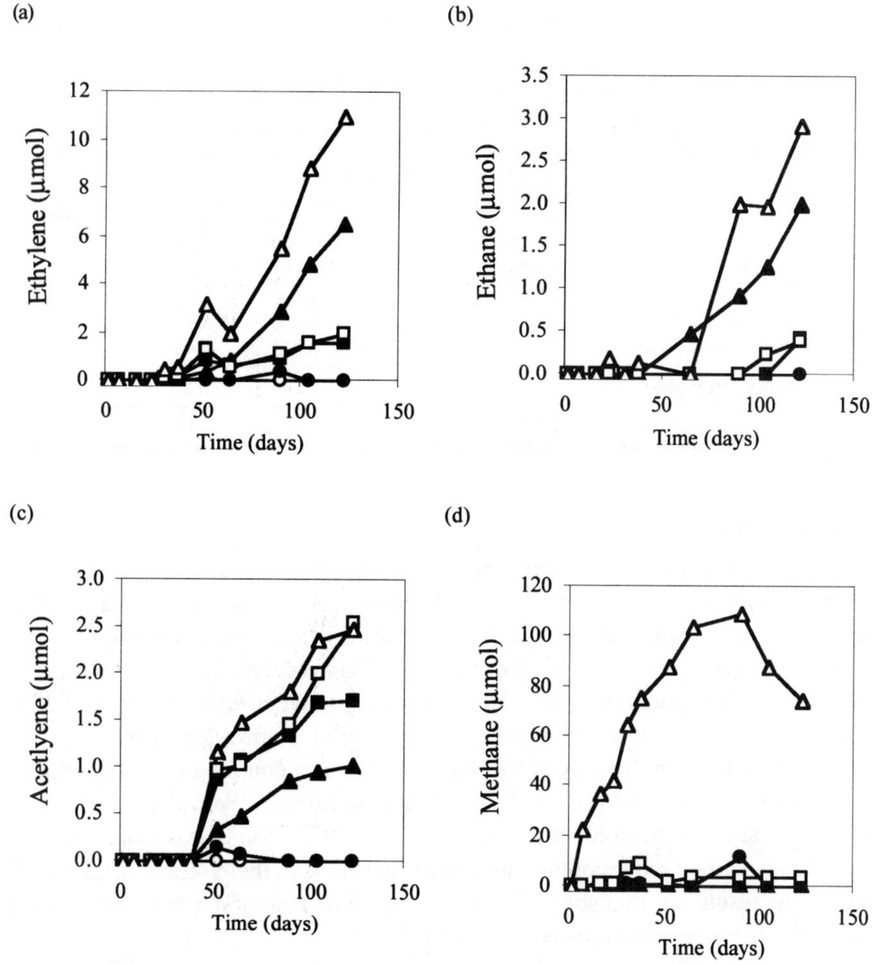
(a)
Ethylene (μmol)
Time (days)
(b)
Ethane (μmol)
Time (days)
(c)
Acetlyene (μmol)
Time (days)
(d)
Methane (μmol)
Time (days)

FIGURE 2. Effects of iron type on anaerobic PCE degradation (O control; ● enrichment only; ■ iron A (5 g/L) only; □ enrichment + iron A (5 g/L); ▲ iron B (5 g/L) only; Δ enrichment + iron B (5 g/L))

Two concentrations of iron (5 g/L and 50 g/L) were chosen to investigate the impact of the amount of iron on PCE degradation. The end product concentration and methane formation in these systems at day 122 (the end of the experiment) are summarized in Table 1. Data in Table 1 show that the amount of added iron had a significant influence on PCE degradation. For both iron A and iron B, increasing the amount of iron increased the proportion of PCE degraded to harmless end products. In the system with enrichment and 50 g/L iron B, all PCE was degraded to ethylene and ethane, while only 55% PCE was completely degraded in the system with enrichment and 5 g/L iron B. In the system with iron A, this proportion was 67% (50 g/L iron A) and 16% (5 g/L iron A), respectively.

TABLE 1. Harmless end products production from PCE degradation and methane formation at day 122 (initial PCE concentration: 30 μmol)

System Configuration	Distribution of Harmless End Products from PCE Degradation			% PCE Degraded to End Products	Methane (μmol)
	Ethylene (μmol)	Ethane (μmol)	Acetylene (μmol)		
Enrichment+iron A (5g/L)	2.0	0.4	2.5	16	3.2
Enrichment+iron A (50g/L)	14.0	2.6	3.4	67	74.5
Enrichment+iron B (5g/L)	11.0	3.0	2.5	55	35.1
Enrichment+iron B (50g/L)	21.6	9.1	0.0	102	121.1

CONCLUSIONS

Batch experiments were performed in this study to evaluate PCE degradation in a system which combined zero-valent iron and a mixed anaerobic culture. The following conclusions could be drawn from the experimental results: (1) in an organic-carbon-poor subsurface environment, anaerobic bacteria could utilize cathodic hydrogen as the electron donor for methane generation and PCE biotransformation, (2) as compared to iron-only or culture-only systems, PCE degradation was significantly enhanced in the systems coupling iron and culture, and (3) the amount and type of zero-valent iron affected both the abiotic PCE reduction and the anaerobic biotransformation of PCE. Iron B was more effective both in the abiotic reduction and in the enhancement of the biotransformation of PCE. The results of this study should be helpful in developing new remediation strategies for sites contaminated with PCE.

DISCLAIMERS

Although this research was funded by the U.S. Environmental Protection Agency, it has not been subjected to the Agency's peer or administrative review and therefore may not necessarily reflect the views of the Agency and no official endorsement may be inferred.

ACKNOWLEDGEMENTS

This research was supported by U.S. Environmental Protection Agency. The work was performed while the author Xiaohong Luo held a National Research Council Associateship Award at USEPA NRMRL/SPRD.

REFERENCES

Gregory, K.B., Mason, M.G., Picken, H.D., Weathers, L.J., and G.F. Parkin. 2000. "Biroaugmentation of Fe(0) for the Remediation of Chlorinated Aliphatic Hydrocarbons", *Environmental Engineering Science, 17*(3):169-181.

Novak, P.J., Deniels, L., and G.F. Parkin. 1998. "Enhanced Dechlorination of Carbon Tetrachloride and Chloroform in the Presence of Elemental Iron and *Methanosarcina barkeri, Methanosarcina thermophila,* or *Methanosaeta concillii.*" *Environ. Sci. Technol., 32(*5):1438-1443.

Sacre, J. A. 1997. *Treatment walls: a status update*. Ground-water Remediation Technologies Analysis Center, TP-97-02, Pittsburgh, PA.

Sewell, G.W., and M. Hightower. 1998. *Cost and Performance Report-In Situ Anaerobic Bioremediation Pinellas Northeast Site, Largo Florida*. US-Department of Energy. April 1998. EPA/600/R-98/115.

Weathers, L.J., Parkin, G.F. and P.J. Alvarez. 1997. "Utilization of Cathodic Hydrogen as Electron Donor for Chloroform Cometabolism by a Mixed, Methanogenic Culture." *Environ. Sci. Technol., 31*(3):880-885.

RDX DEGRADATION WITH BIOAUGMENTED Fe^0 FILINGS: IMPLICATIONS FOR ENHANCED PRB PERFORMANCE

Byung-Taek Oh and Pedro J.J. Alvarez
(The University of Iowa, Civil and Environmental Engineering, Iowa City, Iowa)

ABSTRACT: Hexahydro-1,3,5-trinitro-1,3,5-triazine (RDX) degradation was studied in aquifer microcosms amended with Fe^0 flings. Faster removal occurred in treatments with Fe^0 plus anaerobic sludge, and agitation significantly increased RDX degradation rates. The nitroso byproducts 1,3-dinitro-5-nitroso-1,3,5-triazacyclohexane (MNX), 1,3-dinitroso-5-nitro-1,3,5-triazacyclo-hexane (DNX), and 1,3,5-trinitroso-1,3,5-triazacyclohexane (TNX) were detected in both bioaugmented and abiotic microcosms, although these compounds never accumulated above 5% of the added RDX on a molar basis. [^{14}C]-RDX was synthesized for additional fate studies. Microcosms with both Fe^0 and sludge mineralized RDX faster and to a greater extent than separate treatments, with up to 51% $^{14}CO_2$ recovery after 77 days. A soluble (unidentified) metabolite was found in all microcosms, and this compound accumulated to a much lower extent in combined-treatment reactors than in sets with Fe^0 or sludge alone. Some of the radiolabel was bound to soil and Fe^0 and could not be extracted with CH_3CN. This fraction, which was recovered by combustion with a biological oxidizer, was also found at lower concentrations in combined-treatment reactors. This work suggests that permeable reactive Fe^0 barriers (PRBs) might be an effective approach to intercept and degrade RDX plumes, and that treatment efficiency might be enhanced by biogeochemical interactions through bioaugmentation.

INTRODUCTION

RDX (Hexahydro-1,3,5-trinitro-1,3,5-triazine) is the British code name for Research Department Explosive (Testud et al., 1996). RDX is a suspected carcinogen that represents a major remediation challenge at numerous munitions facilities due to its recalcitrance to biodegradation, low volatility, and high mobility in aquifers. While several *ex situ* physical-chemical and biological processes have been proposed to manage RDX contamination, many of these approaches are prohibitively expensive for groundwater treatment, or are limited by the accumulation of transformation products of equal or even greater toxicity. The need exists for an *in situ* remediation strategy that is easy, cost-effective, less prone to accumulate toxic byproducts, and addresses both chemical and microbiological advantages and constraints.

Encouraging results in laboratory and field experiments have recently stimulated a rapid increase in the use of zero-valent iron (Fe^0) as a reactive material to remove redox-sensitive contaminants from groundwater (Scherer et al., 2000). Semipermeable Fe^0 barriers are particularly attractive for *in situ* remediation in that they conserve energy and water, and through long-term low operating costs, have the potential to be considerably less costly than conventional cleanup methods (Orth and Gillham, 1996). This approach has been mainly used to remove waste chlorinated solvents and redox-sensitive metals such as chromium and uranium (Vidic and Pohland, 1996), but recent studies have reported that Fe^0 can also chemically reduce RDX in contaminated water and soil (Wildman and Alvarez, 2001; Singh et al., 1998).

Up to now, research on Fe^0 systems has focused primarily on abiotic processes. Nevertheless, we recently found that hydrogen gas produced from the reduction of water-derived protons during Fe^0 corrosion (Equation 1) can serve as an electron donor for the biotransformation of reducible contaminants.

$$Fe^0 + 2H_2O \rightarrow Fe^{2+} + 2OH^- + H_2 \qquad (1)$$

In fact, combining Fe^0 with an active methanogenic consortium significantly enhanced both the rate and extent of transformation of chlorinated methanes (Weathers et al., 1997). Further experiments were conducted with pure cultures of methanogens, including hydrogenotrophic species that could grow on H_2 as well as aceticlastic species that could not. These experiments demonstrated that cathodic H_2 could stimulate anaerobic bioremediation of chlorinated solvents, even when H_2 does not serve as growth substrate. In addition, Fe^0 stimulated *M. thermophila* to excrete an extracellular factor with protein-like characteristics that degraded both carbon tetrachloride and chloroform (Novak et al., 1998). We also showed that a similar approach, combining Fe^0 and autotrophic denitrifiers, significantly enhanced the treatment of nitrate-contaminated water by increasing removal rates and improving the end product distribution, favoring N_2 over NH_4^+ (Till et al., 1998). These experiments suggest that some biogeochemical interactions can significantly enhance the efficacy of Fe^0 barriers.

This project investigated the potential benefits of bioaugmenting Fe^0 barriers to enhance the removal of RDX from contaminated groundwater. Emphasis was placed on determining whether combining anaerobic municipal sludge with Fe^0 filings is synergistic in terms of the rates and extent of RDX mineralization. The fate of RDX in combined versus separate treatment systems was also compared.

METHODS

RDX Removal. Microcosms were prepared by adding uncontaminated soil (100 g) from the Iowa Army Ammunitions Plant, anaerobic mineral medium (150 mL, bicarbonate-buffered near pH 7) (Hughes and Parkin, 1996), and RDX (25 mg/L) to 250-mL bottles capped with screw-cap Mininert™ valves. Four reactor sets were prepared in triplicate: no-treatment control (without Master Builder® Fe^0 filings and bacteria addition), sterilized control (poisoned with 350 mg/L $HgCl_2$) plus Fe^0 filings (10 g), anaerobic sludge alone (10%, v/v), and Fe^0 filings plus sludge. The anaerobic sludge was obtained from Iowa City's wastewater treatment plant (6.6 g/L volatile suspended solid (VSS)). Emphasis was placed on comparing the removal efficiency and the fate of RDX in different treatments.

All microcosms were purged for 2 hours with N_2/CO_2 (80/20, v/v) to remove dissolved oxygen, and were incubated quiescently at 30 ± 2°C in the dark in a Coy anaerobic chamber. Aqueous samples were regularly collected with disposable syringes and filtered using a 0.2-μm syringe filter. Changes in the concentration of RDX and its degradation products were monitored.

RDX Mineralization. Experiments were also conducted to determine the potential benefits of an integrated microbial-Fe^0 system for RDX mineralization. [^{14}C]-RDX was synthesized as decribed by Ampleman et al. (1995). Reactors were prepared in triplicate using wide-mouth jars under anaerobic conditions, as described previously. Hot and cold RDX were added to reactors to achieve an initial concentration of about 25 mg/L RDX (113 μM) and 1 μCi of total radioactivity. A 50-mL test tube containing 5 mL of 1 M NaOH was placed inside each jar to trap $^{14}CO_2$ from RDX mineralization. Test tubes were removed and replaced every 7 days. RDX mineralization was determined from trapped $^{14}CO_2$ at each sampling by liquid scintillation counting (LSC). All reactors were incubated quiescently at 30 ± 2°C in the dark in a Coy anaerobic chamber and shaken manually every sampling event.

Analytical methods. HPLC analysis of RDX and its nitroso derivatives 1,3-dinitro-5-nitroso-1,3,5-triazacyclohexane (MNX), 1,3-dinitroso-5-nitro-1,3,5-triazacyclohexane (DNX), and 1,3,5-trinitroso-1,3,5-triazacyclohexane (TNX) was conducted using a Hewlett Packard 1100 Series HPLC equipped with a 250 × 4.6 mm Supelcosil™ LC-18 column. The mobile phase was iscocratic, consisting of deionized water and methanol (4:6, v/v) at a flow rate of 1.0 mL/min. Detection was spectrophotometric at 240 nm. [^{14}C]-RDX and its ^{14}C-metabolites were quantified by HPLC using a radioactivity detector (Radiomatic, Series A-500, Packard Instrument Co.). RDX mineralization was determined from trapped $^{14}CO_2$ by mixing 1 mL of sample with 10 mL of LSC cocktail (Fisher Scientific) and counting on a Beckman LS 6000IC liquid scintillation counter. For N_2O measurements, headspace samples were taken through the Mininert™ valves using a 100-μL gas-tight syringe (Dynatec Precision Sampling Corp.). Nitrous oxide was measured by injecting headspace (100 μL) into a HP 5890 Series II GC. The GC was equipped with a thermal-conductivity detector (TCD). Separation was achieved using an Alltech (Deerefield, IL) packed Haysep Q molecular sieve column.

RESULTS AND DISCUSSION

Microcosm experiments suggest that an integrated microbial-Fe^0 treatment approach can effectively degrade RDX. No RDX removal was observed in no-treatment controls (Figure 1A), corroborating the notion that RDX is a recalcitrant compound. RDX was removed faster in the combined-treatment microcosms (Figure 1D) than in treatments with Fe^0 (Figure 1B) or municipal anaerobic sludge alone (Figure 1C). Agitation significantly increased RDX degradation rates (up to 5 mg/L/day), which suggests that degradation kinetics under quiescent conditions were limited by inadequate contact between RDX and Fe^0 and/or cells.

MNX, DNX and TNX were detected as transient intermediates in both bioaugmented and abiotic microcosms, although these compounds never accumulated above 5% of the added RDX on a molar basis. The detection of these heterocyclic nitroso compounds is not surprising. McCormick et al. (1981)

reported that MNX, DNX, and TNX were major intermediates of RDX transformation by anaerobic sludge, and Singh et al. (1998) also found these byproducts when RDX was treated with Fe^0.

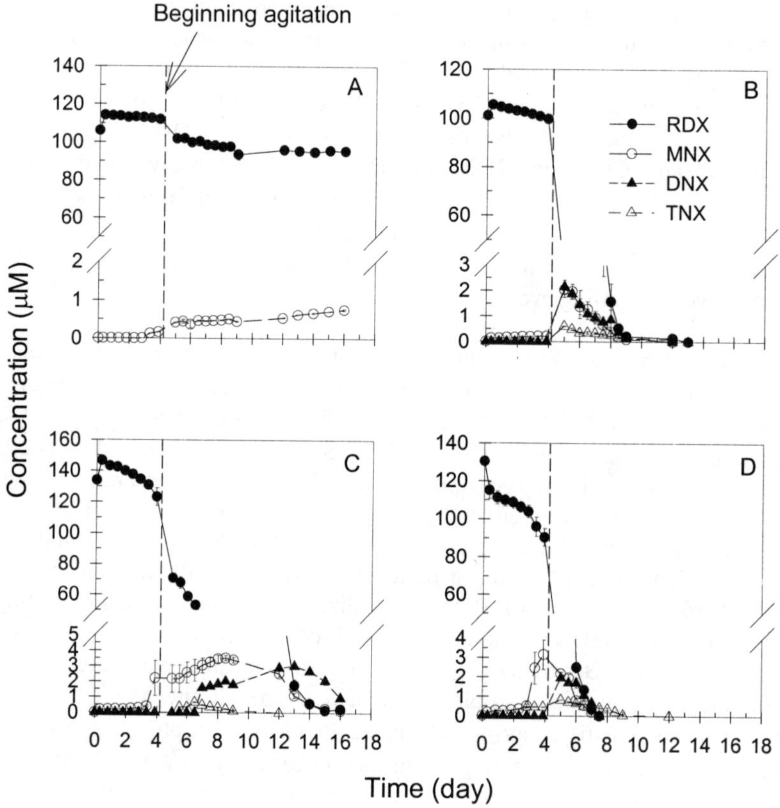
Beginning agitation
A
B
C
D
RDX
MNX
DNX
TNX
Concentration (μM)
Time (day)

FIGURE 1. Changes in RDX, MNX, DNX, and TNX concentrations in unamended (no-treatment) controls (A), and in reactors amended with Fe^0 filings (10% of soil weight) (B), anaerobic sludge (10% v/v) (C), and both (D).

Combined-treatment microcosms were more effective in mineralizing RDX, as shown by $^{14}CO_2$ recovery over 77 days of incubation (i.e., 51% for combined treatment, 22% for Fe^0 alone, and 29% for sludge alone) (Figure 2). This degree of RDX mineralization compares favorably to that reported by Sheremata and Hawari (2000) for the white-rot fungus *Phanerochaete chrysosporium* (52% mineralization after 60 days). Nitrous oxide (N_2O) was also found in the headspace of all microcosms, which corroborates the work of Hawari et al. (2000) who found that N_2O can be an end product of RDX degradation. More N_2O was produced after 28 days in the combined-treatment set (55 μM) than in sets with sludge or Fe^0 alone (20 and 5 μM, respectively), which reinforces the notion that the combined treatment results in greater transformation of RDX to innocuous products.

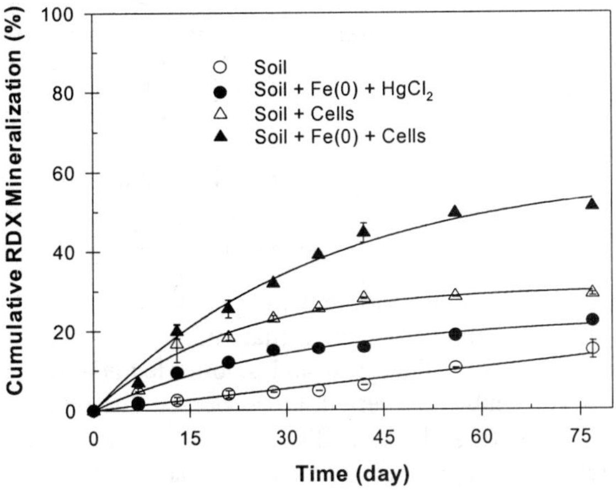
Cumulative RDX Mineralization (%)
100
80
60
40
20
0
Soil
Soil + Fe(0) + HgCl2
Soil + Cells
Soil + Fe(0) + Cells
0
15
30
45
60
75
Time (day)

FIGURE 2. Cumulative mineralization of RDX in reactors amended with Fe^0 filings (10% of soil weight), anaerobic sludge (10% v/v), and both.

RDX mineralization followed first-order kinetics (i.e., $C = C^* [1-e^{-kt}]$, where C = $^{14}CO_2$ concentration at time t, and C* is the asymptotic value). This model is depicted as solid lines in Figure 2. Rate coefficients were estimated by fitting the data to this model (Table 1). Interestingly, the first-order rate coefficient (*k*) for the combined Fe^0 + cells treatment (0.010 day^{-1}) was 30% greater than the sum of the *k* values for the separate Fe^0 and cells treatments (i.e, 0.003 + 0.004 = 0.007 day^{-1}), suggesting that the combination was synergistic with respect to RDX mineralization. This synergism could be due to several factors. Fe^0 corrosion rapidly induces anoxic conditions that favor RDX degradation. The production of cathodic (water-derived) hydrogen by Fe^0 corrosion (Eq.1) would increase the availability of an excellent electron donor to support microbial reduction of RDX and the further degradation of some dead-end products that could accumulate during abiotic reduction by Fe^0. In addition to Fe^0 enhancing bacterial activity, some bacteria (e.g., iron reducers) could also enhance iron reactivity not only by cathodic depolarization, which accelerates corrosion, but also by reductive dissolution and activation of some oxides that passivate the iron surface.

Regarding the fate of [^{14}C]-RDX, we obtained very good mass balances, ranging from 84 to 99% (Table 2). Some of the radiolabel was bound to soil and Fe^0 and could not be extracted with CH_3CN. This bound residue, which was recovered by combustion with a biological oxidizer, was greater for treatments with Fe^0 alone (28% in soil and 14% in Fe^0) than with sludge alone (14% in soil) or with both Fe^0 plus sludge (14% in soil and 12% in Fe^0). Apparently, RDX byproducts can be irreversibly bound with some surface material as an additional pathway to mineralization. Whether this pathway leads to an acceptable treatment endpoint (due to lack of bioavailabilty) remains to be determined.

TABLE 1. Mineralization rate coefficients for different treatments.

Treatment	First-order mineralization rate coefficient (*k*)
No treatments	0.002 ± 0.0001 day^{-1}
Fe^0 alone	0.003 ± 0.0004 day^{-1}
Cells alone	0.004 ± 0.0009 day^{-1}
Fe^0 + Cells	0.010 ± 0.0012 day^{-1}

A soluble ^{14}C-labeled metabolite was detected in all microcosms by HPLC with radio-chromatographic detection (data not shown). This intermediate appears to be common to both biological and abiotic degradation pathways. At the end of the 77-day incubation period, it accounted for about 50% of the radiolabel in treatments with soil alone, 29% with sludge alone, 22% with Fe^0 alone, and 5% with sludge plus Fe^0. We have extracted this compound with methylene chloride and concentrated it by evaporating the solvent. Preliminary GC/MS analysis suggests that this metabolite could be different from any other intermediate that has been reported by others (e.g., Hawari et al., 2000; McCormick et al., 1981). The accumulation of this metabolite suggests that its degradation may be the rate-limiting step in RDX mineralization. To identify this metabolite, we will rely on further GC/MS, LC/MS and NMR analyses.

CONCLUSIONS

Laboratory experiments suggest that permeable reactive Fe^0 barriers might be a viable alternative to intercept and degrade RDX plumes, and that system performance can be enhanced by some biogeochemical interactions through bioaugmentation. This integrated approach is more than a mere juxtaposition of two technologies since it involves a synergistic combination of Fe^0 and bacteria, presumably related to the exploitation of cathodic depolarization and bioremediation as metabolic niches. This approach may also be practical and effective to treat other redox-sensitive groundwater pollutants, such as chlorinated solvents, nitroaromatic compounds, nitrate, hexavalent chromium, hexavalent uranium, and some pesticides. Nevertheless, further studies are needed to delineate better the applicability and limitations of iron-based bioremediation.

ACKNOWLEDGMENTS

This work was funded by Strategic Environmental Research and Development Program (SERDP) (Project No. DACA72-00-P-0057). The authors wish to thank Mr. Craig Just for synthesizing [^{14}C]-RDX and for valuable analytical assistance.

TABLE 2. Distribution of ^{14}C in microcosms amended with Fe(0) filings (10% of soil weight), anaerobic sludge (10%, v/v), none, or both after 77 days of incubation. The initial RDX concentration was 25 mg/L (spiked with 1 μCi ^{14}C-RDX).

			Solid phase (%)				
					Unextractable		
Amendments	Released $^{14}CO_2$ (%)[a]	Aqueous phase (%)	3 mM $CaCl_3$	CH_3CN	Soil	Fe(0)	Total (%)
Soil	15.0 (2.3)[b]	49.6 (7.2)	8.8 (0.4)	3.7 (0.1)	13.6 (3.1)	--	90.7
Soil + Fe(0) + $HgCl_2$	22.3 (0.2)	22.0 (3.9)	9.3 (0.5)	3.6 (0.2)	28.0 (1.3)	13.8 (1.3)	99.0
Soil + Cells	29.1 (0.4)	28.7 (1.2)	8.7 (1.2)	3.0 (0.1)	14.5 (2.9)	--	84.0
Soil + Fe(0) + Cells	51.0 (0.2)	4.6 (0.9)	3.5 (0.2)	2.6 (0.1)	14.3 (0.3)	11.9 (0.5)	87.9

[a] Recovery of ^{14}C is percent of total ^{14}C added.
[b] Parenthetic values indicate sample standard deviations.

REFERENCES

Ampleman G., S. Thiboutot, J. Lavigne, and A. Marois. 1995. "Synthesis of ^{14}C-labeled hexahydro-1,3,5-trinitro-1,3,5-triazine (RDX), 2,4,6-trinitrotoluene (TNT), nitrocellulose (NC) and glycidyl azide polymer (GAP) for use in assessing the biodegradation potential of these energetic compounds". *J. Labeled Compd. Radiopharm. 36*(6): 559-577.

Hawari J., A. Halasz, T. Sheremata, S. Beaudet, C. Groom, L. Paquet, C. Rhofir, G. Ampleman, and S. Thiboutot. 2000. "Characterization of metabolites during biodegradation of hexahydro-1,3,5-trinitro-1,3,5-triazine (RDX) with municipal anaerobic sludge". *Appl. Environ. Microbiol. 66*(6): 2652-2657.

Hughes J.B., and G.F. Parkin. 1996. "Concentration effects of chlorinated aliphatic transformation kinetics". *J. Env. Eng. 122*: 92-98.

McCormick N.G., J.H. Cornell, and A.M. Kaplan. 1981. "Biodegradation of hexahydro-1,3,5-trinitro-1,3,5-triazine". *Appl. Environ. Microbiol. 42*(5): 817-823.

Novak P., L. Daniels, and G. Parkin. 1998. "Rapid dechlorination of carbon tetrachloride and chloroform by extracellular agents in cultures *Methanosarcina thermophila*". *Environ. Sci. Technol. 32*: 3132-3136.

Orth W.S., and R.W. Gillham. 1996. "Dechlorination of trichloroethene in aqueous solution using Fe^0". *Environ. Sci. Technol. 30*: 66-71.

Sheremata T., and J. Hawari. 2000. "Mineralization of RDX by the white rot fungus *Phanerochaete chrysosporium* to carbon dioxide and nitrous oxide". *Environ. Sci. Technol. 34*: 3384-3388.

Scherer M.M., S. Richter, R.L. Valentine, and P.J. Alvarez. 2000. "Chemistry and microbiology of permeable reactive barriers for *in situ* groundwater cleanup". *Crit. Rev. Environ. Sci. Technol. 30*: 363-411.

Singh J., S.D. Comfort, and P.J. Shea. 1998. "Remediating RDX-contaminated water and soil using zero-valent iron". *J. Environ. Qual. 27*: 1240-1245.

Testud F., J.M. Glanclaude, and J. Descotes. 1996. "Acute hexogen poisoning after occupational exposure". *Clinical Toxicol. 34*(1): 109-111.

Till B.A., L.J. Weathers, and P.J.J. Alvarez. 1998. "Fe(0)-supported autotrophic denitrification". *Environ. Sci. Technol. 32*(5): 634-639.

Vidic R.D., and F.G. Pohland. 1996. "*Treatment Walls*". Ground-Water Remediation Technologies Analysis Center, TE-96-01, Pittsburgh, PA.

Weathers L.J., G.F. Parkin, and P.J.J. Alvarez. 1997. "Utilization of cathodic hydrogen as electron donor for chloroform cometabolism by a mixed methanogenic culture". *Environ. Sci. Technol. 31*(3): 880-885.

Wildman M.J., and P.J.J. Alvarez. 2001. "RDX degradation using an integrated Fe(0)-microbial treatment approach". *Wat. Sci. Technol. 43*(2): 25-33.

PROBABALISTIC DESIGN OF A COMBINED PERMEABLE BARRIER AND NATURAL BIODEGRADATION REMEDY

John E. Vidumsky, P.E. and Richard C. Landis
(DuPont, Wilmington, Delaware)

ABSTRACT: A probabilistic model was used to design a combined permeable reactive barrier (PRB) and natural biodegradation remedy for a carbon tetrachloride (CT) plume at a former manufacturing site. A distribution of CT concentrations was used as input to the model, with a most probable concentration of 80 parts per million (ppm). Based on this input concentration and other input parameters, the Monte Carlo analysis showed that, at the 90% confidence level, CT will degrade to below detectable levels in a 6-inch thick PRB using zero valent iron (ZVI) as the reactive media. Model results also predicted the effluent dichloromethane (DCM) concentration to be approximately 8 ppm, and the effluent trichloromethane (TCM) concentration to be 100 parts per billion (ppb) or less. A biodegradation evaluation was then conducted to predict the downgradient biodegradation of the TCM and DCM in the PRB effluent. The biodegradation evaluation indicated that 100 ppb of TCM would be reduced to less than 1 ppb within 350 ft (107 m) downgradient of the PRB. Similarly, 8 ppm of DCM will be reduced to less than 1 ppb in less than 75 ft (23 m) downgradient of the PRB. Based on these results, it was concluded that all groundwater quality objectives would be met prior to reaching the compliance point downgradient of the PRB.

INTRODUCTION

Probabilistic design models are useful and effective tools for optimizing the design of PRBs. These models allow the design engineer to use the full range of available site data and treatability testing results. The model can use these ranges of input parameters to develop a complete distribution of anticipated PRB performance with an associated probability distribution. This analysis can then be used to select a design thickness of ZVI at the desired confidence level, typically in the 85 to 95% range. This approach provides more efficient resource utilization when compared with the more traditional deterministic approach using only worst-case input conditions to calculate a design thickness, or applying arbitrary safety factors.

At many sites, chlorinated solvent biodegradation is occurring naturally to some extent, particularly if the geochemical conditions are favorable and the right population of organisms is present. At some sites, the rate of natural biodegradation is sufficient to achieve groundwater remedial goals prior to reaching the compliance point, typically the property boundary or a point where groundwater discharges to a surface water body such as a river or stream.

At some sites, however, the rate of natural biodegradation is not sufficient, and intervention is required to accelerate the rate of degradation, reduce chlorinated solvent plume concentrations, convert the solvent into a more rapidly degradable form, or any combination of the above. At the subject site, a permeable reactive barrier (PRB) using zero valent iron (ZVI) as the reactive media was selected as the means of intervention. Since 1994, over 30 full-scale PRBs using ZVI as the reactive media have been installed in the United States to treat chlorinated solvents (e.g., chlorinated methanes, ethanes, and ethenes). ZVI effectively degrades these compounds with reaction half-lives ranging from several minutes to several hours. A conceptual model of how a PRB can work in combination with natural attenuation to reduce constituent concentrations in a plume is shown in Figure 1.

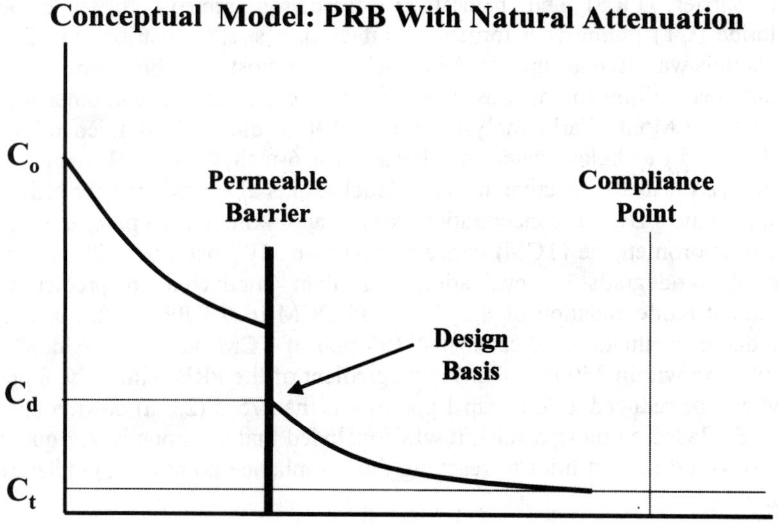
Conceptual Model: PRB With Natural Attenuation
C_o
Permeable
Barrier
Compliance
Point
Design
Basis
C_d
C_t

FIGURE 1. PRB volatile organic compound degradation model coupled with natural attenuation.

Figure 1 shows the concentration of the target constituent (vertical axis) versus distance in the downgradient direction (horizontal axis). C_0 represents the initial concentration in the source area, which decreases in the downgradient direction at the baseline rate of natural biodegradation at the site. C_t is the target concentration that needs to be achieved prior to reaching the downgradient compliance point. It is not necessary to design the PRB to achieve C_t at the downgradient edge of the PRB, because space is available downgradient of the PRB for biodegradation to further reduce the concentration prior to reaching the compliance point. C_d is the PRB design concentration selected to achieve the target concentration at the compliance point, taking downgradient biodegradation into account. At the subject site, the target constituent is carbon tetrachloride (CT), and the compliance point is 1,000 feet or 305 meters downgradient of the PRB.

Objective. The objective of this analysis was to predict the performance of a 6-inch (15.2 centimeter) thick PRB at the subject site through probabilistic modeling. Laboratory column studies were conducted to determine reaction rates and daughter product yield, which were then used as input to the PRB design model. The model predicted the PRB effluent concentrations of CT, TCM, and DCM. The PRB effluent concentrations were then used as input to a biodegradation analysis, which was conducted to determine whether TCM and DCM could be biodegraded to below 1 part per billion (ppb) prior to reaching the compliance point.

PRB DESIGN ANALYSIS

The primary objective in PRB design is to provide adequate residence time in the PRB for the ZVI-driven degradation reactions to proceed to the desired endpoints. The design residence time is a function of reaction rates (as determined by laboratory studies), influent contaminant concentrations, groundwater temperature, and desired contaminant concentration endpoints. Residence time is then translated into a design thickness of the PRB using the groundwater velocity, and the porosity of the iron in the PRB. These parameters were used as input to a probabilistic design model, which served as the basis for predicting the performance of a 6-inch (15.2 cm) thick PRB at the subject site.

CT degrades relatively rapidly in the presence of ZVI with a half-life typically less than 30 minutes (Powell, et al., 1998). CT degrades to methane and carbon dioxide with partial conversion to TCM and DCM. Given sufficient reaction time, CT and TCM can be completely destroyed, yielding DCM as the only chlorinated end product. It is important to note that reaction rates and yield of daughter products are dependent on the specific geochemistry and temperature of the groundwater being treated.

Laboratory Column Studies. In order to measure site-specific reaction rates and daughter product yields, laboratory column studies were conducted using groundwater from the site The half-life for carbon tetrachloride (CT) was found to be 0.2 hours or 12 minutes. The completed reaction will yield approximately 18% DCM on a molar basis, which is equivalent to approximately a 10% conversion on a concentration basis. For example, with an expected influent concentration of 80 parts per million (ppm) of CT, the treated groundwater exiting the PRB will contain nonchlorinated end-products, 100 ppb of TCM, and approximately 8 ppm of DCM. The reaction pathway and site-specific yield coefficients are shown on Figure 2.

The column studies were performed by Envirometal Technologies Inc. of Waterloo, Ontario Canada. The results of the column studies are presented in a confidential report dated October 2000. Site groundwater was fed through a Plexiglass™ column filled with granular iron with a grain size of –30 to +76 US Standard Mesh size (0.2 to 0.6 mm) supplied by Connelly-GPM, Inc. of Chicago, IL. Constituent concentrations were measured as a function of distance along the column after equilibrium had been reached.

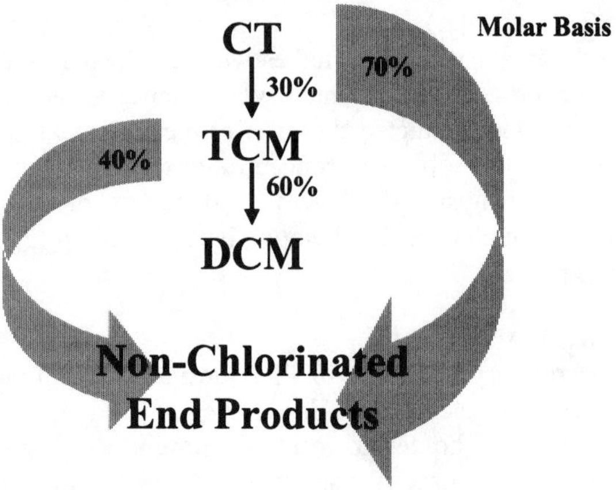
CT
30%
TCM
60%
DCM
70%
40%
Molar Basis
Non-Chlorinated
End Products

FIGURE 2. CT degradation pathway with ZVI.

Model Input Parameters. The first step in the probabilistic design analysis was to input the range of values for each design parameter. These ranges were based on historical and measured data from the site. A Monte Carlo analysis was conducted using 5,000 iterations, resulting in a range of confidence levels for meeting the given treatment goal. A corresponding required PRB thickness was associated with each confidence level.

The input parameters for residence time within the PRB included ranges for groundwater velocity of 0.5 ft/day (0.152 m) to 1.0 ft/day (0.305 m), a soil porosity of 0.35 to 0.45, and a reactive media porosity (iron) of 0.50 to 0.60. The input degradation rates included a range of contaminant half-lives of 0.1 to 0.4 hours for CT, and 1.5 to 3.0 hours for TCM, based on laboratory column tests conducted at 25°C. Other input parameters included a groundwater temperature range of 18 to 21°C and initial concentrations for CT of 40 to 170 ppm (with a most probable value of 80 ppm) and 1 to 3 ppm of TCM.

Modeling Results. Based on the input parameters, the Monte Carlo analysis indicates that, at a 90% confidence level, CT degrades within roughly the first inch (2.54 cm) of the 6-inch (15.2 cm) thick PRB, and that TCM generation and degradation are the controlling factors for the PRB design. The analysis also shows that, at the 90% confidence level for a 6-inch (15.2 cm) PRB thickness, the effluent TCM concentration will be 100 ppb or less as shown in the figures below.

Figure 3 shows the output of a Monte Carlo analysis for degrading CT to 2.5 ppb within a 6-inch (15.2 cm) ZVI PRB. At the 90% confidence level, CT will degrade from the input concentration of 80 ppm down to 2.5 ppb or less within the first inch of a 6-inch (15.2 cm) ZVI PRB. Clearly, the effluent concentration from a 6-inch (15.2 cm) PRB will be below detectable levels.

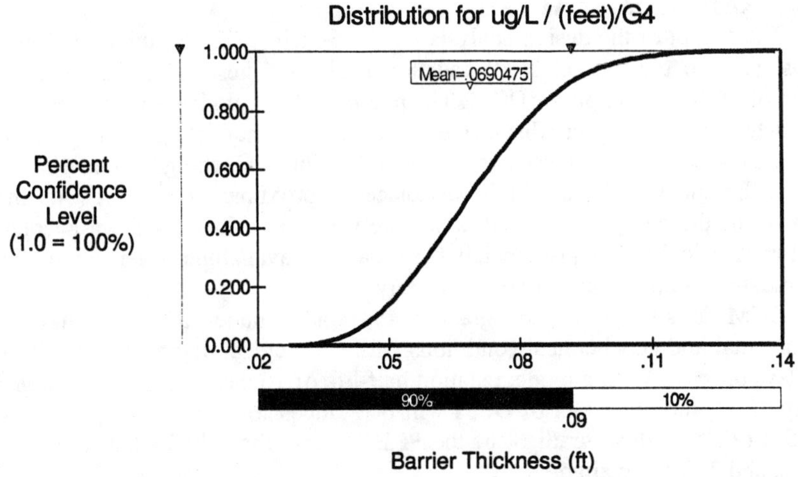
Distribution for ug/L / (feet)/G4
Mean=.0690475
Percent Confidence Level (1.0 = 100%)
1.000
0.800
0.600
0.400
0.200
0.000
.02
.05
.08
.11
.14
90%
10%
.09
Barrier Thickness (ft)

FIGURE 3. Model output for carbon tetrachloride degradation.

Figure 4 shows the output confidence level of a Monte Carlo analysis for degrading TCM to 100 ppb within a 6-inch (15.2 cm) ZVI PRB. Using 6 inches (15.2 cm) of ZVI, there is a 90% confidence level that the effluent TCM concentration will be 100 ppb or less at the specified input concentrations of CT and TCM.

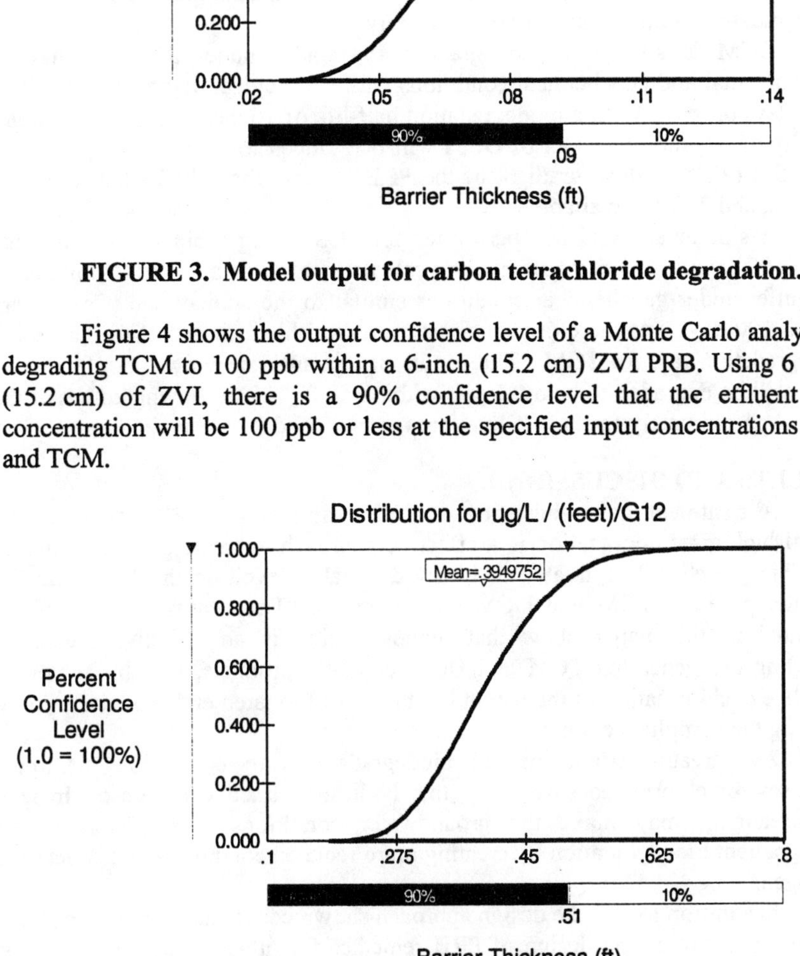
Distribution for ug/L / (feet)/G12
Mean=.3949752
Percent Confidence Level (1.0 = 100%)
1.000
0.800
0.600
0.400
0.200
0.000
.1
.275
.45
.625
.8
90%
10%
.51
Barrier Thickness (ft)

FIGURE 4. Model output for trichloromethane.

BIODEGRADATION ANALYSIS

The output of the design analysis for the 6-inch (15.2 cm) thick PRB was used as input to the natural biodegradation analysis. These input values were 100 ppb of TCM and 8 ppm of DCM. The purpose of the biodegradation analysis was to determine if sufficient distance exists downgradient of the PRB to achieve the remedial objectives for these two compounds prior to reaching the compliance point. At the subject site, the PRB was placed approximately 1,000 ft (305 m) upgradient of the compliance point, a surface water boundary. With an average groundwater velocity of approximately 0.5 ft/day, the available residence time for biodegradation is 2,000 days or approximately 5.5 years.

DCM is known to biodegrade very rapidly under a wide range of environmental and geochemical conditions with a biological degradation half-life of 10 days or less. With a biodegradation half-life of 10 days and a groundwater velocity of 0.5 ft/day, 8 ppm of DCM will be reduced to less than 1 ppb in less than 75 ft (22.9 m) downgradient of the PRB. Approximately 13 half-lives are required, and 200 are available.

A similar analysis was performed for TCM using a biodegradation half-life of 100 days. A 70-day half-life was calculated from data for a TCM plume in an aquifer under geochemical conditions similar to the subject site (Cox, et al., 1998). 100 days was therefore considered a reasonable estimate of the anaerobic biodegradation rate for TCM. The results of this analysis show that 100 ppb of TCM will be degraded to below 1 ppb within 350 ft (107 m) downgradient of the PRB. Approximately seven half-lives are required, and 20 are available.

RESULTS AND DISCUSSION

A treatment train consisting of ZVI treatment in a PRB followed by natural biological degradation is a effective approach to treating a CT plume at the subject site. CT is treated to below detectable levels in the PRB, and the daughter products TCM and DCM are generated. The results of the analysis presented in this paper show that biodegradation is an effective means of degrading the generated TCM and DCM daughter products, thereby achieving complete dechlorination of the parent CT to non-chlorinated end products prior to reaching the compliance point.

ZVI treatment and natural biodegradation are compatible treatment processes for chlorinated solvents in that both are reductive processes. In fact, ZVI treatment may make the groundwater conditions more favorable for downgradient biodegradation by creating more reduced conditions and generating hydrogen.

In addition to CT, the design approach shown conceptually in Figure 1 can be used to optimize the design of PRB remedies for other chlorinated solvents, where daughter products are generated which degrade (in the presence of ZVI) more slowly than the parent compounds. Examples would include tetrachloroethene (PCE) and trichloroethene (TCE). In the case of PCE and TCE, the daughter compounds cis-dichloroethene (cis-DCE) and vinyl chloride (VC) are generated as part of the ZVI driven degradation reactions. VC and cis-DCE can be fully dechlorinated in a ZVI PRB given a sufficient thickness of iron and corresponding residence time. However, significant cost savings may be realized by using the available aquifer space downgradient of the PRB as a natural bioreactor to degrade residual DCE and VC, rather than increasing PRB thickness. This approach is only practicable at sites where the biodegradation rates of these daughter compounds is sufficiently rapid and where sufficient space and residence time is available downgradient of the PRB prior to reaching the compliance point.

REFERENCES

Powell, R. M., R. W. Puls, D. W. Blowes, R. W. Gillham, and D. Schultz. 1998. *Permeable Reactive Barrier Technologies for Contaminant Remediation.* 114 pp. EPA/600/R-98/125 (also available at http://www.epa.gov/ada/reports.html).

Cox, E.E., M. McMaster, and D.W. Major. 1998. "Natural Attenuation of 1,2 Dichloroethane and Chloroform in Groundwater at a Superfund Site." In G.B. Wickramanayake and R.E. Hinchee (Eds.), *Remediation of Chlorinated and Recalcitrant Compounds: Volume 1.* Battelle Press, Columbus, OH.

BIOTIC AND ABIOTIC DECHLORINATION IN IRON-REDUCING AND SULFIDOGENIC ENVIRONMENTS

Peter Adriaens, Michael J. Barcelona, Kim F. Hayes, and ***Michael L. McCormick***
(University of Michigan, Ann Arbor, MI, USA)
Karen L. Skubal
(Case Western Reserve University, Cleveland, OH, USA)

ABSTRACT: The contributions of biotic and abiotic reactions to reductive dechlorination of chlorinated alkanes and alkenes were assessed in model iron and sulfate reducing environments. Comparisons were made on the basis of reaction kinetics and product distributions. In well attenuated iron reducing cultures containing *Geobacter metallireducens* and biogenic magnetite particles produced by this same strain, tetrachloromethane (CT) transformation by the magnetite surfaces (abiotic) was approximately two orders of magnitude faster than the cell mediated reaction (biotic). Moreover, the abiotic reaction could be differentiated from the biotic reaction on the basis of methane and carbon monoxide production. Similarly, under sulfidogenic conditions, formation of "tell tale" products allowed discrimination between biotic and abiotic dechlorination reactions. Ferrous sulfide-mediated dechlorination of tetrachloroethene (PCE) resulted in the production of acetylene, whereas biotic dechlorination by the sulfate reducing bacteria *Desulfovibrio vulgaris subsp. vulgaris* proceeded via hydrogenolysis forming tri- and di-chloroethenes. Even though the relative contributions of biotic and abiotic dechlorination under sulfidogenic conditions has not yet been assessed, the predominance of abiotic dechlorination in the iron reducing system and the ubiquity of biogenic iron and sulfide minerals in the environment suggest that biogenic minerals may play a more significant role in contaminant transformation than previously thought.

INTRODUCTION

Organic matter oxidation in contaminated anaerobic aquifers is commonly coupled to iron- and sulfate reduction, which has been demonstrated to result in the depletion of iron oxide coatings on aquifer solid minerals, and in the precipitation of reduced iron and sulfide minerals (e.g. Lendvay et al., 1998). The presence of iron- and sulfate-reducers, along with mineral reductants consequential to their metabolism, presents the opportunity for biotic and abiotic dechlorination reactions to occur in haloalkane-contaminated aquifers. Under iron-reducing conditions, both *Geobacter metallireducens* and the biogenic solids which form as the result of iron reduction (e.g. magnetite), have been demonstrated to exhibit dechlorination activity against a limited range of haloalkanes, such as hexachloroethane (HCA) or carbon tetrachloride (CT) (McCormick et al., 1998; McCormick and Adriaens, 2000b). Sulfate-reduction in the presence of iron results in the precipitation of ferrous sulfide minerals such as mackinawite or crystalline FeS. Both of which have been shown to abiotically dechlorinate a variety of haloalkanes and alkenes (e.g., Butler and Hayes, 2000). Similarly, biotic dechlorination by sulfate-reducing bacteria has also been demonstrated. For example, limited direct dechlorination of tetrachloromethane and 1,1,1-trichloroethane have been exhibited by *Desulfobacterium autotrophicum* (Egli et al., 1987).

While the individual dechlorination activities of reactive minerals and pure cultures of iron- and sulfate-reducing bacteria have been evaluated separately, their relative contributions in iron and sulfate reducing environments has not been assessed. This paper describes efforts to delineate biotic and abiotic contributions to dechlorination in these environments on the basis of reaction kinetics and product formation in separate and combined iron reducing and sulfate reducing systems.

APPROACH

The generalized approach to address the overarching goal of this study is illustrated in Figure 1, which shows the four objectives associated with the mineral- and cell-mediated dechlorination reactions. Specifically, the following materials sources were used:

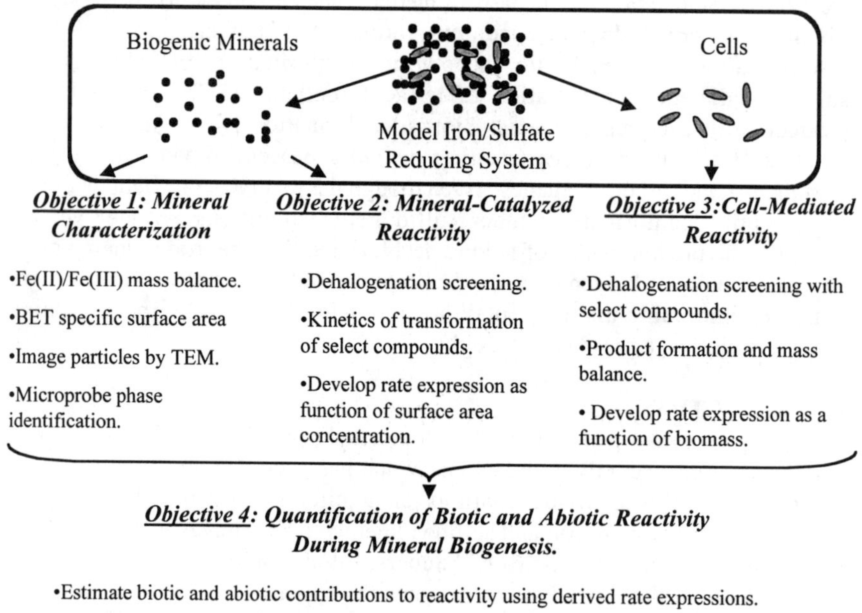
Biogenic Minerals
Model Iron/Sulfate Reducing System
Cells
Objective 1: Mineral Characterization
•Fe(II)/Fe(III) mass balance.
•BET specific surface area
•Image particles by TEM.
•Microprobe phase identification.
Objective 2: Mineral-Catalyzed Reactivity
•Dehalogenation screening.
•Kinetics of transformation of select compounds.
•Develop rate expression as function of surface area concentration.
Objective 3:Cell-Mediated Reactivity
•Dehalogenation screening with select compounds.
•Product formation and mass balance.
• Develop rate expression as a function of biomass.
Objective 4: Quantification of Biotic and Abiotic Reactivity During Mineral Biogenesis.
•Estimate biotic and abiotic contributions to reactivity using derived rate expressions.

FIGURE 1. Experimental approach to assess biotic and mineral reactivity under iron- and sulfate-reducing conditions

(i) Solids: *Geobacter metallireducens* was grown in the presence of hydrous ferric oxide (HFO) as an electron acceptor, acetate as an electron donor, and bicarbonate buffer to generate the magnetite (Fe_3O_4). The solids were magnetically separated and washed to remove residual cell material prior carbon monoxide (CO), 37%; and methane, 9%. In contrast, microbial dechlorination by *G. metallireducens* resulted in the production of 14% chloroform, and 86% cell-bound products, based on carbon and free chloride mass balances. Interestingly, CF was dechlorinated only in the presence of *D. autotrophicum*, and tetrachloroethene (PCE) was dechlorinated only in the presence of FeS. Acetylene and methane (or carbon monoxide) appear to be indicator products for the abiotic transformations of PCE and CT respectively and may allow differentiation between mineral and microbial dechlorination activity in systems where cells and biogenic minerals are both present.

TABLE 1. Comparative results of dehalogenation screening experiment

Alkyl Halide	Synthetic[a] $FeS_{(1-x)}$	Biogenic Fe_3O_4	*D. autotrophicum*[c]	*G. metallireducens*
HCA	PCE	PCE	–	–
1122-TeCA	TCE, acetylene	No reaction[b]	–	–
CT	CF	CF, CO, CH_4	CF, DCM	CF, cell bound products
CF	–	No reaction	DCM	–
TBM	DBM	DBM, CH_4	–	–
PCE	TCE, cDCE, acetylene	No reaction	No reaction	No reaction[d]
TCE	cDCE, acetylene	No reaction	No reaction	–

Abbreviations: (–): not tested; TeCA: tetrachloroethane; TCE: trichloroethene; DCE: dichloroethene; TBM: tribromomethane; DBM: dibromomethane. [a]Butler and Hayes (2000), [b]TCE formed but was indistinguishable from hydrolytic formation of TCE in buffer controls, [c]Data from Egli et al. (1987), [d]Krumholz et al. (1996).

The products of PCE dechlorination by the minerals produced under sulfate-reducing conditions in the presence of various iron sources is shown in Table 2.

TABLE 2. Abiotic dechlorination of PCE under sulfate-reducing conditions

Amendment*	Media Control**	$FeSO_4$ (14.4 μM Fe^{2+})	Fe-citrate (17 g/L)	$Fe(OH)_3$ (10.6 mM Fe)
Product Distribution	TCE CDCE (1,1-DCE)	TCE cDCE	TCE c/t-DCE (1,1-DCE)	TCE cDCE
Yield*** (Σ prod./PCE)	8-17%	5-26%	8-74%	8-74%

* Media: lactate, 20 mM; sulfate, 21.1 mM; iron, see table; MOPSO, 10mM; PCE, 0.5 μM
**MOPSO buffer amended with 0.24 g/L cysteine
*** All yields corrected for dechlorination products observed in the aqueous control

The products were limited to TCE and cDCE, with an additional product tentatively identified as 1,1-DCE. The yield of dechlorinated products as a percent of PCE removed ranged from a low of 5% to a high of 74%. It should be to the dechlorination assays. *Desulfovibrio vulgaris subsp. vulgaris* was grown in the presence of lactate, three different iron sources ($FeSO_4$, HFO, ferric citrate) and media for sulfate reducing bacteria (Egli et al., 1987) to generate biogenic solids which were removed, washed and freeze-dried prior to initiation of the reactivity assays. *D. autotrophicum* did not produce solids under any of the conditions tested.

(ii) Cells: *G. metallireducens* was grown in the presence of ferric citrate and acetate as the electron acceptor and donor respectively. *D. vulgaris* was grown in the presence of SRB media for brackish strains with lactate as the electron donor.

(iii) Mineral biogenesis experiment: On the basis of protein and mineral surface area yields during growth, the protein- and surface area-normalized rate expressions obtained in objectives 2 and 3 were used to estimate the biotic and abiotic contributions to dechlorination activity at several time points during mineral biogenesis.

RESULTS

Due to the complex and extensive matrix of experiments, the results will be illustrative of the type of information obtained under each objective, and will emphasize the results obtained under iron-reducing conditions.

Solids Characterization. During incubation on HFO, microbial iron reduction produced a mixture of siderite, $FeCO_3$, (~11 % m/m) and magnetite, Fe_3O_4, (~89 % m/m). The specific surface area of the solids decreased with time from 300 m^2/g initially to 150 m^2/g in one month, and a further decrease to 65 m^2/g after more than one year. Transmission electron microscopy (TEM) revealed the solids to consist predominantly of nano-scale magnetite particles (3-5 nm diameter) with a few relatively large siderite particles (0.3-1 μm diameter). On the basis of particle sizes and mass fractions, magnetite was estimated to comprise over 98% of the total mineral surface area. Under sulfate-reducing conditions and in the presence of *D. vulgaris*, black solids were formed in all conditions (whether $FeSO_4$, HFO or ferric citrate was used as the iron source). These materials were tentatively identified as poorly ordered mackinawite (FeS_{1-x}) based on previously reported work, but the quantities produced did not allow for extensive spectroscopic characterization. Follow-on experiments will emphasize this objective of the study.

Dechlorination Screening Activity. The dehalogenation activity of microbial and mineral reductants was unevenly distributed with respect to the alkyl halides tested (Table 1). The results show that only tetrachloromethane was dechlorinated in all systems. However, the product distribution and yields differ substantially between incubations. Mineral-mediated dehalogenation tended to result in a better mass balance based on dechlorination products than microbially-mediated reactions, due to the production of radical intermediates which tend to bind to cell material. For example, CT dechlorination in the presence of biogenic Fe_3O_4 resulted in the following product distribution: chloroform (CF), 52%;

noted that dechlorination at these low concentrations of PCE was observed in both aqueous and media controls presumably due to the availability of soluble reducing equivalents. The added presence of solids increased the yield by a factor of up to 4 relative to the media control. Considering the variability of iron in the various amendments, it is not clear which component controls dechlorination activity in these systems.

Dechlorination Kinetics. Based on the results previously reported for iron-reducing systems (McCormick et al., 2000a), the biotic and abiotic pseudo-first order rate constants for CT dechlorination can be estimated by the following surface area- and protein normalized expressions (note: the values below are corrected for head space partitioning effects and therefore differ from those previously reported):

$$k_{\text{abiotic}} = 3.6 \text{ x } 10^{-5} \text{ hr}^{-1} \text{ L m}^{-2} * [\textit{Mineral surface} \text{ m}^2 \text{ L}^{-1}] \quad (1)$$

$$k_{\text{biotic}} = 3.9 \text{ x } 10^{-4} \text{ hr}^{-1} \text{ L mg}^{-1} * [\textit{Protein} \text{ mg L}^{-1}] \quad (2)$$

Using the rate expressions above, the biotic and abiotic dechlorination rates of CT were estimated on the basis of total protein and mineral surface area at several time points during magnetite biogenesis. The cartoon in Figure 2 reflects the reactions occurring in this system during incubation, while the modeled and measured dechlorination rates are shown in Figure 3.

Most of the biogeochemical reactions such as HFO reduction, cell growth, aqueous ferrous iron production and associated redox and pH changes took place within the first 10 days of the experiment. In this system, ferrous iron precipitation between 10 and 30 days was associated with siderite and magnetite formation. Based on long term measurements of the mineral specific surface area and microscopic analysis of the mineral phases, the predominant process after 30 days involves mineral ripening and further crystallization of magnetite.

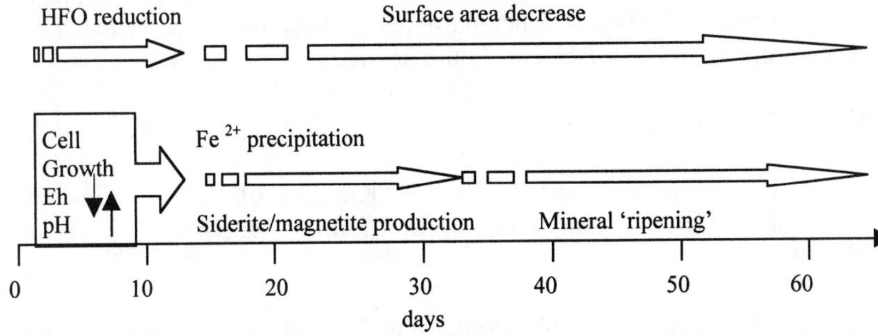
HFO reduction
Surface area decrease
Cell
Growth
Eh
pH
Fe 2+ precipitation
Siderite/magnetite production
Mineral 'ripening'
0
10
20
30
40
50
60
days

FIGURE. 2. Biological, geochemical and mineralogical changes in an active iron-reducing system.

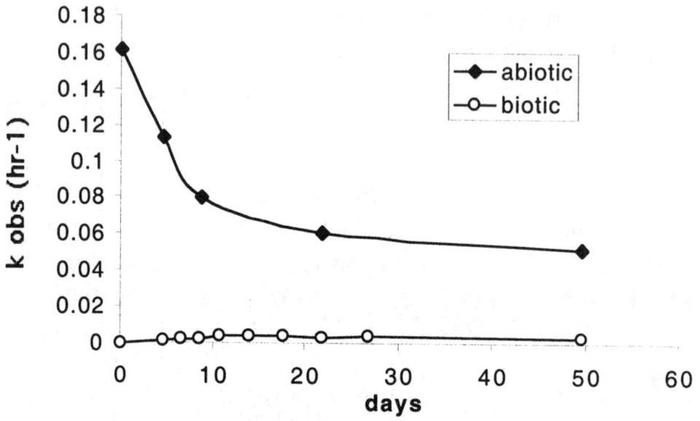
0.18
0.16
0.14
0.12
0.1
0.08
0.06
0.04
0.02
0
k obs (hr-1)
abiotic
biotic
0
10
20
30
40
50
60
days

FIGURE. 3. Carbon tetrachloride (CT) dechlorination dynamics in an active iron-reducing system involving biotic and abiotic components.

CONCLUSIONS

From these data we conclude that during magnetite biogenesis, CT dechlorination is predominantly associated with mineral, rather than with microbial catalysis. Similar experiments with other alkyl halides such as HCA, and under sulfate-reducing conditions will be conducted to assess whether this finding is particular to iron-reducing conditions and CT, or whether this observation is a general characteristic of dechlorination reactions under iron and sulfate-reducing conditions.

ACKNOWLEDGEMENTS

The authors wish to acknowledge the U. S. Environmental Protection Agency STAR Fellowship Program for support of M. L. McCormick, and the Environmental Protection Agency/National Science Foundation/Office of Naval Research Joint Program on Bioremediation for an award to P. Adriaens, M. J. Barcelona and K. F. Hayes. Since these results have not been reviewed by either agency/program, no endorsement should be inferred.

REFERENCES

Butler, E. C., and K. Hayes. 2000. "Kinetics of the Transformation of Halogenated Aliphatic Compounds by Iron Sulfide." *Environ. Sci. Technol. 34*: 422-429.

Egli, C., R. Scholtz, A. M. Cook, and T. Leisinger. 1987. "Anaerobic dechlorination of tetrachloromethane and 1,2-dichloroethane to degradable products by pure cultures of *Desulfobacterium* sp. and *Methanobacterium* sp." *FEMS Microbiol. Lett. 43*:257-261.

Krumholz, L.R., R. Sharp, and S.S. Fishbain. 1996. "A Freshwater Anaerobe Coupling Acetate Oxidation to Tetrachloroethylene Dehalogenation." *Appl. Environ. Microbiol. 62*:4108-4113.

Lendvay, J.M., W.A. Sauck, M.L. McCormick, M.J. Barcelona, D.H. Kampbell, J.T. Wilson, and P. Adriaens. 1998. "Geophysical Characterization, Redox Zonation, and Contaminant Distribution at a Groundwater-Surface Water Interface." *Wat. Resour. Res. 34*: 3545-3559.

McCormick, M.L., Han S. Kim, E.J. Bouwer, and P. Adriaens. 1998. "Abiotic transformation of chlorinated solvents as a consequence of microbial iron reduction: An investigation of the role of biogenic magnetite in mediating reductive dechlorination." In *Proceedings of the 30th Mid-Atlantic Industrial and Hazardous Waste Conference*, pp. 339-348. Technomic Publishing, Lancaster, PA.

McCormick, M.L., Han S. Kim, and P. Adriaens. 2000a. "Transformation of tetrachloromethane in a defined iron reducing culture: Relative contributions of cell and mineral mediated reactions", In *Abstracts of the 219th American Chemical Society National Meeting*, Vol. 40, No.1, pp. 138-141. March 26-30, San Francisco, CA.

McCormick, M.L. and P. Adriaens. 2000b. "Transformation of carbon tetrachloride in a defined iron reducing culture: Product formation via biotic and abiotic pathways". In *Abstracts of the 220th American Chemical Society National Meeting*, Vol. 40, No.2, pp. 357-361. August 20-24, Washington, DC.

ENHANCEMENT OF DISSIMILATORY IRON(III) REDUCTION BY NATURAL ORGANIC MATTER

William Burgos, Richard Royer, Angela Fisher and Richard Unz,
The Pennsylvania State University, University Park, PA 16802

ABSTRACT: Dissimilatory iron(III) reducing bacteria (DIRB) have the ability to directly or indirectly affect a variety of organic and inorganic contaminants. Natural organic matter (NOM) has been shown to increase DIRB activity and hence the in situ bioremediation of target contaminants. NOM has been proposed to enhance Fe(III) reduction by two mechanisms, complexation of biogenic Fe(II) and shuttling of electrons from the bacterium to the Fe(III) oxide surface. The experimental objective of this laboratory study was to compare the relative enhancement of a number of NOMs to two "functional analogs": ferrozine, a Fe(II) complexing agent that can not shuttle electrons, and; anthraquinone-2,6-disulfonate (AQDS), an electron shuttling compound that can not complex Fe(II). The bioreduction of hematite (α-Fe_2O_3) by *Shewanella putrefaciens* CN32 with H_2 as the electron donor was measured after 1 or 5 days with these amendments. Results demonstrated that NOM can complex Fe(II) which enhances the extent of Fe(III) reduction, and shuttle electrons which enhances the initial rate of Fe(III) reduction. NOM-enhanced stimulation of Fe(III) reduction may be a viable strategy for the bioremediation of many contaminants.

INTRODUCTION

Dissimilatory iron(III) reduction is a process of major importance in the biogeochemistry of non-sulfidogenic sediments. The abundance of Fe(III) minerals in the subsurface and their affinity for contaminants through adsorption has made solid-phase Fe(III) oxide reduction a subject of importance with respect to a variety of groundwater contaminants. For example, dissimilatory iron-reducing bacteria (DIRB) can directly oxidize aromatic contaminants (Lovley et al., 1994) and directly reduce a number of inorganic contaminants such as As(V) (Langner and Inskeep, 2000), Tc(VII) (Wildung et al., 2000) and U(VI) (Fredrickson et al., 2000). In addition, biogenic Fe(II) produced by DIRB is a strong, non-specific reductant that can reduce chlorinated solvents (Kim and Picardal, 2000), nitroaromatics (Klausen et al., 1995), As(V), Co(III), Cr(VI), Tc(VII) and U(VI). Thus, the stimulation of DIRB activity could be exploited for the in situ bioremediation of a wide variety of contaminants and contaminant mixtures.

The addition of appropriate electron donors and nutrients can stimulate DIRB abundance and activity for in situ bioremediation. The addition of natural organic matter (NOM) may further stimulate DIRB activity and contaminant reactivity because NOM can serve as an alternate electron acceptor to Fe(III), and bioreduced NOM can act as an "electron shuttle" between the bacterium and Fe(III) or reducible contaminants (Lovley et al., 1998). In addition to serving as an electron shuttle, NOM may also enhance Fe(III) reduction due to its ability to complex biogenic Fe(II). Biogenic Fe(II) has been proposed to "passivate" solid-phase Fe(III) surfaces (Roden and Urrutia, 1999) and interfere with DIRB activity (Urrutia et al., 1998), thus Fe(II) complexation could significantly enhance Fe(III) reduction.

The goal of the present study was to clarify the relative contribution of electron shuttling and Fe(II) complexation as mechanisms by which NOM enhances the bioreduction of ferric oxides.

MATERIALS AND METHODS

The DIRB *Shewanella putrefaciens* strain CN32 was grown aerobically on tryptic soy broth without dextrose (Difco) at 20°C, cells were harvested by centrifugation from a 16-hour-old culture, washed in PIPES-phosphate buffer (50 mM PIPES plus 30 μM phosphate buffer, pH=6.8), and resuspended in deoxygenated PIPES-phosphate buffer in an anaerobic chamber (Coy) under a N_2:H_2 (ca. 97.5:2.5) atmosphere. Cell density was determined by absorbance at 420 nm. The hematite (α-Fe_2O_3, J.T. Baker) used in all experiments had a specific surface area of 9.04 m^2 g^{-1} measured by 5-point BET-N_2 adsorption. Ferrozine (J.T. Baker), a specific Fe(II) iron chelator which forms a 3:1 ferrozine:Fe(II) complex (36.7 mg Fe(II) g^{-1} ferrozine), was added to experimental systems as a dry powder. Anthraquinone-2,6-disulfonate (AQDS, Aldrich) was added volumetrically from a filtered (0.1 μm) concentrated stock solution. NOMs from the International Humic Substances Society (IHSS) tested included: Leonardite Humic Acid (LHA), Soil Humic Acid (SHA), Summit Hill Humic Acid (SHHA), Suwanee River Fulvic Acid (SRFA), Suwanee River Humic Acid (SRHA), and Suwanee River Natural Organic Matter (SRNOM); and Georgetown Natural Organic Matter (GNOM) was provided by Dr. Baohua Gu (Oak Ridge National Laboratory).

Test systems for bioreduction experiments consisted of crimp sealed amber serum bottles (10 mL) containing 5 mL of medium. All preparations were performed in an anaerobic chamber. Sealed test vessels were incubated at 20°C on orbital shakers outside of the anaerobic chamber. The test medium contained 50 mM PIPES-phosphate buffer, 2.0 g L^{-1} hematite (25 mM as Fe) and variable concentrations of the different amendments, and was inoculated to achieve a final cell density of 10^8 cells mL^{-1}. Parallel to all treatments, amendment-free controls were run that contained only the inoculated test medium. All treatments and controls were run in at least triplicate. Uninoculated controls for each amendment type were incubated in quintuplicate for 5 days. In all experiments, serum bottles were sacrificed for iron analysis after 5 days of incubation and, in some experiments, after 1 day of incubation.

Fe(II) was reported as dissolved and acid-extractable. Dissolved Fe(II) was measured with filtered (0.1 μm) sample aliquots added to ferrozine (1 g L^{-1} ferrozine in 50 mM HEPES, pH=8.0) in the anaerobic chamber. The absorbance at 562 nm was determined with a spectrophotometer, results were corrected for dilution and converted to Fe(II) by comparison with standards. Acid-extractable Fe(II) was determined by a 24 h, 0.5 N HCl extraction, then filtered and analyzed as per the dissolved Fe(II) analysis. Solution pH was determined by combination electrode on sample filtrate in the anaerobic chamber. All experiments were performed in a 20°C constant temperature room.

RESULTS AND DISCUSSION

AQDS significantly increased the extent of bioreduction of hematite after either 1 or 5 days of incubation (Figure 1a). Ferrozine did not enhance hematite bioreduction after 1 day although it significantly increased the extent of hematite bioreduction after 5 days (Figure 1b). In contrast to results obtained with AQDS, Fe(II) production in 5-day ferrozine amended cultures was linearly dependent on the concentration of ferrozine (Figure 1b). The enhanced production of Fe(II) versus ferrozine concentration was nearly equal to ferrozine's Fe(II) complexation capacity. Neither AQDS nor ferrozine reduced hematite in uninoculated controls (data not shown).

NOM may enhance ferric oxide bioreduction by complexing Fe(II), thus inhibiting surface passivation, or by acting as an electron shuttle. The goal of this research has been to clarify the relative contribution of each of these proposed functions. Results obtained with LHA are shown as representative of most of the IHSS materials tested (Figure 1c). LHA enhanced the bioreduction of hematite with Fe(II) production being generally linear with LHA concentration. The concentration dependence between the NOMs tested and biogenic Fe(II) production was similar to both AQDS and ferrozine; NOMs enhanced hematite reduction rates during the first day of incubation, and also increased the 5-day extent of reduction in direct proportion to their concentration in the medium. Thus, NOM can act both as an Fe(II) complexant and as an electron shuttle. As an Fe(II) complexant, NOM can prevent Fe(II) sorption to Fe(III) oxide or bacterial cell surfaces and primarily enhance the extent of Fe(III) reduction. As an electron shuttle, NOM can relax the requirement for direct DIRB-oxide contact and/or gain access to more of the oxide surface and primarily enhance the rate of Fe(III) reduction.

To further examine the importance of Fe(II) complexation with respect to the enhancement of Fe(III) bioreduction by NOM, a series of Fe(II) "pre-loading" experiments were performed. All NOMs were added at concentrations that resulted in a Fe(II) complexation capacity of 15 mg L^{-1} as estimated from NOM acidity. Fe(II) was added in sufficient quantity to satisfy this complexation capacity and that of the cell and hematite surfaces (final Fe(II) added was 43 mg L^{-1}). AQDS was included as a positive control for electron shuttling, and an unamended system was tested to assess the affect of Fe(II) on bioreduction in the absence of amendments. An uninoculated control was used to analytically determine the final Fe(II) concentration. All amended systems were compared to the inoculated control using a two tailed t-test ($\alpha = 0.05$). Only the AQDS system was statistically significantly higher than the control (Figure 2). The lack of enhancement by any of the NOMs tested indicated that either: these materials do not shuttle electrons, their electron shuttling function is interfered by Fe(II), the concentrations of NOM used were too low to produce a statistically significant shuttling effect, or the level of Fe(II) in the systems was inhibitory to Fe(III)

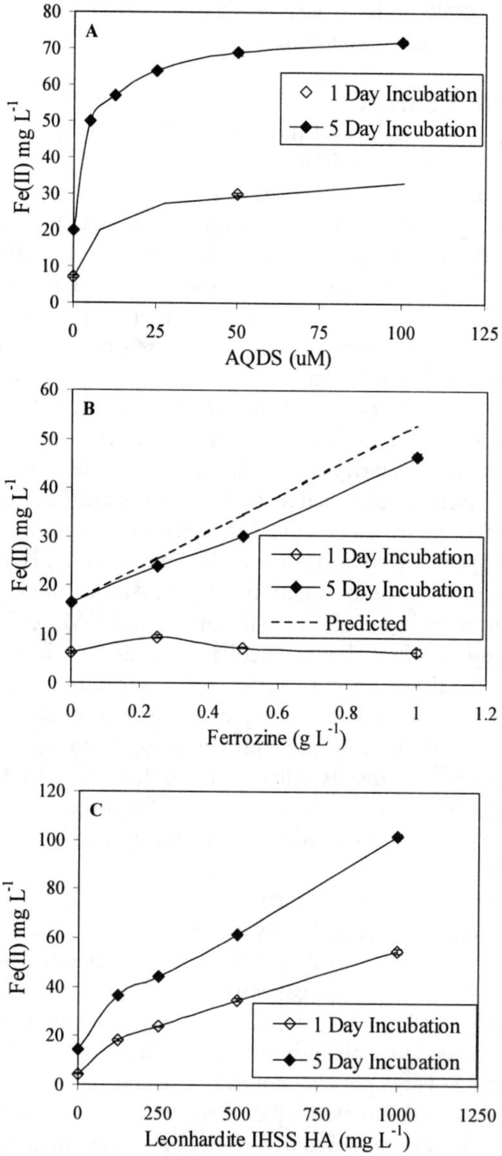
A
Fe(II) mg L-1
AQDS (uM)
1 Day Incubation
5 Day Incubation
B
Fe(II) mg L-1
Ferrozine (g L-1)
1 Day Incubation
5 Day Incubation
Predicted
C
Fe(II) mg L-1
Leonhardite IHSS HA (mg L-1)
1 Day Incubation
5 Day Incubation

FIGURE 1. Production of biogenic acid-extractable Fe(II) (0.5 N HCl) after 1 day or 5 day incubations with variable concentrations of A) AQDS, B) ferrozine and C) IHSS Leonardite humic acid. The predicted line for 5 days shown with ferrozine is based upon the ferrozine-free inoculated control plus the theoretical complexation capacity of 36.7 mg Fe(II) g^{-1} ferrozine. Values are means of triplicate measurements ± one standard deviation.

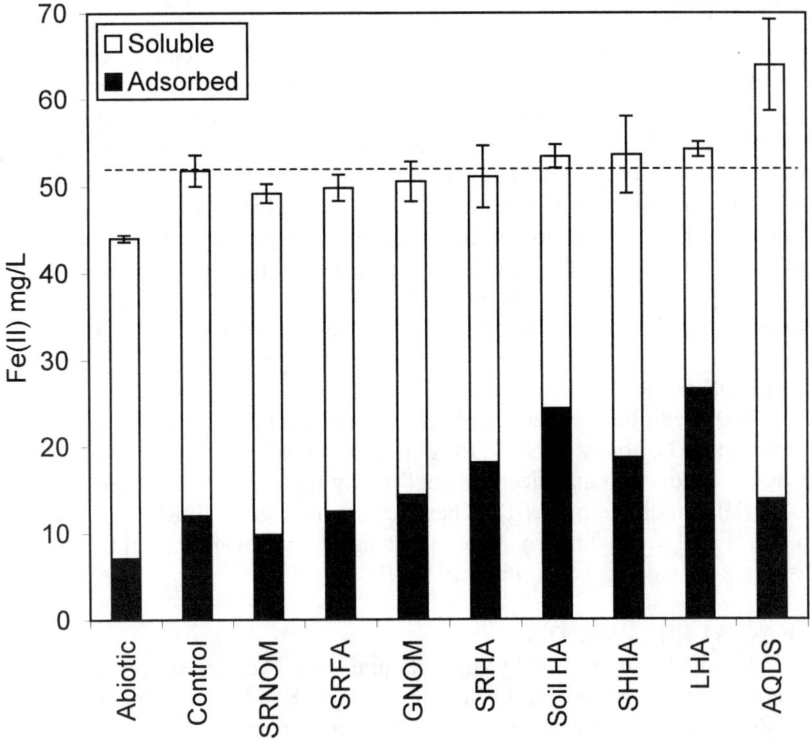
Soluble
Adsorbed
Fe(II) mg/L
70
60
50
40
30
20
10
0
Abiotic
Control
SRNOM
SRFA
GNOM
SRHA
Soil HA
SHHA
LHA
AQDS

FIGURE 2. Measured values of Fe(II) after 5 day incubations with Fe(II) "pre-loaded" into the system. All experimental systems were spiked with 43 mg L^{-1} Fe(II) from a $FeCl_2$ stock solution. All NOM concentrations were selected such that the NOM Fe(II) complexation capacity would be saturated. Values are means of triplicate measurements ± one standard deviation. The dotted line is mean value for the control and the NOM test systems.

reduction. The unamended control showed significant Fe(III) reduction over the uninoculated control but the production of biogenic Fe(II) was quite less in these "pre-loading" experiments as compared to when no initial Fe(II) was present (e.g., ca. 8 mg Fe(II) L^{-1} vs. 16 mg Fe(II) L^{-1}, respectively, Figures 1 and 2). This may be indicative of a process that slows not only the rate of direct bioreduction but also the rate of electron shuttling. While the concentrations of NOM employed were fairly low compared to other experiments (all less than 170 mg L^{-1}) they were very high relative to typical environmental concentrations. It is possible that the formation of NOM-Fe(II) complexes lessened or destroyed the ability of the NOM to shuttle electrons. The ability of AQDS to shuttle electrons was lessened under these "pre-loading" conditions compared to when no initial Fe(II) was present. However, the enhancement of hematite bioreduction by AQDS demonstrates that electron shuttling was still possible as no other enhancement function is attributed to this material.

Comparison of the results of the AQDS, ferrozine and NOM amendments indicate that electron shuttling was occurring in NOM systems as evidenced by their enhancement of iron reduction after 1 day of incubation (Figure 1). These results indicate that electron shuttling by NOM does occur in the absence of substantial free Fe(II) but that this process is severely disrupted by the addition of Fe(II) to the systems. In contrast, electron shuttling by AQDS was able to proceed in the presence of Fe(II) albeit at a somewhat lower level. The partial inhibition of Fe(III) reduction in the unamended control indicates that even direct bioreduction is interfered by Fe(II), although not to the same degree as electron shuttling by NOM. The results of this study indicate that NOM can enhance iron reduction by both electron shuttling and Fe(II) complexation.

CONCLUSIONS

NOM exhibits properties of an electron shuttling compound and a Fe(II) complexant. The ability of NOM to shuttle electrons is inhibited by Fe(II). Direct hematite bioreduction and electron shuttling by the humic acid analog AQDS are also partially inhibited by Fe(II). These results indicate that NOM can effectively enhance Fe(III) reduction by two mechanisms, however, this enhancement is sensitive to elevated concentrations of Fe(II).

ACKNOWLEDGEMENTS

Research supported by the Natural and Accelerated Bioremediation Research Program (NABIR), Office of Biological and Environmental Research (OBER), Office of Energy Research, U.S. Department of Energy (DOE), Grant no. DE-FG02-98ER62691 is gratefully acknowledged. The continued support of Dr. Anna C. Palmisano is greatly appreciated.

REFERENCES

Fredrickson J.K., J.M. Zachara, D.W. Kennedy, M.C. Duff, Y.A. Gorby, S.W. Li, and K.M. Krupka. 2000. "Reduction of U(VI) in goethite (α-FeOOH) suspensions by a dissimilatory metal-reducing bacterium." *Geochim. Cosmochim. Acta 64*: 3085-3098.

Kim, S., and F.W. Picardal. 2000. "Enhanced anaerobic biotransformation of carbon tetrachloride in the presence of reduced iron oxides." *Environ. Toxicol. Chem.*, *18*: 2142-2150.

Klausen, J., S.P. Trober, S.B. Haderlein, and R.P. Schwarzenbach. 1995. "Reduction of substituted nitrobenzenes by Fe(II) in aqueous mineral suspensions." *Environ. Sci. Technol.*, *29*: 2396-2404.

Langner, H.W., and W.P. Inskeep. 2000. "Microbial reduction of arsenate in the presence of ferrihydrite." *Environ Sci. Technol.*, *34*: 3131-3136.

Lovley, D.R., J.C. Woodward, and F.H. Chapelle. 1994. “Stimulated anoxic biodegradation of aromatic hydrocarbons using Fe(III) ligands.” *Nature*, 370: 128-131.

Lovley, D.R., J.L. Fraga, E.L. Blunt-Harris, L.A. Hayes, E.J.P. Phillips, and J.D. Coates. 1998. “Humic substances as a mediator for microbially catalyzed metal reduction.” *Acta Hydrochim. Hydrobiol.*, *26*: 152-157.

Roden, E.E., and M.M. Urrutia. 1999. “Ferrous iron removal promotes microbial reduction of crystalline iron(III) oxides.” *Environ. Sci. Technol.*, *33*: 1847-1853.

Urrutia M.M., E.E. Roden, J.K. Fredrickson, and J.M. Zachara. 1998. “Microbial and geochemical controls on synthetic Fe(III) oxide reduction by *Shewanella alga* strain BrY.” *Geomicrobiol. J. 15*: 269-291.

Wildung, R.E., Y.A. Gorby, K.M. Krupka, N.J. Hess, S.M. Li, A.E. Plymale, J.P. McKinley, and J.K. Fredrickson. 2000. “Effect of electron donor and solution chemistry on products of dissimilatory redcution of technetium by *Shewanella putrefaciens*.” *Appl. Environ. Microbiol.*, *66*: 2451-2460.

A BIOAVAILABLE FERRIC IRON ASSAY AND RELEVANCE TO REDUCTIVE DECHLORINATION

Patrick J. Evans (Camp Dresser & McKee Inc., Bellevue, Washington)
Stephen S. Koenigsberg (Regenesis, San Clemente, California)

ABSTRACT: Reductive dechlorination of chlorinated organic compounds can be promoted through the addition of electron donors to the subsurface. The presence of ferric iron that is *bioavailable* can affect reductive dechlorination by supporting the growth of iron-reducing bacteria. These bacteria can maintain sufficiently low dissolved hydrogen concentrations to prevent reductive dechlorination of *cis*-dichloroethene (c-DCE) for example. Evaluation of data from 13 sites that have undergone pilot testing with Regenesis Hydrogen Release Compound (HRC®) were evaluated for reductive dechlorination of chlorinated ethenes and production of dissolved ferrous iron. Data from nine of the sites supported the hypothesis that bioavailable ferric iron can prevent the reductive dechlorination of c-DCE to vinyl chloride. Data from one of the sites were to the contrary and data from three of the sites were inconclusive. The hypothesis that this "c-DCE slow-down" is mediated by bioavailable ferric iron and that it can be a transient phenomenon is supported by the data. This is important from the standpoint that there is some regulatory concern for the accumulation of c-DCE when parent compounds are actively dechlorinated. An assay was developed that is capable of measuring bioavailable ferric iron in soil. Such an assay may be used to determine the potential for c-DCE accumulation and subsequently engineer total cleanup strategies. These strategies may include sequential reductive-oxidative treatments in which reductively dechlorinated compounds are subsequently oxidized though natural attenuation or engineered methods.

INTRODUCTION & THEORY

Reductive dechlorination of chlorinated organic compounds is recognized as an important remediation tool. The technology involves use of organic or inorganic electron donors that promote dechlorination of electron accepting chlorinated compounds. This process is a component of several remediation strategies. Natural attenuation involving use of natural (e.g., naturally occurring organic carbon) or anthropogenic (e.g., fuel hydrocarbons) sources of electron donors by anaerobic bacteria is one application. Another application is enhanced anaerobic bioremediation which involves injection of electron donors such as molasses, lactic acid, volatile fatty acids, gaseous hydrogen, and commercial products such as Regenesis Hydrogen Release Compound (HRC®). Chemical methods involve the use of inorganic electron donors such as reduced sulfur compounds to promote reductive dechlorination.

A variety of organic and inorganic electron acceptors can be present in addition to chlorinated organic contaminants. Different electron acceptors will be reduced to differing extents depending on thermodynamic and kinetic considerations. Reductive dechlorination typically does not occur under aerobic conditions because oxygen is thermodynamically a more favorable electron acceptor than, for example, trichloroethene (TCE). Additionally, aerobic bacteria have a much greater affinity for molecular hydrogen and will decrease dissolved hydrogen (DH) concentrations to levels that are not conducive to reductive dechlorination for kinetic reasons (Yang and McCarty, 1998).

While the effect of oxygen on reductive dechlorination is widely accepted, the effect of ferric iron (Fe III) is not as well understood. Fe III is an electron acceptor that iron-reducing bacteria reduce to ferrous iron (Fe II) in order to obtain energy. Unlike oxygen, Fe III is present in nature in a variety of forms. Most of these forms are iron oxides and hydroxides that are relatively insoluble in water at typical groundwater pH. These forms vary widely in their ability to be reduced by iron-reducing bacteria. The ability to be used by iron-reducing bacteria is referred to *bioavailability* as defined here:

> *Ferric iron (Fe III) that is capable of being reduced by microorganisms that oxidize another chemical species and derive energy from the electron transfer.*

Many factors can affect Fe III bioavailability. These include solid-phase crystallinity and surface area (Roden and Zachara, 1996). In addition, they include groundwater chemistry such as pH, specific conductivity, divalent cations, electron shuttles such as quinones, and chelators (Evans, 2000). These factors affect the ability of iron-reducing bacteria to reduce Fe III. For example, Table 1 shows the free energy associated with reduction of various electron acceptors including inorganic compounds and chlorinated ethenes (EPA, 1998). Three different forms of Fe III are shown each having different free energies due to differences in crystallinity (or lack thereof) and surface area. Amorphous iron oxides and hydroxides [amorphous $Fe(OH)_3$ and FeOOH] are typically considered to be more bioavailable than crystalline species such as goethite.

TABLE 1. Free energy values for reduction of various electron acceptors.

Electron Acceptor	Half-Cell Reaction Product	Free Energy per Electron Transferred (kJ/mole)
NO_3^-	N_2	-120
O_2	H_2O	-119
Amorphous $Fe(OH)_3$	Fe II	-89.9
FeOOH	Fe II	-62.9
PCE	TCE	-61.8
TCE	c-DCE	-60.6
VC	Ethene	-57.5
c-DCE	VC	-50.7
Crystallized Goethite	Fe II	-49.2
SO_4^-	HS^-	-24
CO_2	CH_4	-16.4

(EPA, 1998)

Table 1 also shows that reductive dechlorination of tetrachloroethene

(PCE), trichloroethene (TCE), *cis*-dichloroethene (c-DCE), and vinyl chloride (VC) is less favorable than reduction of amorphous $Fe(OH)_3$ and FeOOH but more favorable than reduction of crystalline goethite. The difference is most significant for c-DCE. Thus from a thermodynamic perspective, reductive dechlorination of c-DCE to VC may be inhibited by the presence of bioavailable Fe III. The effects of bioavailable Fe III can help to explain a common observation during enhanced anaerobic bioremediation where PCE and TCE are reductively dechlorinated to c-DCE but c-DCE is not reductively dechlorinated to VC. We refer to this phenomenon as the "c-DCE slow-down".

RESULTS AND DISCUSSION

Data from 13 sites across the United States that underwent pilot testing with Regenesis HRC were evaluated. All sites were contaminated with PCE, TCE and/or reductive dechlorination products. The data were classified into four categories as shown in Table 2. Data from 9 of the 13 sites were consistent with the hypothesis that iron reduction inhibited c-DCE reduction and/or VC production. Representative data are shown in Figures 1 and 2. Figure 1 contains data for the NJ1 site in which c-DCE reduction and VC production were not observed. High amounts of dissolved Fe II production were observed possibly inhibiting the reductive dechlorination of c-DCE. Another possible explanation may be that oxidative degradation of c-DCE and/or VC occurred under iron-reducing conditions. Figure 2 contains data for the VA site where high dissolved Fe II production occurred initially but then stopped. TCE and c-DCE reduction and VC production occurred after Fe II production ceased. These data suggest that inhibition of c-DCE reduction was a transient phenomenon that was alleviated once bioavailable ferric iron was consumed. One of the sites (IL, see Table 2) provided data contrary to the hypothesis since VC was produced even though Fe II production occurred. Data from three sites were inconclusive because the data were not internally consistent.

Table 2. Relationship between dissolved Fe II production and reductive dechlorination at 13 sites during HRC pilot testing.

Sites	Observation	Consistent
WI, NJ2, OR, OH, VA	VC production or c-DCE reduction observed after Fe II increase stopped.	Yes
FL2, MI, NY	VC production or c-DCE reduction. Little or no Fe II production.	Yes
NJ1	No VC production or c-DCE reduction. Substantial Fe II production.	Yes
IL	VC production and Fe II production.	No
KS, CA, FL1	Data inconclusive.	—

Because of the observed relationship between dissolved Fe II in groundwater and reductive dechlorination, a better understanding of the effects of bioavailable Fe III is warranted. Quantification of bioavailable Fe III is a key element in this understanding. A field-portable assay has been developed (patent pending) that measures bioavailable Fe III in soil or sediment. The bioavailable ferric iron assay is a bioassay that uses the iron-reducing bacterium *Shewanella alga* BrY to provide a direct measurement of ferric iron bioavailability. The assay medium is shown in Table 3 (patent pending). This medium contains several factors in optimal concentrations to promote iron reduction by *Shewanella alga* BrY. In addition, a phosphate buffer and nitrogen source is provided to promote pH control, growth, and iron reduction. Lactic acid is provided as the electron donor.

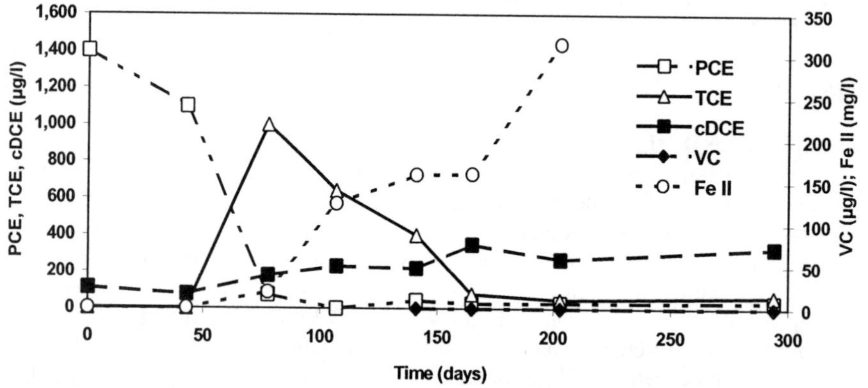
1,600
1,400
1,200
1,000
800
600
400
200
0
PCE, TCE, cDCE (µg/l)
350
300
250
200
150
100
50
0
VC (µg/l); Fe II (mg/l)
0
50
100
150
200
250
300
Time (days)
PCE
TCE
cDCE
VC
Fe II

Figure 1. Site NJ1 results showing that while reduction of PCE to TCE and TCE to c-DCE were observed, reduction of c-DCE to VC was not observed and Fe II production was observed.

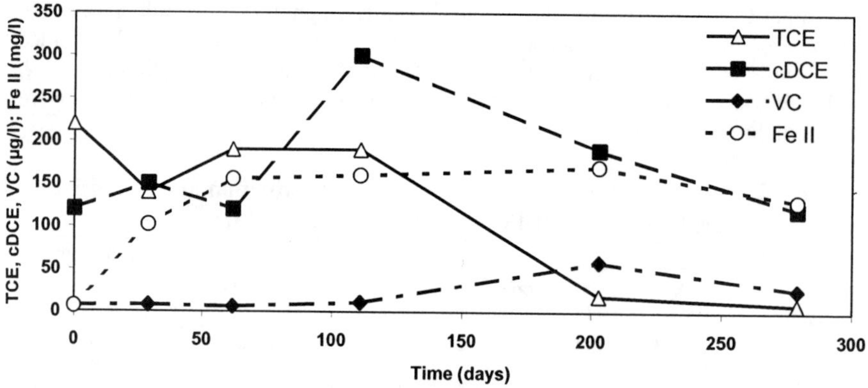
350
300
250
200
150
100
50
0
TCE, cDCE, VC (µg/l); Fe II (mg/l)
0
50
100
150
200
250
300
Time (days)
TCE
cDCE
VC
Fe II

Figure 2. Site VA results showing that TCE reduction to c-DCE and c-DCE reduction to VC was observed after production of dissolved Fe II ceased.

The assay is prepackaged for ease of use. *Shewanella alga* BrY was lyophilized separately (Reagent B) and is best stored under refrigeration or freezing so as to maintain stability over long periods of time. The remaining reagents are combined in a dry state (Reagent A) and can be stored at room temperature. The assay is conducted as shown in Figure 3 by adding 5-gram (g) soil samples to each of two 25-milliliter (ml) tubes. One tube (time thirty or T30) contains Reagent A and the other tube (time zero or T0) does not.

Distilled water plus 10 ml of concentrated HCl is added to tube T0. The tube is mixed for 24 hours during which adsorbed Fe II is dissolved. Then dissolved Fe II is measured using the ferrozine assay or a standard Hach or Chemetrics test kit. This measurement gives the initial or T0 result.

TABLE 3. Bioavailable ferric iron assay composition (patent pending).

Component	Concentration	Units
KH_2PO_4	2.2	g/l
K_2HPO_4	2.4	g/l
NH_4Cl	1.5	g/l
KCl	0.1	g/l
Humic acids	1	g/l
Anthraquinone-1,5-disulfonic acid (AQDS)	0.041	g/l
$CaCl_2$-$2H_2O$	1	g/l
L-Lactic acid, sodium salt	40	MM
Shewanella alga BrY	-	-
Trace metals	64	mg/l
pH	6.5	-

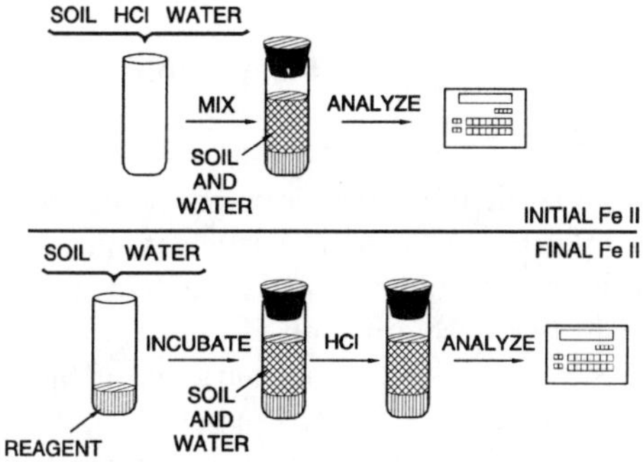
SOIL HCl WATER
MIX
ANALYZE
SOIL AND WATER
INITIAL Fe II
SOIL WATER
FINAL Fe II
INCUBATE
HCl
ANALYZE
SOIL AND WATER
REAGENT

Figure 3. Schematic representation of bioavailable ferric iron assay protocol (patent pending).

Reagent B and distilled water are added to tube T30 that already contains Reagent A. The tube is incubated at room temperature in the dark for 30 days during which *Shewanella alga* BrY reduces any bioavailable Fe III to Fe II. After the incubation period, 10 ml of concentrated HCl is added to tube T30 to dissolve adsorbed Fe II. After 24 hours, dissolved Fe II is measured giving the final or T30 result. The difference between T30 and T0 Fe II is the amount of Fe III that was reduced over the 30-day incubation and is the bioavailable ferric iron. The concentration of bioavailable ferric iron on the soil is calculated from this result.

Figure 4 shows results for a variation of the assay where *Geobacter metalloreducens* GS-15 and acetate were used instead of BrY and lactate.

Amorphous iron oxide was varied and Fe II production was measured. The linearity of the bioassay was good, with a correlation coefficient (r^2) of 0.965. A greater amount of Fe II was produced than expected, as indicated by a slope 1.6. The concentration of the stock amorphous iron oxide (640 mM) was an estimate and was likely the source of this discrepancy. The sensitivity of the bioassay was estimated to be less than 1 mM based on these results. The relative percent deviation defined as the difference between replicates divided by the mean concentration was typically less than 25% and at most 41%.

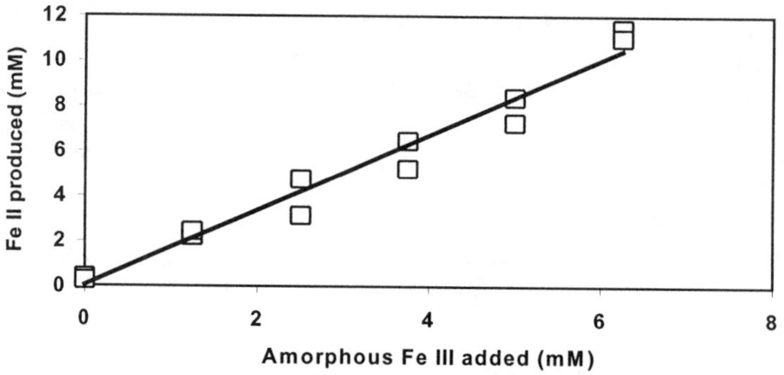
Fe II produced (mM)
12
10
8
6
4
2
0
0
2
4
6
8
Amorphous Fe III added (mM)

Figure 4. Linearity of the bioavailable Fe III assay.

CONCLUSIONS

The effect of bioavailable ferric iron on reductive dechlorination has practical importance in site remediation. Sufficient quantities of bioavailable ferric iron may result in the consumption of limited electron donors, in the form of DH, and prevent reductive dechlorination of c-DCE to VC. This phenomenon is not observed at all sites because the concentration of bioavailable ferric iron can vary significantly from site to site (Evans et al., 1999). Furthermore, the prevention of reductive dechlorination may be transient that ceases once bioavailable ferric iron is consumed (Figure 2). This review of site data has provided initial support of the hypothesis that bioavailable ferric iron is a factor that may cause c-DCE slow-down. Future work to further validate the hypothesis that bioavailable ferric iron can cause c-DCE slow down is recommended.

Measurement of bioavailable ferric iron may be helpful with regard to prediction of c-DCE accumulation during enhanced anaerobic bioremediation. This analyte may also promote understanding of the potential for complete reductive dechlorination of TCE under a monitored natural attenuation scenario. Prediction of this phenomenon has the potential to improve the engineering design of site cleanup. For example, measurement of substantial bioavailable ferric iron concentrations may lead to the following potential courses of action:

- Add sufficient electron donor to consume bioavailable ferric iron.
- Allow for a transient period of c-DCE accumulation followed by reduction.

- Plan on c-DCE accumulation and use an oxidative method for c-DCE degradation to carbon dioxide (e.g., oxygen addition or dependence on oxidative c-DCE biodegradation under iron-reducing conditions).

Accumulation of c-DCE during reductive dechlorination appears to have the potential to be predicted, quantified, and controlled. Measurement of bioavailable ferric iron is one tool that can be used in making engineering decisions is this regard.

ACKNOWLEDGMENTS

The authors wish to thank Ms. Carmen LeBron of the Naval Facilities Engineering Service Center and Lisa Nickens, Mary Trute, and Naht Rammachat of CDM. Funding for Patrick Evans was provided by the U.S. Air Force under the Small Business Innovative Research program and by the U.S. Navy under the Environmental Security Technology Certification Program. New Horizons Diagnostics Corporation (Columbia, Maryland) is acknowledged for producing the bioavailable ferric iron assay.

REFERENCES

EPA. 1998. "Technical Protocol for Evaluating Natural Attenuation of Chlorinated Solvents in Ground Water. p. B3-43.

Evans, P.J., K.A. Jones, C.C. Liu, and D.R. Lovley. 1999. "Development of a Natural Attenuation Test Kit." In: Natural Attenuation of Chlorinated Solvents, Petroleum Hydrocarbons, and Other Organic Compounds (B.C. Alleman and A. Leeson, Eds.) Battelle Press, Columbus. pp. 331-336.

Evans, P.J. 2000. "A Novel Ferric Iron Bioavailability Assay." In: Risk, Regulatory, and Monitoring Considerations (G.B. Wickramanayake et al, eds.) Battelle Press, Columbus. pp. 167-174

Roden, E.E. and J.M. Zachara. 1996. "Microbial Reduction of Crystalline Iron (III) Oxides: Influence of Oxide Surface Area and Potential for Cell Growth." *Environ. Sci. Technol. 30*:1618-1628

Yang, Y, and P.L. McCarty. 1998. "Competition for Hydrogen within a Chlorinated Solvent Dehalogenating Anaerobic Mixed Culture." *Environ. Sci. Technol. 32*: 3591-3597

2001 AUTHOR INDEX

This index contains names, affiliations, and volume/page citations for all authors who contributed to the ten-volume proceedings of the Sixth International In Situ and On-Site Bioremediation Symposium (San Diego, California, June 4-7, 2001). Ordering information is provided on the back cover of this book.

The citations reference the ten volumes as follows:

6(1): Magar, V.S., J.T. Gibbs, K.T. O'Reilly, M.R. Hyman, and A. Leeson (Eds.), *Bioremediation of MTBE, Alcohols, and Ethers*. Battelle Press, Columbus, OH, 2001. 249 pp.

6(2): Leeson, A., M.E. Kelley, H.S. Rifai, and V.S. Magar (Eds.), *Natural Attenuation of Environmental Contaminants*. Battelle Press, Columbus, OH, 2001. 307 pp.

6(3): Magar, V.S., G. Johnson, S.K. Ong, and A. Leeson (Eds.), *Bioremediation of Energetics, Phenolics, and Polycyclic Aromatic Hydrocarbons*. Battelle Press, Columbus, OH, 2001. 313 pp.

6(4): Magar, V.S., T.M. Vogel, C.M. Aelion, and A. Leeson (Eds.), *Innovative Methods in Support of Bioremediation*. Battelle Press, Columbus, OH, 2001. 197 pp.

6(5): Leeson, A., E.A. Foote, M.K. Banks, and V.S. Magar (Eds.), *Phytoremediation, Wetlands, and Sediments*. Battelle Press, Columbus, OH, 2001. 383 pp.

6(6): Magar, V.S., F.M. von Fahnestock, and A. Leeson (Eds.), *Ex Situ Biological Treatment Technologies*. Battelle Press, Columbus, OH, 2001. 423 pp.

6(7): Magar, V.S., D.E. Fennell, J.J. Morse, B.C. Alleman, and A. Leeson (Eds.), *Anaerobic Degradation of Chlorinated Solvents*. Battelle Press, Columbus, OH, 2001. 387 pp.

6(8): Leeson, A., B.C. Alleman, P.J. Alvarez, and V.S. Magar (Eds.), *Bioaugmentation, Biobarriers, and Biogeochemistry*. Battelle Press, Columbus, OH, 2001. 255 pp.

6(9): Leeson, A., B.M. Peyton, J.L. Means, and V.S. Magar (Eds.), *Bioremediation of Inorganic Compounds*. Battelle Press, Columbus, OH, 2001. 377 pp.

6(10): Leeson, A., P.C. Johnson, R.E. Hinchee, L. Semprini, and V.S. Magar (Eds.), *In Situ Aeration and Aerobic Remediation*. Battelle Press, Columbus, OH, 2001. 391 pp.

Aagaard, Per (University of Oslo/NORWAY) *6*(2):181
Aarnink, Pedro J.P. (Tauw BV/THE NETHERLANDS) *6*(10):253
Abbott, James E. (Battelle/USA) *6*(5):231, 237
Accashian, John V. (Camp Dresser & McKee, Inc./USA) *6*(7):133
Adams, Daniel J. (Camp Dresser & McKee, Inc./USA) *6*(8):53
Adams, Jack (Applied Biosciences Corporation/USA) *6*(9):331
Adriaens, Peter (University of Michigan/USA) *6*(8):19, 193
Adrian, Neal R. (U.S. Army Corps of Engineers/USA) *6*(6):133
Agrawal, Abinash (Wright State University/USA) *6*(5):95
Aiken, Brian S. (Parsons Engineering Science/USA) *6*(2): *6*5, 189

Aitchison, Eric (Ecolotree, Inc./USA) *6*(5):121
Al-Awadhi, Nader (Kuwait Institute for Scientific Research/KUWAIT) *6*(6):249
Alblas, B. (Logisticon Water Treatment/THE NETHERLANDS) *6*(8):11
Albores, A. (CINVESTAV-IPN/MEXICO) *6*(6):219
Al-Daher, Reyad (Kuwait Institute for Scientific Research/KUWAIT) *6*(6):249
Al-Fayyomi, Ihsan A. (Metcalf & Eddy, Inc./USA) *6*(7):173
Al-Hakak, A. (McGill University/CANADA) *6*(9):139
Allen, Harry L. (U.S. EPA/USA) *6*(3):259
Allen, Jeffrey (University of Cincinnati/USA) *6*(9):9
Allen, Mark H. (Dames & Moore/USA) *6*(10):95
Allende, J.L. (Universidad Complutense/SPAIN) *6*(4):29
Alonso, R. (Universidad Politecnica/SPAIN) *6*(6):377
Alphenaar, Arne (TAUW bv/THE NETHERLANDS) *6*(7):297
Alvarez, Pedro J. J. (University of Iowa/USA) *6*(1):195; *6*(3):1; *6*(8):147, 175
Alvestad, Kimberly R. (Earth Tech/USA) *6*(3):17
Ambert, Jack (Battelle Europe/SWITZERLAND) *6*(6):241
Amezcua-Vega, Claudia (CINVESTAV-IPN/MEXICO) *6*(3):243
Amy, Penny (University of Nevada Las Vegas/USA) *6*(9):257
Andersen, Peter F. (GeoTrans, Inc./USA) *6*(10):163
Anderson, Bruce (Plan Real AG/AUSTRALIA) *6*(2):223
Anderson, Jack W. (RMT, Inc./USA) *6*(10):201
Anderson, Todd (Texas Tech University/USA) *6*(9):273
Andreotti, Giorgio (ENI Sop.A.) *6*(5):41
Andretta, Massimo (Centro Ricerche Ambientali Montecatini/ITALY) *6*(4):131
Andrews, Eric (Environmental Management, Inc./USA) *6*(10):23
Andrews, John (SHN Consulting Engineers & Geologists, Inc./USA) *6*(3):83
Archibald, Brent B. (Exxon Mobil Environmental Remediation/USA) *6*(8):87
Archibold, Errol (Spelman College/USA) *6*(9):53
Aresta, Michele (Universita di Catania/ITALY) *6*(3):149
Arias, Marianela (PDVSA Intevep/VENEZUELA) *6*(6):257
Atagana, Harrison I. (Mangosuthu Technikon/REP OF SOUTH AFRICA) *6*(6):101
Atta, Amena (U.S. Air Force/USA) *6*(2):73
Ausma, Sandra (University of Guelph/CANADA) *6*(6):185
Autenrieth, Robin L. (Texas A&M University/USA) *6*(5): 17, 25
Aziz, Carol E. (Groundwater Services, Inc./USA) *6*(7):19; *6*(8):73
Azizian, Mohammad (Oregon State University/USA) *6*(10): 145, 155

Babel, Wolfgang (UFZ Center for Environmental Research/GERMANY) *6*(4):81
Bae, Bumhan (Kyungwon University/REPUBLIC OF KOREA) *6*(6):51
Baek, Seung S. (Kyonggi University/REPUBLIC OF KOREA) *6*(1):161
Bagchi, Rajesh (University of Cincinnati/USA) *6*(5):243, 253, 261
Baiden, Laurin (Clemson University/USA) *6*(7):109
Bakker, C. (IWACO/THE NETHERLANDS) *6*(7):141
Balasoiu, Cristina (École Polytechnique de Montreal/CANADA) *6*(9):129
Balba, M. Talaat (Conestoga-Rovers & Associates/USA) *6*(1):99; *6*(6):249; *6*(10):131

Banerjee, Pinaki (Harza Engineering Company, Inc./USA) *6*(7):157
Bankston, Jamie L. (Camp Dresser and McKee Inc./USA) *6*(5):33
Barbé, Pascal (Centre National de Recherche sur les Sites et Sols Pollués/FRANCE) *6*(2):129
Barcelona, Michael J. (University of Michigan/USA) *6*(8):19, 193
Barczewski, Baldur (Universitat Stuttgart/GERMANY) *6*(2):137
Barker, James F. (University of Waterloo/CANADA) *6*(8):95
Barnes, Paul W. (Earth Tech, Inc./USA) *6*(3): 17, 25
Basel, Michael D. (Montgomery Watson Harza/USA) *6*(10):41
Baskunov, Boris B. (Russian Academy of Sciences/RUSSIA) *6*(3):75
Bastiaens, Leen (VITO/BELGIUM) *6*(4):35; *6*(9):87
Batista, Jacimaria (University of Nevada Las Vegas/USA) *6*(9): 257, 265
Bautista-Margulis, Raul G. (Centro de Investigacion en Materiales Avanzados/MEXICO) *6*(6):361
Becker, Paul W. (Exxon Mobil Refining & Supply/USA) *6*(8):87
Beckett, Ronald (Monash University/AUSTRALIA) *6*(4):1
Beckwith, Walt (Solutions Industrial & Environmental Services/USA) *6*(7):249
Beguin, Pierre (Institut Pasteur/FRANCE) *6*(1):153
Behera, N. (Sambalpur University/INDIA) *6*(9):173
Bell, Nigel (Imperial College London/UK) *6*(10):123
Bell, Mike (Coats North America/USA) *6*(7):213
Beller, Harry R. (Lawrence Livermore National Laboratory/USA) *6*(1):195
Belloso, Claudio (Facultad Catolica de Quimica e Ingenieria/ARGENTINA) *6*(6): 235, 303
Benner, S. G. (Stanford University/USA) *6*(9):71
Bensch, Jeffrey C. (GeoTrans, Inc/USA) *6*(7):221
Béron, Patrick (Université du Québec à Montréal/CANADA) *6*(3):165
Berry, Duane F. (Virginia Polytechnic Institute & State University/USA) *6*(2):105
Betts, W. Bernard (Cell Analysis Ltd./UK) *6*(6):27
Billings, Bradford G. (Brad) (Billings & Associates, Inc./USA) *6*(1):115
Bingler, Linda (Battelle Sequim/USA) *6*(5):231, 237
Birkle, M. (Fraunhofer Institute/GERMANY) *6*(2):137
Bitter, Paul (URS Corporation./USA) *6*(2):261
Bittoni, A. (EniTecnologie/ITALY) *6*(6):173
Bjerg, Poul L (Technical University of Denmark/DENMARK) *6*(2):11
Blanchet, Denis (Institut Français du Pétrole/FRANCE) *6*(3):227
Bleckmann, Charles A. (Air Force Institute of Technology/USA) *6*(2):173
Blokzijl, R. (DHV Environment and Infrastructure/THE NETHERLANDS) *6*(8):11
Blowes, David (University of Waterloo/CANADA) *6*(9):71
Bluestone, Simon (Montgomery Watson/ITALY) *6*(10):41
Boben, Carolyn (Williams/USA) *6*(1):175
Böckle, Karin (Technologiezentrum Wasser/GERMANY) *6*(8):105
Boender, H. (Logisticon Water Treatment/THE NETHERLANDS) *6*(8):11
Böhler, Anja (BioPlanta GmbH/GERMANY) *6*(3):67
Bonner, James S. (Texas A&M University/USA) *6*(5):17, 25
Bononi, Vera Lucia Ramos (Instituto de Botânica/BRAZIL) *6*(3):99
Bonsack, Laurence T. (Aerojet/USA) *6*(9):297
Borazjani, Abdolhamid (Mississippi State University/USA) *6*(5):329; *6*(6):279

Borden, Robert C. (Solutions Industrial & Environmental Services/USA) *6*(7):249
Bornholm, Jon (U.S. EPA/USA) *6*(6):81
Bosco, Francesca (Politecnico di Torino/ITALY) *6*(3):211
Bosma, Tom N.P. (TNO Environment/THE NETHERLANDS) *6*(7):61
Bourquin, Al W. (Camp Dresser & McKee Inc./USA) *6*(5):33; *6*(6):81; *6*(7):133,
Bouwer, Edward J. (Johns Hopkins University/USA) *6*(2):19
Bowman, Robert S. (New Mexico Institute of Mining & Technology/USA) *6*(8):131
Boyd, Sian (CEFAS Laboratory/UK) *6*(10):337
Boyd-Kaygi, Patricia (Harding ESE/USA) *6*(10):231
Boyle, Susan L. (Haley & Aldrich, Inc./USA) *6*(7):27, 281
Brady, Warren D. (IT Corporation/USA) *6*(9):215
Breedveld, Gijs (University of Oslo/NORWAY) *6*(2):181
Bregante, M. (Istituto di Cibernetica e Biofisica/ITALY) *6*(5):157
Brenner, Richard C. (U.S. EPA/USA) *6*(5):231, 237
Breteler, Hans (Oostwaardhoeve Co./THE NETHERLANDS) *6*(6):59
Bricka, Mark R. (U.S. Army Corps of Engineers/USA) *6*(9):241
Brickell, James L. (Earth Tech, Inc./USA) *6*(10):65
Brigmon, Robin L. (Westinghouse Savannah River Co/USA) *6*(7):109
Britto, Ronnie (EnSafe, Inc./USA) *6*(9):315
Brossmer, Christoph (Degussa Corporation/USA) *6*(10):73
Brown, Bill (Dunham Environmental Services/USA) *6*(6):35
Brown, Kandi L. (IT Corporation/USA) *6*(1):51
Brown, Richard A. (ERM, Inc./USA) *6*(7):45, 213
Brown, Stephen (Queen's University/CANADA) *6*(2):121
Brown, Susan (National Water Research Institute/CANADA) *6*(7):321, 333, 341
Brubaker, Gaylen (ThermoRetec North Carolina Corp./USA) *6*(7):1
Bruce, Cristin (Arizona State University/USA) *6*(8):61
Bruce, Neil C. (University of Cambridge/UK) *6*(5):69
Buchanan, Gregory (Tait Environmental Management, Inc./USA) *6*(10):267
Bucke, Christopher (University of Westminster/UK) *6*(3):75
Bulloch, Gordon (BAE Systems Properties Ltd./UK) *6*(6):119
Burckle, John (U.S. EPA/USA) *6*(9):9
Burden, David S. (U.S. EPA/USA) *6*(2):163
Burdick, Jeffrey S. (ARCADIS Geraghty & Mills/USA) *6*(7):53
Burgos, William (The Pennsylvania State University/USA) *6*(8):201
Burken, Joel G. (University of Missouri-Rolla/USA) *6*(5):113, 199
Burkett, Sharon E. (ENVIRON International Corp./USA) *6*(7):189
Burnell, Daniel K. (GeoTrans, Inc./USA) *6*(2):163
Burns, David A. (ERM, Inc./USA) *6*(7):213
Burton, Christy D. (Battelle/USA) *6*(1):137; *6*(10):193
Buscheck, Timothy E. (Chevron Research & Technology Co/USA) *6*(1): 35, 203
Buss, James A. (RMT, Inc./USA) *6*(2):97
Butler, Adrian P. (Imperial College London/UK) *6*(10):123
Butler, Jenny (Battelle/USA) *6*(7):13
Büyüksönmez, Fatih (San Diego State University/USA) *6*(10):301

Caccavo, Frank (Whitworth College/USA) *6*(8):1
Callender, James S. (Rockwell Automation/USA) *6*(7):133
Calva-Calva, G. (CINVESTAV-IPN/MEXICO) *6*(6):219
Camper, Anne K. (Montana State University/USA) *6*(7):117

Camrud, Doug (Terracon/USA) *6*(10):15
Canty, Marietta C. (MSE Technology Applications/USA) *6*(9):35
Carman, Kevin R. (Louisiana State University/USA) *6*(5):305
Carrera, Paolo (Ambiente S.p.A./ITALY) *6*(6):227
Carson, David A. (U.S. EPA/USA) *6*(2):247
Carvalho, Cristina (Clemson University/USA) *6*(7):109
Case, Nichole L. (Haley & Aldrich, Inc./USA) *6*(7):27, 281
Castelli, Francesco (Universita di Catania/ITALY) *6*(3):149
Cha, Daniel K. (University of Delaware/USA) *6*(6):149
Chaney, Rufus L. (U.S. Department of Agriculture/USA) *6*(5):77
Chang, Ching-Chia (National Chung Hsing University/TAIWAN) *6*(10):217
Chang, Soon-Woong (Kyonggi University/REPUBLIC OF KOREA) *6*(1):161
Chang, Wook (University of Maryland/USA) *6*(3):205
Chapuis, R. P. (École Polytechnique de Montréal/CANADA) *6*(4):139
Charrois, Jeffrey W.A. (Komex International, Ltd./CANADA) *6*(4):7
Chatham, James (BP Exploration/USA) *6*(2):261
Chekol, Tesema (University of Maryland/USA) *6*(5):77
Chen, Abraham S.C. (Battelle/USA) *6*(10):245
Chen, Chi-Ruey (Florida International University/USA) *6*(10):187
Chen, Zhu (The University of New Mexico/USA) *6*(9):155
Cherry, Jonathan C. (Kennecott Utah Copper Corp/USA) *6*(9):323
Child, Peter (Investigative Science Inc./CANADA) *6*(2):27
Chino, Hiroyuki (Obayashi Corporation/JAPAN) *6*(6):249
Chirnside, Anastasia E.M. (University of Delaware/USA) *6*(6):9
Chiu, Pei C. (University of Delaware/USA) *6*(6):149
Cho, Kyung-Suk (Ewha University/REPUBLIC OF KOREA) *6*(6):51
Choung, Youn-kyoo (Yonsei University/REPUBLIC OF KOREA) *6*(6):51
Clement, Bernard (École Polytechnique de Montréal/CANADA) *6*(9):27
Clemons, Gary (CDM Federal Programs Corp./USA) *6*(6):81
Cocos, Ioana A. (École Polytechnique de Montréal/CANADA) *6*(9):27
Cocucci, M. (Universita' degli Studi di Milano/ITALY) *6*(5):157
Coelho, Rodrigo O. (CSD-GEOLOCK/BRAZIL) *6*(1):27
Collet, Berto (TAUW bv/THE NETHERLANDS) *6*(10):253
Compton, Joanne C. (REACT Environmental Engineers/USA) *6*(3):25
Connell, Doug (Barr Engineering Company/USA) *6*(5):105
Connor, Michael A. (University of Melbourne/AUSTRALIA) *6*(10):329
Cook, Jim (Beazer East, Inc./USA) *6*(2):239
Cooke, Larry (NOVA Chemicals Corporation/USA) *6*(4):117
Coons, Darlene (Conestoga-Rovers & Associates/USA) *6*(1):99; *6*(10):131
Costley, Shauna C. (University of Natal/REP OF SOUTH AFRICA) *6*(9):79
Cota, Jennine L. (ARCADIS Geraghty & Miller, Inc./USA) *6*(7):149
Covell, James R. (EG&G Technical Services, Inc./USA) *6*(10):49
Cowan, James D. (Ensafe Inc./USA) *6*(9):315
Cox, Evan E. (GeoSyntec Consultants/CANADA) *6*(8):27, *6*(9):297
Cox, Jennifer (Clemson University/USA) *6*(7):109
Craig, Shannon (Beazer East, Inc./USA) *6*(2):239
Crawford, Donald L. (University of Idaho/USA) *6*(3):91; *6*(9):147

Crecelius, Eric (Battelle/USA) *6*(5): 231, 237
Crotwell, Terry (Solutions Industrial & Environmental Services/USA) *6*(7):249
Cui, Yanshan (Chinese Academy of Sciences/CHINA) *6*(9):113
Cunningham, Al B. (Montana State University/USA) *6*(7):117; *6*(8):1
Cunningham, Jeffrey A. (Stanford University/USA) *6*(7):95
Cutright, Teresa J. (The University of Akron/USA) *6*(3):235

da Silva, Marcio Luis Busi (University of Iowa/USA) *6*(1):195
Daly, Daniel J. (Energy & Environmental Research Center/USA) *6*(5):129
Daniel, Fabien (AEA Technology Environment/UK) *6*(10):337
Daniels, Gary (GeoTrans/USA) *6*(8):19
Das, K.C. (University of Georgia/USA) *6*(9):289
Davel, Jan L. (University of Cincinnati/USA) *6*(6):133
Davis, Gregory A. (Microbial Insights Inc./USA) *6*(2):97
Davis, Jeffrey L. (U.S. Army/USA) *6*(3): 43, 51
Davis, John W. (The Dow Chemical Company/USA) *6*(2):89
Davis-Hoover, Wendy J. (U.S. EPA/USA) *6*(2):247
De'Ath, Anna M. (Cranfield University/UK) *6*(6):329
Dean, Sean (Camp Dresser & McKee. Inc/USA) *6*(7):133
DeBacker, Dennis (Battelle/USA) *6*(10):145
DeHghi, Benny (Honeywell International Inc./USA) *6*(2):39;*6*(10):283
de Jong, Jentsje (TAUW BV/THE NETHERLANDS) *6*(10):253
Del Vecchio, Michael (Envirogen, Inc./USA) *6*(9):281
Delille, Daniel (CNRS/FRANCE) *6*(2):57
DeLong, George (AIMTech/USA) *6*(7):321, 333, 341
Demers, Gregg (ERM/USA) *6*(7):45
De Mot, Rene (Catholic University of Leuven/BELGIUM) *6*(4):35
Deobald, Lee A. (University of Idaho/USA) *6*(9):147
Deschênes, Louise (École Polytechnique de Montréal/CANADA) *6*(3):115; *6*(9):129
Dey, William S. (Illinois State Geological Survey/USA) *6*(9):179
Díaz-Cervantes, Dolores (CINVESTAV-IPN/MEXICO) *6*(6):369
Dick, Vincent B. (Haley & Aldrich, Inc./USA) *6*(7):27, 281
Diehl, Danielle (The University of New Mexico/USA) *6*(9):155
Diehl, Susan V. (Mississippi State University/USA) *6*(5):329
Diels, Ludo (VITO/BELGIUM) *6*(9):87
DiGregorio, Salvatore (University della Calabria/ITALY) *6*(4):131
Di Gregorio, Simona (Universita degli Studi di Verona/ITALY) *6*(3):267
Dijkhuis, Edwin (Bioclear/THE NETHERLANDS) *6*(5):289
Di Leo, Cristina (EniTecnologie/ITALY) *6*(6):173
Dimitriou-Christidis, Petros (Texas A& M University) *6*(5):17
Dixon, Robert (Montgomery Watson/ITALY) *6*(10):41
Dobbs, Gregory M. (United Technologies Research Center/USA) *6*(7):69
Doherty, Amy T. (GZA GeoEnvironmental, Inc./USA) *6*(7):165
Dolan, Mark E. (Oregon State University/USA) *6*(10):145, 155, 179
Dollhopf, Michael (Michigan State University/USA) *6*(8):19
Dondi, Giovanni (Water & Soil Remediation S.r.l./ITALY) *6*(6):179
Dong, Yiting (Chinese Academy of Sciences/CHINA) *6*(9):113
Dooley, Maureen A. (Regenesis/USA) *6*(7):197
Dottridge, Jane (Komex Europe Ltd./UK) *6*(4):17
Dowd, John (University of Georgia/USA) *6*(9):289
Doughty, Herb (U.S. Navy/USA) *6*(10):1

Doze, Jacco (RIZA/THE NETHERLANDS) *6*(5):289
Dragich, Brian (California Polytechnic State University/USA) *6*(2):1
Drake, John T. (Camp Dresser & McKee Inc./USA) *6*(7):273
Dries, Victor (Flemish Public Waste Agency/BELGIUM) *6*(7):87
Du, Yan-Hung (National Chung Hsing University/TAIWAN) *6*(6):353
Dudal, Yves (École Polytechnique de Montréal/CANADA) *6*(3):115
Duffey, J. Tom (Camp Dresser & McKee Inc./USA) *6*(5):33
Duffy, Baxter E. (Inland Pollution Services, Inc./USA) *6*(7):313
Duijn, Rik (Oostwaardhoeve Co./THE NETHERLANDS) *6*(6):59
Durant, Neal D. (GeoTrans, Inc./USA) *6*(2):19, 163
Durell, Gregory (Battelle Ocean Sciences/USA) *6*(5):231
Dworatzek, S. (University of Toronto/CANADA) *6*(8):27
Dwyer, Daryl F. (University of Minnesota/USA) *6*(3):219
Dzantor, E. K. (University of Maryland/USA) *6*(5):77

Ebner, R. (GMF/GERMANY) *6*(2):137
Ederer, Martina (University of Idaho/USA) *6*(9):147
Edgar, Michael (Camp Dresser & McKee Inc./USA) *6*(7):133
Edwards, Elizabeth A. (University of Toronto/CANADA) *6*(8):27
Edwards, Grant C. (University of Guelph/CANADA) *6*(6):185
Eggen, Trine (Jordforsk Centre for Soil and Environmental Research/NORWAY) *6*(6):157
Eggert, Tim (CDM Federal Programs Corp./USA) *6*(6):81
Elberson, Margaret A. (DuPont Co./USA) *6*(8):43
Elliott, Mark (Virginia Polytechnic Institute & State University/USA) *6*(5):1
Ellis, David E. (Dupont Company/USA) *6*(8):43
Ellwood, Derek C. (University of Southampton/UK) *6*(9):61
Else, Terri (University of Nevada Las Vegas/USA) *6*(9):257
Elväng, Annelie M. (Stockholm University/SWEDEN) *6*(3):133
England, Kevin P. (USA) *6*(5):105
Ertas, Tuba Turan (San Diego State University/USA) *6*(10):301
Escalon, Lynn (U.S. Army Corps of Engineers/USA) *6*(3):51
Esparza-Garcia, Fernando (CINVESTAV-IPN/MEXICO) *6*(6):219
Evans, Christine S. (University of Westminster/UK) *6*(3):75
Evans, Patrick J. (Camp Dresser & McKee, Inc./USA) *6*(2):113, 199; *6*(8):209

Fabiani, Fabio (EniTecnologie S.p.A./ITALY) *6*(6):173
Fadullon, Frances Steinacker (CH2M Hill/USA) *6*(3):107
Fang, Min (University of Massachusetts/USA) *6*(6):73
Faris, Bart (New Mexico Environmental Department/USA) *6*(9):223
Farone, William A. (Applied Power Concepts, Inc./USA) *6*(7):103
Fathepure, Babu Z. (Oklahoma State University/USA) *6*(8):19
Faust, Charles (GeoTrans, Inc./USA) *6*(2):163
Fayolle, Françoise (Institut Français du Pétrole/FRANCE) *6*(1):153
Feldhake, David (University of Cincinnati/USA) *6*(2):247
Felt, Deborah (Applied Research Associates, Inc./USA) *6*(7):125
Feng, Terry H. (Parsons Engineering Science, Inc./USA) *6*(2):39; *6*(10):283
Fenwick, Caroline (Aberdeen University/UK) *6*(2):223
Fernandez, Jose M. (University of Iowa/USA) *6*(1):195
Fernández-Sanchez, J. Manuel (CINVESTAV-IPN/MEXICO) *6*(6):369

Ferrer, E. (Universidad Complutense de Madrid/SPAIN) *6*(4):29
Ferrera-Cerrato, Ronald (Colegio de Postgraduados/MEXICO) *6*(6):219
Fiacco, R. Joseph (Environmental Resources Management) *6*(7):45
Fields, Jim (University of Georgia/USA) *6*(9):289
Fields, Keith A. (Battelle/USA) *6*(10):1
Fikac, Paul J. (Jacobs Engineering Group, Inc./USA) *6*(6):35
Fischer, Nick M. (Aquifer Technology/USA) *6*(8):157, *6*(10):15
Fisher, Angela (The Pennsylvania State University/USA) *6*(8):201
Fisher, Jonathan (Environment Agency/UK) *6*(4):17
Fitch, Mark W. (University of Missouri-Rolla/USA) *6*(5):199
Fleckenstein, Janice V. (USA) *6*(6):89
Fleischmann, Paul (ZEBRA Environmental Corp./USA) *6*(10):139
Fletcher, John S. (University of Oklahoma/USA) *6*(5):61
Foget, Michael K. (SHN Consulting Engineers & Geologists, Inc./USA) *6*(3):83
Foley, K.L. (U.S. Army Engineer Research & Development Center/USA) *6*(5):9
Follner, Christina G. (University of Leipzig/GERMANY) *6*(4):81
Fontenot, Martin M. (Syngenta Crop Protection, Inc./USA) *6*(6):35
Foote, Eric A. (Battelle/USA) *6*(1):137; *6*(7):13
Ford, James (Investigative Science Inc./CANADA) *6*(2):27
Forman, Sarah R. (URS Corporation/USA) *6*(7):321, 333, 341
Fortman, Tim J. (Battelle Marine Sciences Laboratory/USA) *6*(3):157
Francendese, Leo (U.S. EPA/USA) *6*(3):259
Francis, M. McD. (NOVA Research & Technology Center/CANADA) *6*(4):117; *6*(5):53,
François, Alan (Institut Français du Pétrole/FRANCE) *6*(1):153
Frankenberger, William T. (University of California/USA) *6*(9):249
Freedman, David L. (Clemson University/USA) *6*(7):109
French, Christopher E. (University of Cambridge/UK) *6*(5):69
Friese, Kurt (UFZ Center for Environmental Research/GERMANY) *6*(9):43
Frisbie, Andrew J. (Purdue University/USA) *6*(3):125
Frisch, Sam (Envirogen Inc./USA) *6*(9):281
Frömmichen, René (UFZ Centre for Environmental Research/GERMANY) *6*(9):43
Fuierer, Alana M. (New Mexico Institute of Mining & Technology/USA) *6*(8):131
Fujii, Kensuke (Obayashi Corporation/JAPAN) *6*(10):239
Fujii, Shigeo (Kyoto University/JAPAN) *6*(4):149
Furuki, Masakazu (Hyogo Prefectural Institute of Environmental Science/JAPAN) *6*(5):321

Gallagher, John R. (University of North Dakota/USA) *6*(5):129; *6*(6):141
Gambale, Franco (Istituto di Cibernetica e Biofisica/ITALY) *6*(5):157
Gambrell, Robert P. (Louisiana State University/USA) *6*(5):305
Gandhi, Sumeet (University of Iowa/USA) *6*(8):147
Garbi, C. (Universidad Complutense de Madrid/SPAIN) *6*(4):29; *6*(6):377
García-Arrazola, Roeb (CINVESTAV-IPN/MEXICO) *6*(6):369
García-Barajas, Rubén Joel (ESIQIE-IPN/MEXICO) *6*(6):369
Garrett, Kevin (Harding ESE/USA) *6*(7):205
Garry, Erica (Spelman College/USA) *6*(9):53
Gavaskar, Arun R. (Battelle/USA) *6*(7):13
Gavinelli, Marco (Ambiente S.p.A./ITALY) *6*(6):227
Gebhard, Michael (GeoTrans/USA) *6*(8):19

Gec, Bob (Degussa Canada Ltd./CANADA) *6*(10):73

Gehre, Matthias (UFZ - Centre for Environmental Research/GERMANY) *6*(4):99

Gemoets, Johan (VITO/BELGIUM) *6*(4):35; *6*(9):87

Gent, David B. (U.S. Army Corps of Engineers/USA) *6*(9):241

Gentry, E. E. (Science Applications International Corporation/USA) *6*(8):27

Georgiev, Plamen S. (University of Mining & Geology/BULGARIA) *6*(9):97

Gerday, Charles (Université de Liège/BELGIUM) *6*(2):57

Gerlach, Robin (Montana State University/USA) *6*(8):1

Gerritse, Jan (TNO Environmental Sciences/THE NETHERLANDS) *6*(2):231; *6*(7):61

Gerth, André (BioPlanta GmbH/GERMANY) *6*(3):67; *6*(5):173

Ghosh, Upal (Stanford University/USA) *6*(3):189; *6*(6):89

Ghoshal, Subhasis (McGill University/CANADA) *6*(9):139

Gibbs, James T. (Battelle/USA) *6*(1):137

Gibello, A. (Universidad Complutense/SPAIN) *6*(4):29

Giblin, Tara (University of California/USA) *6*(9):249

Gilbertson, Amanda W. (University of Missouri-Rolla/USA) *6*(5):199

Gillespie, Rick D. (Regenesis/USA) *6*(1):107

Gillespie, Terry J. (University of Guelph/CANADA) *6*(6):185

Glover, L. Anne (Aberdeen University /UK) *6*(2):223

Goedbloed, Peter (Oostwaardhoeve Co./THE NETHERLANDS) *6*(6):59

Golovleva, Ludmila A. (Russian Academy of Sciences/RUSSIA) *6*(3):75

Goltz, Mark N. (Air Force Institute of Technology/USA) *6*(2):173

Gong, Weiliang (The University of New Mexico/USA) *6*(9):155

Gossett, James M. (Cornell University/USA) *6*(4):125

Govind, Rakesh (University of Cincinnati/USA) *6*(5):269; *6*(8):35; *6*(9):1, 9, 17

Gozan, Misri (Water Technology Center/GERMANY) *6*(8):105

Grainger, David (IT Corporation/USA) *6*(1):51; *6*(2):73

Grandi, Beatrice (Water & Soil Remediation S.r.l./ITALY) *6*(6):179

Granley, Brad A. (Leggette, Brashears, & Graham/USA) *6*(10):259

Grant, Russell J. (University of York/UK) *6*(6):27

Graves, Duane (IT Corporation/USA) *6*(2):253; *6*(4):109; *6*(9):215

Green, Chad E. (University of California/USA) *6*(10):311

Green, Donald J. (USAG Aberdeen Proving Ground/USA) *6*(7):321, 333, 341

Green, Robert (Alcoa/USA) *6*(6):89

Green, Roger B. (Waste Management, Inc./USA) *6*(2):247; *6*(6):127

Gregory, Kelvin B. (University of Iowa/USA) *6*(3):1

Griswold, Jim (Construction Analysis & Management, Inc./USA) *6*(1):115

Groen, Jacobus (Vrije Universiteit/THE NETHERLANDS) *6*(4):91

Groenendijk, Gijsbert Jan (Hoek Loos bv/THE NETHERLANDS) *6*(7):297

Grotenhuis, Tim (Wageningen Agricultural University/THE NETHERLANDS) *6*(5):289

Groudev, Stoyan N. (University of Mining & Geology/BULGARIA) *6*(9):97

Guarini, William J. (Envirogen, Inc./USA) *6*(9):281

Guieysse, Benoît (Lund University/SWEDEN) *6*(3):181

Guiot, Serge R. (Biotechnology Research Institute/CANADA) *6*(3):165

Gunsch, Claudia (Clemson University/USA) *6*(7):109

Gurol, Mirat (San Diego State University/USA) *6*(10):301

Ha, Jeonghyub (University of Maryland/USA) *6*(10):57
Haak, Daniel (RMT, Inc./USA) *6*(10):201
Haas, Patrick E. (Mitretek Systems/USA) *6*(7):19, 241, 249; *6*(8):73
Haasnoot, C. (Logisticon Water Treatment/THE NETHERLANDS) *6*(8):11
Habe, Hiroshi (The University of Tokyo/JAPAN) *6*(4):51; *6*(6):111
Haeseler, Frank (Institut Français du Pétrole/FRANCE) *6*(3):227
Haff, James (Meritor Automotive, Inc./USA) *6*(7):173
Haines, John R. (U.S. EPA/USA) *6*(9):17
Håkansson, Torbjörn (Lund University/SWEDEN) *6*(9):123
Halfpenny-Mitchell, Laurie (University of Guelph/CANADA) *6*(6):185
Hall, Billy (Newfields, Inc./USA) *6*(5):189
Hampton, Mark M. (Groundwater Services/USA) *6*(8):73
Hannick, Nerissa K. (University of Cambridge/UK) *6*(5):69
Hannigan, Mary (Mississippi State University) *6*(5):329; *6*(6):279
Hannon, LaToya (Spelman College/USA) *6*(9):53
Hansen, Hans C. L. (Hedeselskabet /DENMARK) *6*(2):11
Hansen, Lance D. (U.S. Army Corps of Engineers/USA) *6*(3):9, 43, 51; *6*(4):59; *6*(6):43; *6*(7):125; *6*(10):115
Haraguchi, Makoo (Sumitomo Marine Research Institute/JAPAN) *6*(10):345
Hardisty, Paul E. (Komex Europe, Ltd./ENGLAND) *6*(4):17
Harmon, Stephen M. (U.S. EPA/USA) *6*(9):17
Harms, Hauke (Swiss Federal Institute of Technology/SWITZERLAND) *6*(3):251
Harmsen, Joop (Alterra, Wageningen University and Research Center/THE NETHERLANDS) *6*(5):137, 279; *6*(6):1, 59
Harper, Greg (TetraTech EM Inc./USA) *6*(3):259
Harrington-Baker, Mary Ann (MSE, Inc./USA) *6*(9):35
Harris, Benjamin Cord (Texas A&M University/USA) *6*(5):17, 25
Harris, James C. (U.S. EPA/USA) *6*(6):287, 295
Harris, Todd (Mason and Hanger Corporation/USA) *6*(3):35
Harrison, Patton B. (American Airlines/USA) *6*(1):121
Harrison, Susan T.L. (University of Cape Town/REP OF SOUTH AFRICA) *6*(6):339
Hart, Barry (Monash University/AUSTRALIA) *6*(4):1
Hartzell, Kristen E. (Battelle/USA) *6*(1):137; *6*(10):193
Harwood, Christine L. (Michael Baker Corporation/USA) *6*(2):155
Hassett, David J. (Energy & Environmental Research Center/USA) *6*(5):129
Hater, Gary R. (Waste Management Inc./USA) *6*(2):247
Hausmann, Tom S. (Battelle Marine Sciences Laboratory/USA) *6*(3):157
Hawari, Jalal (National Research Council of Canada/CANADA) *6*(9):139
Hayes, Adam J. (Triple Point Engineers, Inc./USA) *6*(1):183
Hayes, Dawn M. (U.S. Navy/USA) *6*(3):107
Hayes, Kim F. (University of Michigan/USA) *6*(8):193
Haynes, R.J. (University of Natal/REP OF SOUTH AFRICA) *6*(6):101
Heaston, Mark S. (Earth Tech/USA) *6*(3):17, 25
Hecox, Gary R. (University of Kansas/USA) *6*(4):109
Heebink, Loreal V. (Energy & Environmental Research Center/USA) *6*(5):129
Heine, Robert (EFX Systems, Inc./USA) *6*(8):19
Heintz, Caryl (Texas Tech University/USA) *6*(3):9

Hendrickson, Edwin R. (DuPont Co./USA) *6*(8):27, 43
Hendriks, Willem (Witteveen+Bos Consulting Engineers/THE NETHERLANDS) *6*(5):289
Henkler, Rolf D. (ICI Paints/UK) *6*(2):223
Henny, Cynthia (University of Maine/USA) *6*(8):139
Henry, Bruce M. (Parsons Engineering Science, Inc/USA) *6*(7):241
Henssen, Maurice J.C. (Bioclear Environmental Biotechnology/THE NETHERLANDS) *6*(8):11
Herson, Diane S. (University of Delaware/USA) *6*(6):9
Hesnawi, Rafik M. (University of Manitoba/CANADA) *6*(6):165
Hetland, Melanie D. (Energy & Environmental Research Center/USA) *6*(5):129
Hickey, Robert F. (EFX Systems, Inc./USA) *6*(8):19
Hicks, Patrick H. (ARCADIS/USA) *6*(1):107
Hiebert, Randy (MSE Technology Applications, Inc./USA) *6*(8):79
Higashi, Teruo (University of Tsukuba/JAPAN) *6*(9):187
Higgins, Mathew J. (Bucknell University/USA) *6*(2):105
Higinbotham, James H. (ExxonMobil Environmental Remediation/USA) *6*(8):87
Hines, April (Spelman College/USA) *6*(9):53
Hinshalwood, Gordon (Delta Environmental Consultants, Inc./USA) *6*(1):43
Hirano, Hiroyuki (The University of Tokyo/JAPAN) *6*(6):111
Hirashima, Shouji (Yakult Pharmaceutical Industry/JAPAN) *6*(10):345
Hirsch, Steve (Environmental Protection Agency/USA) *6*(5):207
Hiwatari, Takehiko (National Institute for Environmental Studies/JAPAN) *6*(5):321
Hoag, Rob (Conestoga-Rovers & Associates/USA) *6*(1):99
Hoelen, Thomas P. (Stanford University/USA) *6*(7):95
Hoeppel, Ronald E. (U.S. Navy/USA) *6*(10):245
Hoffmann, Johannes (Hochtief Umwelt GmbH/GERMANY) *6*(6):227
Hoffmann, Robert E. (Chevron Canada Resources/CANADA) *6*(6):193
Höfte, Monica (Ghent University/BELGIUM) *6*(5):223
Holder, Edith L. (University of Cincinnati/USA) *6*(2):247
Holm, Thomas R. (Illinois State Water Survey/USA) *6*(9):179
Holman, Hoi-Ying (Lawrence Berkeley National Laboratory/USA) *6*(4):67
Holoman, Tracey R. Pulliam (University of Maryland/USA) *6*(3):205
Hopper, Troy (URS Corporation/USA) *6*(2):239
Hornett, Ryan (NOVA Chemicals Corporation/USA) *6*(4):117
Hosangadi, Vitthal S. (Foster Wheeler Environmental Corp./USA) *6*(9):249
Hough, Benjamin (Tetra Tech EM, Inc./USA) *6*(10):293
Hozumi, Toyoharu (Oppenheimer Biotechnology/JAPAN) *6*(10):345
Huang, Chin-I (National Chung Hsing University/TAIWAN) *6*(10):217
Huang, Chin-Pao (University of Delaware/USA) *6*(6):9, 149
Huang, Hui-Bin (DuPont Co./USA) *6*(8):43
Huang, Junqi (Air Force Institute of Technology/USA) *6*(2):173
Huang, Wei (University of Sheffield/UK) *6*(2):207
Hubach, Cor (DHV Noord Nederland/THE NETHERLANDS) *6*(8):11
Huesemann, Michael H. (Battelle/USA) *6*(3):157
Hughes, Joseph B. (Rice University/USA) *6*(5):85; *6*(7):19
Hulsen, Kris (University of Ghent/BELGIUM) *6*(5):223
Hunt, Jonathan (Clemson University/USA) *6*(7):109

Hunter, William J. (U.S. Dept of Agriculture/USA) *6*(9):209, 309
Hwang, Sangchul (University of Akron/USA) *6*(3):235
Hyman, Michael R. (North Carolina State University/USA) *6*(1): 83, 145

Ibeanusi, Victor M. (Spelman College/USA) *6*(9):53
Ickes, Jennifer (Battelle/USA) *6*(5):231, 237
Ide, Kazuki (Obayashi Corporation Ltd./JAPAN) *6*(6):111; *6*(10):239
Igarashi, Tsuyoshi (Nippon Institute of Technology/JAPAN) *6*(5):321
Infante, Carmen (PDVSA Intevep/VENEZUELA) *6*(6):257
Ingram, Sherry (IT Corporation/USA) *6*(4):109
Ishikawa, Yoji (Obayashi Corporation/JAPAN) *6*(6):249; *6*(10):239

Jackson, W. Andrew (Texas Tech University/USA) *6*(5):207, 313; *6*(9):273
Jacobs, Alan K. (EnSafe, Inc./USA) *6*(9):315
Jacques, Margaret E. (Rowan University/USA) *6*(5):215
Jahan, Kauser (Rowan University/USA) *6*(5):215
James, Garth (MSE Inc./USA) *6*(8):79
Jansson, Janet K. (Södertörn University College/SWEDEN) *6*(3):133
Japenga, Jan (Alterra/THE NETHERLANDS) *6*(5):137
Jauregui, Juan (Universidad Nacional Autonoma de Mexico/MEXICO) *6*(6):17
Jensen, James N. (State University of New York at Buffalo/USA) *6*(6):89
eon, Mi-Ae (Texas Tech University/USA) *6*(9):273
Jerger, Douglas E. (IT Corporation/USA) 6(3):35
Jernberg, Cecilia (Södertörn University College/SWEDEN) *6*(3):133
Jindal, Ranjna (Suranaree University of Technology/THAILAND) *6*(4):149
Johnson, Dimitra (Southern University at New Orleans/USA) *6*(5):151
Johnson, Glenn (University of Utah/USA) *6*(5):231
Johnson, Paul C. (Arizona State University/USA) *6*(1):11; *6*(8):61
Johnson, Richard L. (Oregon Graduate Institute/USA) *6*(10):293
Jones, Antony (Komex H_2O Science, Inc./USA) *6*(2):223; *6*(3):173; *6*(10):123
Jones, Clay (University of New Mexico/USA) *6*(9):223
Jones, Triana N. (University of Maryland/USA) *6*(3):205
Jonker, Hendrikus (Vrije Universiteit/THE NETHERLANDS) *6*(4):91
Ju, Lu-Kwang (The University of Akron/USA) *6*(6):319

Kaludjerski, Milica (San Diego State University/USA) *6*(10):301
Kamashwaran, S. Ramanathen (University of Idaho/USA) *6*(3):91
Kambhampati, Murty S. (Southern University at New Orleans/USA) *6*(5):145, 151
Kamimura, Daisuke (Gunma University/JAPAN) *6*(8):113
Kang, James J. (URS Corporation/USA) *6*(1):121; *6*(10):223
Kappelmeyer, Uwe (UFZ Centre for Environmental Research/GERMANY) 6(5):337
Karamanev, Dimitre G. (University of Western Ontario/CANADA) *6*(10):171
Karlson, Ulrich (National Environmental Research Institute) *6*(3):141
Kastner, James R. (University of Georgia/USA) *6*(9):289
Kästner, Matthias (UFZ Centre for Environmental Research/GERMANY) *6*(4):99; *6*(5):337
Katz, Lynn E. (University of Texas/USA) *6*(8):139
Kavanaugh, Rathi G. (University of Cincinnati/USA) *6*(2):247

Kawahara, Fred (U.S. EPA/USA) *6*(9):9
Kawakami, Tsuyoshi (University of Tsukuba/JAPAN) *6*(9):187
Keefer, Donald A. (Illinois State Geological Survey/USA) *6*(9):179
Keith, Nathaniel (Texas A&M University/USA) *6*(5):25
Kelly, Laureen S. (Montana Department of Environmental Quality/USA) *6*(6):287
Kempisty, David M. (U.S. Air Force/USA) *6*(10):145, 155
Kerfoot, William B. (K-V Associates, Inc./USA) *6*(10):33
Keuning, S. (Bioclear Environmental Technology/THE NETHERLANDS) *6*(8):11
Khan, Tariq A. (Groundwater Services, Inc./USA) *6*(7):19
Khodadoust, Amid P. (University of Cincinnati/USA) *6*(5):243, 253, 261
Kieft, Thomas L. (New Mexico Institute of Mining and Technology/USA) *6*(8):131
Kiessig, Gunter (WISMUT GmbH/GERMANY) *6*(5):173; *6*(9):155
Kilbride, Rebecca (CEFAS Laboratory/UK) *6*(10):337
Kim, Jae Young (Seoul National University/REPUBLIC OF KOREA) *6*(9):195
Kim, Jay (University of Cincinnati/USA) *6*(6):133
Kim, Kijung (The Pennsylvania State University/USA) *6*(9):303
Kim, Tae Young (Ewha University/REPUBLIC OF KOREA) *6*(6):51
Kinsall, Barry L. (Oak Ridge National Laboratory/USA) *6*(4):73
Kirschenmann, Kyle (IT Corp/USA) *6*(4):109
Klaas, Norbert (University of Stuttgart/GERMANY) *6*(2):137
Klecka, Gary M. (The Dow Chemical Company/USA) *6*(2):89
Klein, Katrina (GeoTrans, Inc./USA) *6*(2):163
Klens, Julia L. (IT Corporation/USA) *6*(2):253; *6*(9):215
Knotek-Smith, Heather M. (University of Idaho/USA) *6*(9):147
Koch, Stacey A. (RMT, Inc./USA) *6*(7):181
Koenen, Brent A. (U.S. Army Engineer Research & Development Center/USA) *6*(5):9
Koenigsberg, Stephen S. (Regenesis Bioremediation Products/USA) *6*(7):197, 257; *6*(8):209; *6*(10):9, 87
Kohata, Kunio (National Institute for Environmental Studies/JAPAN) *6*(5):321
Kohler, Keisha (ThermoRetec Corporation/USA) *6*(7):1
Kolhatkar, Ravindra V. (BP Corporation/USA) *6*(1):35, 43
Komlos, John (Montana State University/USA) *6*(7):117
Komnitsas, Kostas (National Technical University of Athens/GREECE) *6*(9):97
Kono, Masakazu (Oppenheimer Biotechnology/JAPAN) *6*(10):345
Koons, Brad W. (Leggette, Brashears & Graham, Inc./USA) *6*(1):175
Koschal, Gerard (PNG Environmental/USA) *6*(1):203
Koschorreck, Matthias (UFZ Centre for Environmental Research/GERMANY) *6*(9):43
Koshikawa, Hiroshi (National Institute for Environmental Studies/JAPAN) *6*(5):321
Kramers, Jan D. (University of Bern/SWITZERLAND) *6*(4):91
Krooneman, Jannneke (Bioclear Environmental Biotechnology/THE NETHERLANDS) *6*(7):141
Kruk, Taras B. (URS Corporation/USA) *6*(10):223
Kuhwald, Jerry (NOVA Chemicals Corporation/CANADA) *6*(5):53
Kuschk, Peter (UFZ Centre for Environmental Research Leipzig/GERMANY) *6*(5):337

Laboudigue, Agnes (Centre National de Recherche sur les Sites et Sols Pollués/FRANCE) *6*(2):129
LaFlamme, Brian (Engineering Management Support, Inc./USA) *6*(10):231
Lafontaine, Chantal (École Polytechnique de Montréal/CANADA) *6*(10):171
Laha, Shonali (Florida International University/USA) *6*(10):187
Laing, M.D. (University of Natal/REP OF SOUTH AFRICA) *6*(9):79
Lamar, Richard (EarthFax Development Corp/USA) *6*(6):263
Lamarche, Philippe (Royal Military College of Canada/CANADA) *6*(8):95
Lamb, Steven R. (GZA GeoEnvironmental, Inc./USA) *6*(7):165
Landis, Richard C. (E.I. du Pont de Nemours & Company/USA) *6*(8):185
Lang, Beth (United Technologies Corp./USA) *6*(10):41
Langenhoff, Alette (TNO Institute of Environmental Science/THE NETHERLANDS) *6*(7):141
LaPat-Polasko, Laurie T. (Parsons Engineering Science, Inc./USA) *6*(2):65, 189
Lapus, Kevin (Regenesis/USA) *6*(7):257; *6*(10):9
LaRiviere, Daniel (Texas A&M University/USA) *6*(5):17, 25
Larsen, Lars C. (Hedeselskabet/DENMARK) *6*(2):11
Larson, John R. (TranSystems Corporation/USA) *6*(7):229
Larson, Richard A. (University of Illinois at Urbana-Champaign/USA) *6*(5):181
Lauzon, Francois (Dept of National Defence/CANADA) *6*(8):95
Leavitt, Maureen E. (Newfields Inc./USA) *6*(1):51; *6*(5):189
Lebron, Carmen A. (U.S. Navy/USA) *6*(7):95
Lee, B. J. (Science Applications International Corporation) *6*(8):27
Lee, Brady D. (Idaho National Engineering & Environmental Laboratory/USA) *6*(7):77
Lee, Chi Mei (National Chung Hsing University/TAIWAN) *6*(6):353
Lee, Eun-Ju (Louisiana State University/USA) *6*(5):313
Lee, Kenneth (Fisheries & Oceans Canada/CANADA) *6*(10):337
Lee, Michael D. (Terra Systems, Inc./USA) *6*(7):213, 249
Lee, Ming-Kuo (Auburn University/USA) *6*(9):105
Lee, Patrick (Queen's University/CANADA) *6*(2):121
Lee, Seung-Bong (University of Washington/USA) *6*(10):211
Lee, Si-Jin (Kyonggi University/REPUBLIC OF KOREA) *6*(1):161
Lee, Sung-Jae (ChoongAng University/REPUBLIC OF KOREA) *6*(6):51
Leeson, Andrea (Battelle/USA) *6*(10):1, 145, 155, 193
Lehman, Stewart E. (California Polytechnic State University/USA) *6*(2):1
Lei, Li (University of Cincinnati/USA) *6*(5):243, 261
Leigh, Daniel P. (IT Corporation/USA) *6*(3):35
Leigh, Mary Beth (University of Oklahoma/USA) *6*(5):61
Lendvay, John (University of San Francisco/USA) *6*(8):19
Lenzo, Frank C. (ARCADIS Geraghty & Miller/USA) *6*(7):53
Leon, Nidya (PDVSA Intevep/VENEZUELA) *6*(6):257
Leong, Sylvia (Crescent Heights High School/CANADA) *6*(5):53
Leontievsky, Alexey A. (Russian Academy of Sciences/RUSSIA) *6*(3):75
Lerner, David N. (University of Sheffield/UK) *6*(1):59; *6*(2):207
Lesage, Suzanne (National Water Research Institute/CANADA) *6*(7):321, 333, 341

Leslie, Jolyn C. (Camp Dresser & McKee, Inc./USA) *6*(2):113
Lewis, Ronald F. (U.S. EPA/USA) *6*(5):253, 261
Li, Dong X. (USA) *6*(7):205
Li, Guanghe (Tsinghua University/CHINA) *6*(7):61
Li, Tong (Tetra Tech EM Inc./USA) *6*(10):293
Librando, Vito (Universita di Catania/ITALY) *6*(3):149
Lieberman, M. Tony (Solutions Industrial & Environmental Services/USA) *6*(7):249
Lin, Cindy (Conestoga-Rovers & Associates/USA) *6*(1):99; *6*(10):131
Lipson, David S. (Blasland, Bouck & Lee, Inc./USA) *6*(10):319
Liu, Jian (University of Nevada Las Vegas/USA) *6*(9):265
Liu, Xiumei (Shandong Agricultural University/ CHINA) *6*(9):113
Livingstone, Stephen (Franz Environmental Inc./CANADA) *6*(6):211
Lizzari, Daniela (Universita degli Studi di Verona/ITALY) *6*(3):267
Llewellyn, Tim (URS/USA) *6*(7):321, 333, 341
Lobo, C. (El Encin IMIA/SPAIN) *6*(4):29
Loeffler, Frank E. (Georgia Institute of Technology/USA) *6*(8):19
Logan, Bruce E. (The Pennsylvania State University/USA) *6*(9):303
Long, Gilbert M. (Camp Dresser & McKee Inc./USA) *6*(6):287
Longoni, Giovanni (Montgomery Watson/ITALY) *6*(10):41
Lorbeer, Helmut (Technical University of Dresden/GERMANY) *6*(8):105
Lors, Christine (Centre National de Recherche sur les Sites et Sols Pollués /FRANCE) *6*(2):129
Lorton, Diane M. (King's College London/UK) *6*(2):223; *6*(3):173
Losi, Mark E. (Foster Wheeler Environ. Corp./USA) *6*(9):249
Loucks, Mark (U.S. Air Force/USA) *6*(2):261
Lu, Chih-Jen (National Chung Hsing University/TAIWAN) *6*(6):353; *6*(10):217
Lu, Xiaoxia (Tsinghua University/CHINA) *6*(7):61
Lubenow, Brian (University of Delaware/USA) *6*(6):149
Lucas, Mary (Parsons Engineering Science, Inc./USA) *6*(10):283
Lundgren, Tommy S. (Sydkraft SAKAB AB/SWEDEN) *6*(6):127
Lundstedt, Staffan (Umeå University/SWEDEN) *6*(3):181
Luo, Xiaohong (NRC Research Associate/USA) *6*(8):167
Luthy, Richard G. (Stanford University/USA) *6*(3):189
Lutze, Werner (University of New Mexico/USA) *6*(9):155
Luu, Y.-S. (Queen's University/CANADA) *6*(2):121
Lynch, Regina M. (Battelle/USA) *6*(10):155

Macek, Thomáš (Institute of Chemical Technology/Czech Republic) *6*(5):61
MacEwen, Scott J. (CH2M Hill/USA) *6*(3):107
Machado, Kátia M. G. (Fund. Centro Tecnológico de Minas Gerais/BRAZIL) *6*(3):99
Maciel, Helena Alves (Aberdeen University/UK) *6*(1):1
Mack, E. Erin (E.I. du Pont de Nemours & Co./USA) *6*(2):81; *6*(8):43
Macková, Martina (Institute of Chemical Technology/Czech Republic) *6*(5):61
Macnaughton, Sarah J. (AEA Technology/UK) *6*(5):305; *6*(10):337
Macomber, Jeff R. (University of Cincinnati/USA) *6*(6):133
Macrae, Jean (University of Maine/USA) *6*(8):139
Madden, Patrick C. (Engineering Consultant/USA) *6*(8):87
Madsen, Clint (Terracon/USA) *6*(8):157; *6*(10):15
Magar, Victor S. (Battelle/USA) *6*(1):137; *6*(5):231, 237; *6*(10):145, 155

Mage, Roland (Battelle Europe/SWITZERLAND) *6*(6):241; *6*(10):109
Magistrelli, P. (Istituto di Cibernetica e Biofisica/ITALY) *6*(5):157
Maierle, Michael S. (ARCADIS Geraghty & Miller, Inc./USA) *6*(7):149
Major, C. Lee (Jr.) (University of Michigan/USA) *6*(8):19
Major, David W. (GeoSyntec Consultants/CANADA) *6*(8):27
Maki, Hideaki (National Institute for Environmental Studies/JAPAN) *6*(5):321
Makkar, Randhir S. (University of Illinois-Chicago/USA) *6*(5):297
Malcolm, Dave (BAE Systems Properties Ltd./UK) *6*(6):119
Manabe, Takehiko (Hyogo Prefectural Fisheries Research Institute/JAPAN) *6*(10):345
Maner, P.M. (Equilon Enterprises, LLC/USA) *6*(1):11
Maner, Paul (Shell Development Company/USA) *6*(8):61
Manrique-Ramírez, Emilio Javier (SYMCA, S.A. de C.V./MEXICO) *6*(6):369
Marchal, Rémy (Institut Français du Pétrole/FRANCE) *6*(1):153
Maresco, Vincent (Groundwater & Environmental Srvcs/USA) *6*(10):101
Marnette, Emile C. (TAUW BV/THE NETHERLANDS) *6*(7):297
Marshall, Timothy R. (URS Corporation/USA) *6*(2):49
Martella, L. (Istituto di Cibernetica e Biofisica/ITALY) *6*(5):157
Martin, C. (Universidad Politecnica/SPAIN) *6*(4):29
Martin, Jennifer P. (Idaho National Engineering & Environmental Laboratory/USA) *6*(7):265
Martin, John F. (U.S. EPA/USA) *6*(2):247
Martin, Margarita (Universidad Complutense de Madrid/SPAIN) *6*(4):29; *6*(6):377
Martinez-Inigo, M.J. (El Encin IMIA/SPAIN) *6*(4):29
Martino, Lou (Argonne National Laboratory/USA) *6*(5):207
Mascarenas, Tom (Environmental Chemistry/USA) *6*(8):157
Mason, Jeremy (King's College London/UK) *6*(2):223; *6*(3):173; *6*(10):123
Massella, Oscar (Universita degli Studi di Verona/ITALY) *6*(3):267
Matheus, Dacio R. (Instituto de Botânica/BRAZIL) *6*(3):99
Matos, Tania (University of Puerto Rico at Rio Piedras/USA) *6*(9):179
Matsubara, Takashi (Obayashi Corporation/JAPAN) *6*(6):249
Mattiasson, Bo (Lund University/SWEDEN) *6*(3):181; *6*(6):65; *6*(9):123
McCall, Sarah (Battelle/USA) *6*(10):155, 245
McCarthy, Kevin (Battelle Duxbury Operations/USA) *6*(5):9
McCartney, Daryl M. (University of Manitoba/CANADA) *6*(6):165
McCormick, Michael L. (The University of Michigan/USA) *6*(8):193
McDonald, Thomas J. (Texas A&M University) *6*(5):17
McElligott, Mike (U.S. Air Force/USA) *6*(1):51
McGill, William B. (University of Northern British Columbia/CANADA) *6*(4):7
McIntosh, Heather (U.S. Army/USA) *6*(7):321, 333
McLinn, Eugene L. (RMT, Inc./USA) *6*(5):121
McLoughlin, Patrick W. (Microseeps Inc./USA) *6*(1):35
McMaster, Michaye (GeoSyntec Consultants/CANADA) *6*(8):27, 43, *6*(9):297
McMillen, Sara J. (Chevron Research & Technology Company/USA) *6*(6):193
Meckenstock, Rainer U. (University of Tübingen/GERMANY) *6*(4):99
Mehnert, Edward (Illinois State Geological Survey/USA) *6*(9):179

Meigio, Jodette L. (Idaho National Engineering & Environmental Laboratory/USA) *6*(7):77

Meijer, Harro A.J. (University of Groningen/THE NETHERLANDS) *6*(4):91

Meijerink, E. (Province of Drenthe/THE NETHERLANDS) *6*(8):11

Merino-Castro, Glicina (Inst Technol y de Estudios Superiores/MEXICO) *6*(6):377

Messier, J.P. (U.S. Coast Guard/USA) *6*(1):107

Meyer, Michael (Environmental Resources Management/BELGIUM) *6*(7):87

Meylan, S. (Queen's University/CANADA) *6*(2):121

Miles, Victor (Duracell Inc./USA) *6*(7):87

Millar, Kelly (National Water Research Institute/CANADA) *6*(7):321, 333, 341

Miller, Michael E. (Camp Dresser & McKee, Inc./USA) *6*(7):273

Miller, Thomas Ferrell (Lockheed Martin/USA) *6*(3):259

Mills, Heath J. (Georgia Institute of Technology/USA) *6*(9):165

Millward, Rod N. (Louisiana State University/USA) *6*(5):305

Mishra, Pramod Chandra (Sambalpur University/INDIA) *6*(9):173

Mitchell, David (AEA Technology Environment/UK) *6*(10):337

Mitraka, Maria (Serres/GREECE) *6*(6):89

Mocciaro, PierFilippo (Ambiente S.p.A./ITALY) *6*(6):227

Moeri, Ernesto N. (CSD-GEOKLOCK/BRAZIL) *6*(1):27

Moir, Michael (Chevron Research & Technology Co./USA) *6*(1):83

Molinari, Mauro (AgipPetroli S.p.A/ITALY) *6*(6):173

Mollea, C. (Politecnico di Torino/ITALY) *6*(3):211

Mollhagen, Tony (Texas Tech University/USA) *6*(3):9

Monot, Frédéric (Institut Français du Pétrole/FRANCE) *6*(1):153

Moon, Hee Sun (Seoul National University/REPUBLIC OF KOREA) *6*(9):195

Moosa, Shehnaaz (University of Cape Town/REP OF SOUTH AFRICA) *6*(6):339

Morasch, Barbara (University Konstanz/GERMANY) *6*(4):99

Moreno, Joanna (URS Corporation/USA) *6*(2):239

Morgan, Scott (URS - Dames & Moore/USA) *6*(7):321

Morrill, Pamela J. (Camp, Dresser, & McKee, Inc./USA) *6*(2):113

Morris, Damon (ThermoRetec Corporation/USA) *6*(7):1

Mortimer, Marylove (Mississippi State University/USA) *6*(5):329

Mortimer, Wendy (Bell Canada/CANADA) *6*(2):27; *6*(6):185, 203, 211,

Mossing, Christian (Hedeselskabet/DENMARK) *6*(2):11

Mossmann, Jean-Remi (Centre National de Recherche sur les Sites et Sols Pollués/FRANCE) *6*(2):129

Moteleb, Moustafa A. (University of Cincinnati/USA) *6*(6):133

Mowder, Carol S. (URS/USA) *6*(7):321, 333, 341

Moyer, Ellen E. (ENSR International./USA) *6*(1):75

Mravik, Susan C. (U.S. EPA/USA) *6*(1):167

Mueller, James G. (URS Corporation/USA) *6*(2):239

Müller, Axel (Water Technology Center/GERMANY) *6*(8):105

Müller, Beate (Umweltschutz Nord GmbH/GERMANY) *6*(4):131

Müller, Klaus (Battelle Europe/SWITZERLAND) *6*(5):41; *6*(6):241

Muniz, Herminio (Hart Crowser Inc./USA) *6*(10):9

Murphy, Sean M. (Komex International Ltd./CANADA) *6*(4):7

Murray, Cliff (United States Army Corps of Engineers/USA) *6*(9):281

Murray, Gordon Bruce (Stella-Jones Inc./CANADA) *6*(3):197

Murray, Willard A. (Harding ESE/USA) *6*(7):197
Mutch, Robert D. (Brown and Caldwell/USA) *6*(2):145
Mutti, Francois (Water & Soil Remediation S.r.l./ITALY) *6*(6):179
Myasoedova, Nina M. (Russian Academy of Sciences/RUSSIA) *6*(3):75

Nadolishny, Alex (Nedatek, Inc./USA) *6*(10):139
Nagle, David P. (University of Oklahoma/USA) *6*(5):61
Nam, Kyoungphile (Seoul National University/REPUBLIC OF KOREA) *6*(9):195
Narayanaswamy, Karthik (Parsons Engineering Science/USA) *6*(2):65
Nelson, Mark D. (Delta Environmental Consultants, Inc./USA) *6*(1):175
Nelson, Yarrow (California Polytechnic State University/USA) *6*(10):311
Nemati, M. (University of Cape Town/REP OF SOUTH AFRICA) *6*(6):339
Nestler, Catherine C. (Applied Research Associates, Inc./USA) *6*(4):59, *6*(6):43
Nevárez-Moorillón, G.V. (UACH/MEXICO) *6*(6):361
Neville, Scott L. (Aerojet General Corp./USA) *6*(9):297
Newell, Charles J. (Groundwater Services, Inc./USA) *6*(7):19
Nieman, Karl (Utah State University/USA) *6*(4):67
Niemeyer, Thomas (Hochtief Umwelt Gmbh/GERMANY) *6*(6):227
Nies, Loring (Purdue University/USA) *6*(3):125
Nipshagen, Adri A.M. (IWACO/THE NETHERLANDS) *6*(7):141
Nishino, Shirley (U.S. Air Force/USA) *6*(3):59
Nivens, David E. (University of Tennessee/USA) *6*(4):45
Noffsinger, David (Westinghouse Savannah River Company/USA) *6*(10):163
Noguchi, Takuya (Nippon Institute of Technology/JAPAN) *6*(5):321
Nojiri, Hideaki (The University of Tokyo/JAPAN) *6*(4):51; *6*(6):111
Noland, Scott (NESCO Inc./USA) *6*(10):73
Nolen, C. Hunter (Camp Dresser & McKee/USA) *6*(6):287
Norris, Robert D. (Eckenfelder/Brown and Caldwell/USA) *6*(2):145; *6*(7):35
North, Robert W. (Environ Corporation./USA) *6*(7):189
Novak, John T. (Virginia Polytechnic Institute & State University/USA) *6*(2):105; *6*(5):1
Novick, Norman (Exxon/Mobil Oil Corp/USA) *6*(1):35
Nuttall, H. Eric (The University of New Mexico/USA) *6*(9): 155, 223
Nuyens, Dirk (Environmental Resources Management/BELGIUM) *6*(7):87; *6*(9):87
Nzengung, Valentine A. (University of Georgia/USA) *6*(9):289

Ochs, L. Donald (Regenesis/USA) *6*(10):139
O'Connell, Joseph E. (Environmental Resolutions, Inc./USA) *6*(1):91
Odle, Bill (Newfields, Inc./USA) *6*(5):189
O'Donnell, Ingrid (BAE Systems Properties, Ltd./UK) *6*(6):119
Ogden, Richard (BAE Systems Properties Ltd./UK) *6*(6):119
Oh, Byung-Taek (The University of Iowa/USA) *6*(8):147, 175
Oh, Seok-Young (University of Delaware/USA) *6*(6):149
Omori, Toshio (The University of Tokyo/JAPAN) *6*(4):51; *6*(6):111
O'Neal, Brenda (ARA/USA) *6*(3):43
Oppenheimer, Carl H. (Oppenheimer Biotechnology/USA) *6*(10):345
O'Regan, Gerald (Chevron Products Company/USA) *6*(1):203
O'Reilly, Kirk T. (Chevron Research & Technology Co/USA) *6*(1):83, 145, 203
Oshio, Takahiro (University of Tsukuba/JAPAN) *6*(9):187

Ozdemiroglu, Ece (EFTEC Ltd./UK) *6*(4):17

Padovani, Marco (Centro Ricerche Ambientali/ITALY) *6*(4):131
Paganetto, A. (Istituto di Cibernetica e Biofisica/ITALY) *6*(5):157
Pahr, Michelle R. (ARCADIS Geraghty & Miller/USA) *6*(1):107
Pal, Nirupam (California Polytechnic State University/USA) *6*(2):1
Palmer, Tracy (Applied Power Concepts, Inc./USA) *6*(7):103
Palumbo, Anthony V. (Oak Ridge National Laboratory/USA) *6*(4):73; *6*(9):165
Panciera, Matthew A. (University of Connecticut/USA) *6*(7):69
Pancras, Tessa (Wageningen University/THE NETHERLANDS) *6*(5):289
Pardue, John H. (Louisiana State University/USA) *6*(5): 207, 313; *6*(9):273
Park, Kyoohong (ChoongAng University/REPUBLIC OF KOREA) *6*(6):51
Parkin, Gene F. (University of Iowa/USA) *6*(3):1
Paspaliaris, Ioannis (National Technical University of Athens/GREECE) *6*(9):97
Paton, Graeme I. (Aberdeen University/UK) *6*(1):1
Patrick, John (University of Reading/UK) *6*(10):337
Payne, Frederick C. (ARCADIS Geraghty & Miller/USA) *6*(7):53
Payne, Jo Ann (DuPont Co./USA) *6*(8):43
Peabody, Jack G. (Regenesis/USA) *6*(10):95
Peacock, Aaron D. (University of Tennessee/USA) *6*(4):73; *6*(5):305
Peargin, Tom R. (Chevron Research & Technology Co/USA) *6*(1):67
Peeples, James A. (Metcalf & Eddy, Inc./USA) *6*(7):173
Pehlivan, Mehmet (Tait Environmental Management, Inc./USA) *6*(10):267, 275
Pelletier, Emilien (ISMER/CANADA) *6*(2):57
Pennie, Kimberley A. (Stella-Jones, Inc./CANADA) *6*(3):197
Peramaki, Matthew P. (Leggette, Brashears, & Graham, Inc./USA) *6*(10):259
Perey, Jennie R. (University of Delaware/USA) *6*(6):149
Perez-Vargas, Josefina (CINVESTAV-IPN/MEXICO) *6*(6):219
Perina, Tomas (IT Corporation/USA) *6*(1):51; *6*(2):73
Perlis, Shira R. (Rowan University/USA) *6*(5):215
Perlmutter, Michael W. (EnSafe, Inc./USA) *6*(9):315
Perrier, Michel (École Polytechnique de Montréal/CANADA) *6*(4):139
Perry, L.B. (U.S. Army Engineer Research & Development Center/USA) *6*(5):9
Persico, John L. (Blasland, Bouck & Lee, Inc./USA) *6*(10):319
Peschong, Bradley J. (Leggette, Brashears & Graham, Inc./USA) *6*(1):175
Peters, Dave (URS/USA) *6*(7):333
Peterson, Lance N. (North Wind Environmental, Inc./USA) *6*(7):265
Petrovskis, Erik A. (Geotrans Inc./USA) *6*(8):19
Peven-McCarthy, Carole (Battelle Ocean Sciences/USA) *6*(5):231
Pfiffner, Susan M. (University of Tennessee/USA) *6*(4):73
Phelps, Tommy J. (Oak Ridge National Laboratory/USA) *6*(4):73
Pickett, Tim M. (Applied Biosciences Corporation/USA) *6*(9):331
Pickle, D.W. (Equilon Enterprises LLC/USA) *6*(8):61
Pierre, Stephane (École Polytechnique de Montréal/CANADA) *6*(10):171
Pijls, Charles G.J.M. (TAUW BV/THE NETHERLANDS) *6*(10):253
Pirkle, Robert J. (Microseeps, Inc./USA) *6*(1):35
Pisarik, Michael F. (New Fields/USA) *6*(1):121

Piveteau, Pascal (Institut Français du Pétrole/FRANCE) *6*(1):153
Place, Matthew (Battelle/USA) *6*(10):245
Plata, Nadia (Battelle Europe/SWITZERLAND) *6*(5):41
Poggi-Varaldo, Hector M. (CINVESTAV-IPN/MEXICO) *6*(3):243; *6*(6):219
Pohlmann, Dirk C. (IT Corporation/USA) *6*(2):253
Pokethitiyook, Prayad (Mahidol University/THAILAND) *6*(10):329
Polk, Jonna (U.S. Army Corps of Engineers/USA) *6*(9):281
Pope, Daniel F. (Dynamac Corp/USA) *6*(1):129
Porta, Augusto (Battelle Europe/SWITZERLAND) *6*(5):41; *6*(6):241; *6*(10):109
Portier, Ralph J. (Louisiana State University/USA) *6*(5):305
Powers, Leigh (Georgia Institute of Technology/USA) *6*(9):165
Prandi, Alberto (Water & Soil Remediation S.r.1/ITALY) *6*(6):179
Prasad, M.N.V. (University of Hyderabad/INDIA) *6*(5):165
Price, Steven (Camp Dresser & McKee, Inc./USA) *6*(9):303
Priester, Lamar E. (Priester & Associates/USA) *6*(10):65
Pritchard, P. H. (Hap) (U.S. Navy/USA) *6*(7):125
Profit, Michael D. (CDM Federal Programs Corporation/USA) *6*(6):81
Prosnansky, Michal (Gunma University/JAPAN) *6*(9):201
Pruden, Amy (University of Cincinnati/USA) *6*(1):19
Ptacek, Carol J. (University of Waterloo/CANADA) *6*(9):71

Radosevich, Mark (University of Delaware/USA) *6*(6):9
Radtke, Corey (INEEL/USA) *6*(3):9
Raetz, Richard M. (Global Remediation Technologies, Inc./USA) *6*(6):311
Rainwater, Ken (Texas Tech University/USA) *6*(3):9
Ramani, Mukundan (University of Cincinnati/USA) *6*(5):269
Raming, Julie B. (Georgia-Pacific Corp./USA) *6*(1):183
Ramírez, N. E. (ECOPETROL-ICP/COLOMBIA) *6*(6):319
Ramsay, Bruce A. (Polyferm Canada Inc./CANADA) *6*(2):121; *6*(10):171
Ramsay, Juliana A. (Queen's University/CANADA) *6*(2):121; *6*(10):171
Rao, Prasanna (University of Cincinnati/USA) *6*(9):1
Ratzke, Hans-Peter (Umweltschutz Nord GMBH/GERMANY) *6*(4):131
Reardon, Kenneth F. (Colorado State University/USA) *6*(8):53
Rectanus, Heather V. (Virginia Polytechnic Institute & State University/USA) *6*(2):105
Reed, Thomas A. (URS Corporation/USA) *6*(8):157; *6*(10):15, 95
Rees, Hubert (CEFAS Laboratory/UK) *6*(10):337
Rehm, Bernd W. (RMT, Inc./USA) *6*(2):97; *6*(10):201
Reinecke, Stefan (Franz Environmental Inc./CANADA) *6*(6):211
Reinhard, Martin (Stanford University/USA) *6*(7):95
Reisinger, H. James (Integrated Science & Technology Inc/USA) *6*(1):183
Rek, Dorota (IT Corporation/USA) *6*(2):73
Reynolds, Charles M. (U.S. Army Engineer Research & Development Center/USA) *6*(5):9
Reynolds, Daniel E. (Air Force Institute of Technology/USA) *6*(2):173
Rice, John M. (RMT, Inc./USA) *6*(7):181
Richard, Don E. (Barr Engineering Company/USA) *6*(3):219; *6*(5):105
Richardson, Ian (Conestoga-Rovers & Associates/USA) *6*(10):131
Richnow, Hans H. (UFZ-Centre for Environmental Research/GERMANY) *6*(4):99

Rijnaarts, Huub H.M. (TNO Institute of Environmental Science/THE NETHERLANDS) *6*(2):231
Ringelberg, David B. (U.S. Army Corps of Engineers/USA) *6*(5):9; *6*(6):43; *6*(10):115
Ríos-Leal, E. (CINVESTAV-IPN/MEXICO) *6*(3):243
Ripp, Steven (University of Tennessee/USA) *6*(4):45
Ritter, Michael (URS Corporation/USA) *6*(2):239
Ritter, William F. (University of Delaware/USA) *6*(6):9
Riva, Vanessa (Parsons Engineering Science, Inc./USA) *6*(2):39
Rivas-Lucero, B.A. (Centro de Investigacion en Materiales Avanzados/MEXICO) *6*(6):361
Rivetta, A. (Universita degli Studi di Milano/ITALY) *6*(5):157
Robb, Joseph (ENSR International/USA) *6*(1):75
Robertiello, Andrea (EniTecnologie S.p.A./ITALY) *6*(6):173
Robertson, K. (Queen's University/CANADA) *6*(2):121
Robinson, David (ERM, Inc./USA) *6*(7):45
Robinson, Sandra L. (Virginia Polytechnic Institute & State University/USA) *6*(5):1
Rockne, Karl J. (University of Illinois-Chicago/USA) *6*(5):297
Rodríguez-Vázquez, Refugio (CINVESTAV-IPN/MEXICO) *6*(3):243; *6*(6):219, 369
Römkens, Paul (Alterra/THE NETHERLANDS) *6*(5):137
Rongo, Rocco (University della Calabria/ITALY) *6*(4):131
Roorda, Marcus L. (Rowan University/USA) *6*(5):215
Rosser, Susan J. (University of Cambridge/UK) *6*(5):69
Rowland, Martin A. (Lockheed-Martin Michoud Space Systems/USA) *6*(7):1
Royer, Richard (The Pennsylvania State University/USA) *6*(8):201
Ruggeri, Bernardo (Politecnico di Torino/ITALY) *6*(3):211
Ruiz, Graciela M. (University of Iowa/USA) *6*(1):195
Rupassara, S. Indumathie (University of Illinois at Urbana-Champaign/USA) *6*(5):181

Sacchi, G.A. (Universita degli Studi di Milano/ITALY) *6*(5):157
Sahagun, Tracy (U.S. Marine Corps./USA) *6*(10):1
Sakakibara, Yutaka (Waseda University/JAPAN) *6*(8):113; *6*(9):201
Sakamoto, T. (Queen's University/CANADA) *6*(10):171
Salam, Munazza (Crescent Heights High School/CANADA) *6*(5):53
Salanitro, Joseph P. (Equilon Enterprises, LLC/USA) *6*(1):11; *6*(8):61
Salvador, Maria Cristina (CSD-GEOKLOCK/BRAZIL) *6*(1):27
Samson, Réjean (École Polytechnique de Montréal/CANADA) *6*(3):115; *6*(4):139; *6*(9):27
San Felipe, Zenaida (Monash University/AUSTRALIA) *6*(4):1
Sánchez, F.N. (ECOPETROL-ICP/COLOMBIA) *6*(6):319
Sánchez, Gisela (PDVSA Intevep/VENEZUELA) *6*(6):257
Sánchez, Luis (PDVSA Intevep/VENEZUELA) *6*(6):257
Sanchez, M. (Universidad Complutense de Madrid/SPAIN) *6*(4):29; *6*(6):377
Sandefur, Craig A. (Regenesis/USA) *6*(7):257; *6*(10):87
Sanford, Robert A. (University of Illinois at Urbana-Champaign/USA) *6*(9):179
Santangelo-Dreiling, Theresa (Colorado Dept. of Transportation/USA) *6*(10):231
Saran, Jennifer (Kennecott Utah Copper Corp./USA) *6*(9):323
Sarpietro, M.G. (Universita di Catania/ITALY) *6*(3):149

Sartoros, Catherine (Université du Québec à Montréal/CANADA) *6*(3):165
Saucedo-Terán, R.A. (Centro de Investigacion en Materiales Avanzados/MEXICO) *6*(6):361
Saunders, James A. (Auburn University/USA) *6*(9):105
Sayler, Gary S. (University of Tennessee/USA) *6*(4):45
Scalzi, Michael M. (Innovative Environmental Technologies, Inc./USA) *6*(10):23
Scarborough, Shirley (IT Corporation/USA) *6*(2):253
Schaffner, I. Richard (GZA GeoEnvironmental, Inc./USA) *6*(7):165
Scharp, Richard A. (U.S. EPA/USA) *6*(9):9
Schell, Heico (Water Technology Center/GERMANY) *6*(8):105
Scherer, Michelle M. (The University of Iowa/USA) *6*(3):1
Schipper, Mark (Groundwater Services) *6*(8):73
Schmelling, Stephen (U.S. EPA/USA) *6*(1):129
Schnoor, Jerald L. (University of Iowa/USA) *6*(8):147
Schoefs, Olivier (École Polytechnique de Montréal/CANADA) *6*(4):139
Schratzberger, Michaela (CEFAS Laboratory/UK) *6*(10):337
Schulze, Susanne (Water Technology Center/GERMANY) *6*(2):137
Schuur, Jessica H. (Lund University/SWEDEN) *6*(6):65
Scrocchi, Susan (Conestoga-Rovers & Associates/USA) *6*(1):99; *6*(10):131
Sczechowski, Jeff (California Polytechnic State University/USA) *6*(10):311
Seagren, Eric A. (University of Maryland/USA) *6*(10):57
Sedran, Marie A. (University of Cincinnati/USA) *6*(1):19
Seifert, Dorte (Technical University of Denmark/DENMARK) *6*(2):11
Semer, Robin (Harza Engineering Company, Inc./USA) *6*(7):157
Semprini, Lewis (Oregon State University/USA) *6*(10):145, 155, 179
Seracuse, Joe (Harding ESE/USA) *6*(7):205
Serra, Roberto (Centro Ricerche Ambientali/ITALY) *6*(4):131
Sewell, Guy W. (U.S. EPA/USA) *6*(1):167; *6*(7):125; *6*(8):167
Sharma, Pawan (Camp Dresser & McKee Inc./USA) *6*(7):305
Sharp, Robert R. (Manhattan College/USA) *6*(7):117
Shay, Devin T. (Groundwater & Environmental Services, Inc./USA) *6*(10):101
Shelley, Michael L. (Air Force Institute of Technology/USA) *6*(5):95
Shen, Hai (Dynamac Corporation/USA) *6*(1): 129, 167
Sherman, Neil (Louisiana-Pacific Corporation/USA) *6*(3):83
Sherwood Lollar, Barbara (University of Toronto/CANADA) *6*(4):91, 109
Shi, Jing (EFX Systems, Inc./USA) *6*(8):19
Shields, Adrian R.G. (Komex Europe/UK) *6*(10):123
Shiffer, Shawn (University of Illinois/USA) *6*(9):179
Shin, Won Sik (Lousiana State University/USA) *6*(5):313
Shiohara, Kei (Mississippi State University/USA) *6*(6):279
Shirazi, Fatemeh R. (Stratum Engineering Inc./USA) *6*(8):121
Shoemaker, Christine (Cornell University/USA) *6*(4):125
Sibbett, Bruce (IT Corporation/USA) *6*(2):73
Silver, Cannon F. (Parsons Engineering Science, Inc./USA) *6*(10):283
Silverman, Thomas S. (RMT, Inc./USA) *6*(10):201
Simon, Michelle A. (U.S. EPA/USA) *6*(10):293
Sims, Gerald K. (USDA-ARS/USA) *6*(5):181
Sims, Ronald C. (Utah State University/USA) *6*(4):67; *6*(6):1
Sincock, M. Jennifer (ENVIRON International Corp./USA) *6*(7):189

Sittler, Steven P. (Advanced Pollution Technologists, Ltd./USA) *6*(2):215
Skladany, George J. (ERM, Inc./USA) *6*(7):45, 213
Skubal, Karen L. (Case Western Reserve University/USA) *6*(8):193
Slenders, Hans (TNO-MEP/THE NETHERLANDS) *6*(7):289
Slomczynski, David J. (University of Cincinnati/USA) *6*(2):247
Slusser, Thomas J. (Wright State University/USA) *6*(5):95
Smallbeck, Donald R. (Harding Lawson/USA) *6*(10):231
Smets, Barth F. (University of Connecticut/USA) *6*(7):69
Smith, Christy (North Carolina State University/USA) *6*(1):145
Smith, Colin C. (University of Sheffield/UK) *6*(2):207
Smith, John R. (Alcoa Inc./USA) *6*(6):89
Smith, Jonathan (The Environment Agency/UK) *6*(4):17
Smith, Steve (King's College London/UK) *6*(2):223; *6*(3):173; *6*(10):123
Smyth, David J.A. (University of Waterloo/CANADA) *6*(9):71
Sobecky, Patricia (Georgia Institute of Technology/USA) *6*(9):165
Sola, Adrianna (Spelman College/USA) *6*(9):53
Sordini, E. (EniTechnologie/ITALY) *6*(6):173
Sorensen, James A. (University of North Dakota/USA) *6*(6):141
Sorenson, Kent S. (Idaho National Engineering and Environmental Laboratory./USA) *6*(7):265
South, Daniel (Harding ESE/USA) *6*(7):205
Spain, Jim (U.S. Air Force/USA) *6*(3):59; *6*(7):125
Spasova, Irena Ilieva (University of Mining & Geology/BULGARIA) *6*(9):97
Spataro, William (University della Calabria/ITALY) *6*(4):131
Spinnler, Gerard E. (Equilon Enterprises, LLC/USA) *6*(1):11; *6*(8):61
Springael, Dirk (VITO/BELGIUM) *6*(4):35
Srinivasan, P. (GeoTrans, Inc./USA) *6*(2):163
Stansbery, Anita (California Polytechnic State University/USA) *6*(10):311
Starr, Mark G. (DuPont Co./USA) *6*(8):43
Stehmeier, Lester G. (NOVA Research Technology Centre/CANADA) *6*(4):117; *6*(5):53
Stensel, H. David (University of Washington/USA) *6*(10):211
Stordahl, Darrel M. (Camp Dresser & McKee Inc./USA) *6*(6):287
Stout, Scott (Battelle/USA) *6*(5):237
Strand, Stuart E. (University of Washington/USA) *6*(10):211
Stratton, Glenn (Nova Scotia Agricultural College/CANADA) *6*(3):197
Strybel, Dan (IT Corporation/USA) *6*(9):215
Stuetz, R.M. (Cranfield University/UK) *6*(6):329
Suarez, B. (ECOPETROL-ICP/COLOMBIA) *6*(6):319
Suidan, Makram T. (University of Cincinnati/USA) *6*(1):19; *6*(5):243, 253, 261; *6*(6):133,
Suthersan, Suthan S. (ARCADIS Geraghty & Miller/USA) *6*(7):53
Suzuki, Masahiro (Nippon Institute of Technology/JAPAN) *6*(5):321
Sveum, Per (Deconterra AS/NORWAY) *6*(6):157
Swallow, Ian (BAE Systems Properties Ltd./UK) *6*(6):119
Swann, Benjamin M. (Camp Dresser & McKee Inc./USA) *6*(7):305
Swannell, Richard P.J. (AEA Technology Environment/UK) *6*(10):337

Tabak, Henry H. (U.S. EPA/USA) *6*(5):243, 253, 261, 269; *6*(9):1, 17
Takai, Koji (Fuji Packing/JAPAN) *6*(10):345

Talley, Jeffrey W. (University of Notre Dame/USA) *6*(3):189; *6*(4):59; *6*(6):43; *6*(7):125; *6*(10):115
Tao, Shu (Peking University/CHINA) *6*(7):61
Taylor, Christine D. (North Carolina State University/USA) *6*(1):83
Ter Meer, Jeroen (TNO Institute of Environmental Science/THE NETHERLANDS) *6*(2):231; *6*(7):289
Tétreault, Michel (Royal Military College of Canada/CANADA) *6*(8):95
Tharpe, D.L. (Equilon Enterprises LLC/USA) *6*(8):61
Theeuwen, J. (Grontmij BV/THE NETHERLANDS) *6*(7):289
Thomas, Hartmut (WASAG DECON GMbH/GERMANY) *6*(3):67
Thomas, Mark (EG&G Technical Services, Inc./USA) *6*(10):49
Thomas, Paul R. (Thomas Consultants, Inc./USA) *6*(5):189
Thomas, Robert C. (University of Georgia/USA) *6*(9):105
Thomson, Michelle M. (URS Corporation/USA) *6*(2):81
Thornton, Steven F. (University of Sheffield/UK) *6*(1):59, *6*(2):207
Tian, C. (University of Cincinnati/USA) *6*(8):35
Tiedje, James M. (Michigan State University/USA) *6*(7):125; *6*(8):19
Tiehm, Andreas (Water Technology Center/GERMANY) *6*(2):137; *6*(8):105
Tietje, David (Foster Wheeler Environmental Corportation/USA) *6*(9):249
Timmins, Brian (Oregon State University/USA) *6*(10):179
Togna, A. Paul (Envirogen Inc/USA) *6*(9):281
Tolbert, David E.(U.S. Army/USA) *6*(9):281
Tonnaer, Haimo (TAUW BV/THE NETHERLANDS) *6*(7):297; *6*(10):253
Toth, Brad (Harding ESE/USA) *6*(10):231
Tovanabootr, Adisorn (Oregon State University/USA) *6*(10):145
Travis, Bryan (Los Alamos National Laboratory/USA) *6*(10):163
Trudnowski, John M. (MSE Technology Applications, Inc./USA) *6*(9):35
Truax, Dennis D. (Mississippi State University/USA) *6*(9):241
Trute, Mary M. (Camp Dresser & McKee, Inc./USA) *6*(2):113
Tsuji, Hirokazu (Obayashi Corporation Ltd./JAPAN) *6*(6):111, 249; *6*(10):239
Tsutsumi, Hiroaki (Prefectural University of Kumamoto/JAPAN) *6*(10):345
Turner, Tim (CDM Federal Programs Corp./USA) *6*(6):81
Turner, Xandra (International Biochemicals Group/USA) *6*(10):23
Tyner, Larry (IT Corporation/USA) *6*(1):51; *6*(2):73

Ugolini, Nick (U.S. Navy/USA) *6*(10):65
Uhler, Richard (Battelle/USA) *6*(5):237
Unz, Richard F. (The Pennsylvania State University/USA) *6*(8):201
Utgikar, Vivek P. (U.S. EPA/USA) *6*(9):17

Valderrama, Brenda (Universidad Nacional Autónoma de México/MEXICO) *6*(6):17
Vallini, Giovanni (Universita degli Studi di Verona/ITALY) *6*(3):267
van Bavel, Bert (Umeå University/SWEDEN) *6*(3):181
van Breukelen, Boris M. (Vrije University/THE NETHERLANDS) *6*(4):91
VanBroekhoven, K. (Catholic University of Leuven/BELGIUM) *6*(4):35
Vandecasteele, Jean-Paul (Institut Français du Pétrole/FRANCE) *6*(3):227
VanDelft, Frank (NOVA Chemicals/CANADA) *6*(5):53
van der Gun, Johan (BodemBeheer bv/THE NETHERLANDS) *6*(5):289

van der Werf, A. W. (Bioclear Environmental Technology/THE NETHERLANDS) *6*(8):11

van Eekert, Miriam (TNO Environmental Sciences /THE NETHERLANDS) *6*(2):231; *6*(7):289

Van Hout, Amy H. (IT Corporation/USA) *6*(3):35

Van Keulen, E. (DHV Environment and Infrastructure/THE NETHERLANDS) *6*(8):11

Vargas, M.C. (ECOPETROL-ICP/COLOMBIA) *6*(6):319

Vazquez-Duhalt, Rafael (Universidad Nacional Autónoma de México/MEXICO) *6*(6):17

Venosa, Albert (U.S. EPA/USA) *6*(1):19

Verhaagen, P. (Grontmij BV/THE NETHERLANDS) *6*(7):289

Verheij, T. (DAF/THE NETHERLANDS) *6*(7):289

Vidumsky, John E. (E.I. du Pont de Nemours & Company/USA) *6*(2):81; *6*(8):185

Villani, Marco (Centro Ricerche Ambientali/ITALY) *6*(4):131

Vinnai, Louise (Investigative Science Inc./CANADA) *6*(2):27

Visscher, Gerolf (Province of Groningen/THE NETHERLANDS) *6*(7):141

Voegeli, Vincent (TranSystems Corporation/USA) *6*(7):229

Vogt, Bob (Louisiana-Pacific Corporation/USA) *6*(3):83

Volkering, Frank (TAUW bv/THE NETHERLANDS) *6*(4):91

von Arb, Michelle (University of Iowa) *6*(3):1

Vondracek, James E. (Ashland Inc./USA) *6*(5):121

Vos, Johan (VITO/BELGIUM) *6*(9):87

Voscott, Hoa T. (Camp Dresser & McKee, Inc./USA) *6*(7):305

Vough, Lester R. (University of Maryland/USA) *6*(5):77

Waisner, Scott A. (TA Environmental, Inc./USA) *6*(4):59; *6*(10):115

Walecka-Hutchison, Claudia M. (University of Arizona/USA) *6*(9):231

Wall, Caroline (CEFAS Laboratory/UK) *6*(10):337

Wallace, Steve (Lattice Property Holdings Plc./UK) *6*(4):17

Wallis, F.M. (University of Natal/REP OF SOUTH AFRICA) *6*(6):101; *6*(9):79

Walton, Michelle R. (Idaho National Engineering & Environmental Laboratory/USA) *6*(7):77

Walworth, James L. (University of Arizona/USA) *6*(9):231

Wan, C.K. (Hong Kong Baptist University/CHINA) *6*(6):73

Wang, Chuanyue (Rice University/USA) *6*(5):85

Wang, Qingren (Chinese Academy of Sciences/CHINA [PRC]) *6*(9):113

Wani, Altaf (Applied Research Associates, Inc./USA) *6*(10):115

Wanty, Duane A. (The Gillette Company/USA) *6*(7):87

Warburton, Joseph M. (Parsons Engineering Science/USA) *6*(7):173

Watanabe, Masataka (National Institute for Environmental Studies/JAPAN) *6*(5):321

Watson, James H.P. (University of Southampton/UK) *6*(9):61

Wealthall, Gary P. (University of Sheffield/UK) *6*(1):59

Weathers, Lenly J. (Tennessee Technological University/USA) *6*(8):139

Weaver, Dallas E. (Scientific Hatcheries/USA) *6*(1):91

Weaverling, Paul (Harding ESE/USA) *6*(10):231

Weber, A. Scott (State University of New York at Buffalo/USA) *6*(6):89

Weeber, Philip A. (Geotrans/USA) *6*(10):163

Wendt-Potthoff, Katrin (UFZ Centre for Environmental Research/GERMANY) *6*(9):43

Werner, Peter (Technical University of Dresden/GERMANY) *6*(3):227; *6*(8):105

West, Robert J. (The Dow Chemical Company/USA) *6*(2):89
Westerberg, Karolina (Stockholm University/SWEDEN) *6*(3):133
Weston, Alan F. (Conestoga-Rovers & Associates/USA) *6*(1):99; *6*(10):131
Westray, Mark (ThermoRetec Corp/USA) *6*(7):1
Wheater, H.S. (Imperial College of Science and Technology/UK) *6*(10):123
White, David C. (University of Tennessee/USA) *6*(4):73; *6*(5):305
White, Richard (EarthFax Engineering Inc/USA) *6*(6):263
Whitmer, Jill M. (GeoSyntec Consultants/USA) *6*(9):105
Wick, Lukas Y. (Swiss Federal Institute of Technology/SWITZERLAND) *6*(3):251
Wickramanayake, Godage B. (Battelle/USA) *6*(10):1
Widada, Jaka (The University of Tokyo/JAPAN) *6*(4):51
Widdowson, Mark A. (Virginia Polytechnic Institute & State University/USA) *6*(2):105; *6*(5):1
Wieck, James M. (GZA GeoEnvironmental, Inc./USA) *6*(7):165
Wiedemeier, Todd H. (Parsons Engineering Science, Inc./USA) *6*(7):241
Wiessner, Arndt (UFZ - Centre for Environmental Research/GERMANY) *6*(5):337
Wilken, Jon (Harding ESE/USA) *6*(10):231
Williams, Lakesha (Southern University at New Orleans/USA) *6*(5):145
Williamson, Travis (Battelle/USA) *6*(10):245
Willis, Matthew B. (Cornell University/USA) *6*(4):125
Willumsen, Pia Arentsen (National Environmental Research Institute/DENMARK) *6*(3):141
Wilson, Barbara H. (Dynamac Corporation/USA) *6*(1):129
Wilson, Gregory J. (University of Cincinnati/USA) *6*(1):19
Wilson, John T. (U.S. EPA/USA) *6*(1):43, 167
Wiseman, Lee (Camp Dresser & McKee Inc./USA) *6*(7):133
Wisniewski, H.L. (Equilon Enterprises LLC/USA) *6*(8):61
Witt, Michael E. (The Dow Chemical Company/USA) *6*(2):89
Wong, Edwina K. (University of Guelph/CANADA) *6*(6):185
Wong, J.W.C. (Hong Kong Baptist University/CHINA) *6*(6):73
Wood, Thomas K. (University of Connecticut/USA) *6*(5):199
Wrobel, John (U.S. Army/USA) *6*(5):207

Xella, Claudio (Water & Soil Remediation S.r.l./ITALY) *6*(6):179
Xing, Jian (Global Remediation Technologies, Inc./USA) *6*(6):311

Yamamoto, Isao (Sumitomo Marine Research Institute/JAPAN) *6*(10):345
Yamazaki, Fumio (Hyogo Prefectural Institute of Environmental Science/JAPAN) *6*(5):321
Yang, Jeff (URS Corporation/USA) *6*(2):239
Yerushalmi, Laleh (Biotechnology Research Institute/CANADA) *6*(3):165
Yoon, Woong-Sang (Sam) (Battelle/USA) *6*(7):13
Yoshida, Takako (The University of Tokyo/JAPAN) *6*(4):51; *6*(6):111
Yotsumoto, Mizuyo (Obayashi Corporation Ltd./JAPAN) *6*(6):111
Young, Harold C. (Air Force Institute of Technology/USA) *6*(2):173

Zagury, Gérald J. (École Polytechnique de Montréal/CANADA) *6*(9): 27, 129
Zahiraleslamzadeh, Zahra (FMC Corporation/USA) *6*(7):221
Zaluski, Marek H. (MSE Technology Applications/USA) *6*(9):35
Zappi, Mark E. (Mississippi State University/USA) *6*(9):241

Zelennikova, Olga (University of Connecticut/USA) *6*(7):69
Zhang, Chuanlun L. (University of Missouri/USA) *6*(9):165
Zhang, Wei (Cornell University/USA) *6*(4):125
Zhang, Zhong (University of Nevada Las Vegas/USA) *6*(9):257
Zheng, Zuoping (University of Oslo/NORWAY) *6*(2):181
Zocca, Chiara (Universita degli Studi di Verona/ITALY) *6*(3):267
Zwick, Thomas C. (Battelle/USA) *6*(10):1

KEYWORD INDEX

This index contains keyword terms assigned to the articles in the ten-volume proceedings of the Sixth International In Situ and On-Site Bioremediation Symposium (San Diego, California, June 4-7, 2001). Ordering information is provided on the back cover of this book.

In assigning the terms that appear in this index, no attempt was made to reference all subjects addressed. Instead, terms were assigned to each article to reflect the primary topics covered by that article. Authors' suggestions were taken into consideration and expanded or revised as necessary. The citations reference the ten volumes as follows:

6(1): Magar, V.S., J.T. Gibbs, K.T. O'Reilly, M.R. Hyman, and A. Leeson (Eds.), *Bioremediation of MTBE, Alcohols, and Ethers*. Battelle Press, Columbus, OH, 2001. 249 pp.

6(2): Leeson, A., M.E. Kelley, H.S. Rifai, and V.S. Magar (Eds.), *Natural Attenuation of Environmental Contaminants*. Battelle Press, Columbus, OH, 2001. 307 pp.

6(3): Magar, V.S., G. Johnson, S.K. Ong, and A. Leeson (Eds.), *Bioremediation of Energetics, Phenolics, and Polycyclic Aromatic Hydrocarbons*. Battelle Press, Columbus, OH, 2001. 313 pp.

6(4): Magar, V.S., T.M. Vogel, C.M. Aelion, and A. Leeson (Eds.), *Innovative Methods in Support of Bioremediation*. Battelle Press, Columbus, OH, 2001. 197 pp.

6(5): Leeson, A., E.A. Foote, M.K. Banks, and V.S. Magar (Eds.), *Phytoremediation, Wetlands, and Sediments*. Battelle Press, Columbus, OH, 2001. 383 pp.

6(6): Magar, V.S., F.M. von Fahnestock, and A. Leeson (Eds.), *Ex Situ Biological Treatment Technologies*. Battelle Press, Columbus, OH, 2001. 423 pp.

6(7): Magar, V.S., D.E. Fennell, J.J. Morse, B.C. Alleman, and A. Leeson (Eds.), *Anaerobic Degradation of Chlorinated Solvents*. Battelle Press, Columbus, OH, 2001. 387 pp.

6(8): Leeson, A., B.C. Alleman, P.J. Alvarez, and V.S. Magar (Eds.), *Bioaugmentation, Biobarriers, and Biogeochemistry*. Battelle Press, Columbus, OH, 2001. 255 pp.

6(9): Leeson, A., B.M. Peyton, J.L. Means, and V.S. Magar (Eds.), *Bioremediation of Inorganic Compounds*. Battelle Press, Columbus, OH, 2001. 377 pp.

6(10): Leeson, A., P.C. Johnson, R.E. Hinchee, L. Semprini, and V.S. Magar (Eds.), *In Situ Aeration and Aerobic Remediation*. Battelle Press, Columbus, OH, 2001. 391 pp.

A

abiotic/biotic dechlorination ***6*(8)**:193
acenaphthene ***6*(5)**:253
acetate as electron donor ***6*(3)**:51; ***6*(9)**:297
acetone ***6*(2)**:49
acid mine drainage, (*see also* mine tailings) ***6*(9)**:1, 9, 27, 35, 43, 53
acrylic vessel ***6*(5)**:321
actinomycetes ***6*(10)**:211
activated carbon biomass carrier ***6*(6)**:311; ***6*(8)**:113

activated carbon ***6*(8)**:105
adsorption ***6*(3)**:243; ***6*(5)**:253; ***6*(6)**:377; ***6*(7)**:77; ***6*(8)**:131; ***6*(9)**:86
advanced oxidation ***6*(1)**:121; ***6*(10)**:33
aerated submerged ***6*(10)**:329
aeration ***6*(6)**:203
anaerobic/aerobic treatment ***6*(6)**:361; ***6*(7)**:229
age dating ***6*(5)**:231, 237
air sparging ***6*(1)**:115, 175; ***6*(2)**:239; ***6*(9)**:215; ***6*(10)**:1, 9, 41, 49, 65, 101, 115, 123, 163, 223
alachlor ***6*(6)**:9
algae ***6*(5)**:181
alkaline phosphatase ***6*(9)**:165
alkane degradation ***6*(5)**:313
alkylaromatic compounds ***6*(6)**:173
alkylbenzene ***6*(2)**:19
alkylphenolethoxylate ***6*(5)**:215
Amaranthaceae ***6*(5)**:165
Ames test ***6*(6)**:249
ammonia ***6*(1)**:175; ***6*(5)**:337
amphipod toxicity test ***6*(5)**:321
anaerobic ***6*(1)**:35, 43; ***6*(3)**:91; 205; ***6*(5)**:17, 25, 261, 297, 313; ***6*(6)**:133; ***6*(7)**:249, 297; ***6*(9)**:147, 303
anaerobic biodegradation ***6*(1)**:137; ***6*(5)**:1; ***6*(8)**:167
anaerobic bioventing ***6*(3)**:9
anaerobic petroleum degradation ***6*(5)**:25
anaerobic sparging ***6*(7)**:297
aniline ***6*(6)**:149
Antarctica ***6*(2)**:57
anthracene ***6*(3)**:165, 251; ***6*(6)**:73
aquatic plants ***6*(5)**:181
arid-region soils ***6*(9)**:231
aromatic dyes ***6*(6)**:369
arsenic ***6*(2)**:239, 261; ***6*(5)**:173; ***6*(9)**: 97, 129
atrazine ***6*(5)**:181; ***6*(6)**:9
azoaromatic compounds ***6*(6)**:149
Azomonas ***6*(6)**:219

B

bacterial transport ***6*(8)**:1
barrier technologies ***6*(1)**:11; ***6*(3)**:165; ***6*(7)**:289; ***6*(8)**:61, 79, 87, 105, 121; ***6*(9)**:27, 71, 195, 209, 309
basidiomycete ***6*(6)**:101
benthic ***6*(10)**:337
benzene ***6*(1)**:1, 67, 75, 145, 167, 203; ***6*(4)**:91,117; ***6*(8)**:87; ***6*(10)**:123
benzene, toluene, ethylbenzene, and xylenes (BTEX) ***6*(1)**:43, 51, 59, 107, 129, 167, 195; ***6*(2)**:11, 19, 137, 215, 223, 270; ***6*(4)**:99; ***6*(5)**:33; ***6*(7)**:133; ***6*(8)**:105; ***6*(10)**: 1, 23, 49, 65, 95, 123, 131
benzo(a)pyrene ***6*(3)**:149; ***6*(6)**:101
benzo(e)pyrene ***6*(3)**:149
BER, *see* biofilm-electrode reactor
bioassays ***6*(3)**:219
bioaugmentation ***6*(1)**:11; ***6*(3)**:133; ***6*(4)**:59; ***6*(6)**:9, 43, 111; ***6*(7)**:125; ***6*(8)**:1, 11, 19, 27, 43, 53, 61, 147, 175
bioavailability ***6*(3)**:115, 157, 173, 189, 51; ***6*(4)**:7; ***6*(5)**:253, 279, 289; ***6*(6)**:1
bioavailable FeIII assay ***6*(8)**:209
biobarrier ***6*(1)**:11; ***6*(3)**:165; ***6*(7)**:289; ***6*(8)**:61, 79, 105, 121; ***6*(9)**:27, 71, 209, 309
BIOCHLOR model ***6*(2)**:155
biocide ***6*(7)**:321, 333
biodegradability ***6*(6)**:193
biodegradation ***6*(1)**:19,153; ***6*(3)**:165, 181, 205, 235; ***6*(10)**:187
biofilm ***6*(3)**:251; ***6*(4)**:149; ***6*(8)**:79; ***6*(9)**:201, 303
biofilm-electrode reactor (BER) ***6*(9)**:201
biofiltration ***6*(4)**:149
biofouling ***6*(7)**:321, 333
bioindicators ***6*(1)**:1; ***6*(3)**:173; ***6*(5)**:223
biological carbon regeneration ***6*(8)**:105
bioluminescence ***6*(1)**:1; ***6*(3)**:173; ***6*(4)**:45
biopile ***6*(6)**:81, 127, 141, 227, 249, 287
bioreactors ***6*(1)**:91; ***6*(6)**:361; ***6*(8)**:11, 35; ***6*(9)**:1, 265, 281, 303, 315; ***6*(10)**:171, 211
biorecovery of metals ***6*(9)**:9
bioreporters ***6*(4)**:45
biosensors ***6*(1)**:1
bioslurping ***6*(10)**:245, 253, 267, 275
bioslurry and bioslurry reactors ***6*(3)**:189; ***6*(6)**:51, 65
biosparging ***6*(10)**:115, 163
biostabilization ***6*(6)**:89
biostimulation ***6*(6)**:43
biosurfactant ***6*(3)**:243; ***6*(7)**:53
bioventing ***6*(10)**:109, 115, 131
biphasic reactor ***6*(3)**:181

biological oxygen demand (BOD) **6(10)**:311
BTEX, *see* benzene, toluene, ethylbenzene, and xylenes
Burkholderia cepacia **6(1)**:153; **6(7)**:117; **6(8)**:53
butane **6(1)**:137, 161
butyrate **6(7)**:289

C

cadmium **6(3)**:91; **6(9)**:79, 147
carAa, see carbazole 1,9a-dioxygenase gene
carbazole-degrading bacterium **6(6)**:111
carbazole 1,9a-dioxygenase gene (*carAa*) **6(4)**:51
Carbokalk **6(9)**:43
carbon isotope **6(4)**:91, 99, 109, 117; **6(10)**:115
carbon tetrachloride (CT) **6(2)**:81, 89; **6(5)**:113; **6(7)**:241; **6(8)**:185, 193
cesium-137 **6(5)**:231
CF, *see* chloroform
charged coupled device camera **6(2)**:207
chelators addition (EDGA, EDTA) **6(5)**:129, 137, 145, 151; **6(9)**:123, 147
chemical oxidation **6(7)**:45
chicken manure **6(9)**:289
chlorinated ethenes **6(7)**:27, 61, 69, 109; **6(10)**:163, 201, 231
chlorinated solvents **6(2)**:145; **6(7)**:all; **6(8)**:19; **6(10)**:231
chlorobenzene **6(8)**:105
chloroethane **6(2)**:113; **6(7)**:133, 249
chloroform (CF) **6(2)**:81; **6(8)**:193
chloromethanes **6(8)**:185
chlorophenol **6(3)**:75, 133
chlorophyll fluorescence **6(5)**:223
chromated copper arsenate **6(9)**:129
chromium (Cr[VI]) **6(8)**:139, 147; **6(9)**:129, 139, 315
chrysene **6(6)**:101
citrate and citric acid **6(5)**:137; **6(7)**:289
cleanup levels **6(6)**:1
coextraction method **6(4)**:51
Coke Facility waste **6(2)**:129
combined chemical toxicity (*see also* toxicity) **6(5)**:305
cometabolic air sparging **6(10)**:145, 155, 223
cometabolism **6(1)**:137, 145, 153, 161; **6(2)**:19; **6(6)**:81, 141; **6(7)**:117; **6(10)**:145, 155, 163, 171, 179, 193, 201, 211, 217, 223, 231; 239
competitive inhibition **6(2)**:19
composting **6(3)**:83; **6(5)**:129, **6(6)**:73, 119, 165, 257; **6(7)**:141
constructed wetlands **6(5)**:173, 329
contaminant aging **6(3)**:157, 197
contaminant transport **6(3)**:115
copper **6(9)**:79, 129
cosolvent effects **6(1)**:175, 195, 203, 243
cosolvent extraction **6(7)**:125
cost analyses and economics of environmental restoration **6(1)**:129; **6(4)**:17; **6(8)**:121; **6(9)**:331; **6(10)**:65, 211
Cr(VI), *see* chromium
creosote **6(3)**:259; **6(4)**:59; **6(5)**:1, 237, 329; **6(6)**:81, 101, 141, 295
cresols **6(10)**:123
crude oil **6(5)**:313; **6(6)**:193, 249; **6(10)**:329
CT, *see* carbon tetrachloride
cyanide **6(9)**:331
cytochrome P-450 **6(6)**:17

D

2,4-DAT, *see* diaminotoluene
DCA, *see* dichloroethane
1,1-DCA, *see* 1,1-dichloroethane
1,2-DCA, *see* 1,2-dichloroethane
DCE, *see* dichloroethene
1,1-DCE, *see* 1,1-dichloroethene
1,2-DCE, *see* 1,2-dichloroethene
c-DCE, *see* cis-dichloroethene
DCM, *see* dichloromethane
DDT, *see also* dioxins *and* pesticides **6(6)**:157
2,4-DNT, *see* dinitrotoluene
dechlorination kinetics **6(2)**:105; **6(7)**:61
dechlorination **6(2)**:231; **6(3)**:125; **6(5)**:95; **6(7)**:13, 61, 165, 173, 333; **6(8)**:19, 27, 43
DEE, *see* diethyl ether
Dehalococcoides ethenogenes **6(8)**:19, 43
dehalogenation **6(8)**:167
denaturing gradient gel electrophoresis (DGGE) **6(1)**:19; **6(4)**:35

denitrification ***6*(2)**:19; ***6*(4)**:149; ***6*(5)**:17, 261; ***6*(8)**:95; ***6*(9)**:179, 187, 195, 201, 209, 223, 309
dense, nonaqueous-phase liquid (DNAPL) ***6*(7)**:13, 19, 35, 181; ***6*(10)**:319
depletion rate ***6*(1)**:67
desorption ***6*(3)**:235, 243; ***6*(5)**:253; ***6*(6)**:377; ***6*(7)**:53, 77; ***6*(8)**:131
DGGE, *see* denaturing gradient gel electrophoresis
DHPA, *see* dihydroxyphenylacetate
dialysis sampler ***6*(5)**:207
diaminotoluene (2,4-DAT) ***6*(6)**:149
dibenzofuran-degrading bacterium ***6*(6)**:111
dibenzo-p-dioxin ***6*(6)**:111
dibenzothiophene ***6*(3)**:267
dichlorodiethyl ether ***6*(10)**:301
dichloroethane (DCA) ***6*(2)**:39; ***6*(7)**:289
1,1-dichloroethane (1,1-DCA; 1,2-DCA) ***6*(2)**:113; ***6*(5)**:207; ***6*(7)**:133, 165
1,2-dichloroethane (1,2-DCA) ***6*(5)**:207
dichloroethene, dichloroethylene ***6*(2)**:97, 155; ***6*(4)**:125; ***6*(5)**:105,113; ***6*(7)**:157, 197
cis-dichloroethene, *cis*-dichloroethylene (*c*-DCE) ***6*(2)**:39, 65, 73; 105, 173; ***6*(5)**:33, 95, 207; ***6*(7)**:1, 13, 61, 133, 141, 149, 165, 173, 181, 189, 205, 213, 221, 249, 273, 281, 289, 297, 305; ***6*(8)**:11, 19, 27, 43, 73, 105, 157, 209; ***6*(10)**:41, 145, 155, 179, 201
1,1-dichloroethene, 1,1-dichloroethylene (1,1-DCE) ***6*(2)**:39; ***6*(7)**:165, 229; ***6*(8)**:157; ***6*(10)**:231
1,2-dichloroethene and 1,2-dichloroethylene (1,2-DCE) ***6*(2)**:113
dichloromethane (DCM) ***6*(2)**:81; ***6*(8)**:185
diesel fuel ***6*(1)**:175; ***6*(2)**:57; ***6*(5)**:305; ***6*(6)**:81, 141, 165; ***6*(10)**:9
diesel-range organics (DRO) ***6*(10)**:9
diethyl ether (DEE) ***6*(1)**:19
dihydroxyphenylacetate (DHPA) ***6*(4)**:29
diisopropyl ether (DIPE) ***6*(1)**:19, 161
1,3-dinitro-5-nitroso-1,3,5-triazacyclohexane (MNX) (*see also* explosives *and* energetics) ***6*(3)**:51; ***6*(8)**:175
dinitrotoluene (2,4-DNT) ***6*(3)**:25, 59; ***6*(6)**:127, 149
dioxins ***6*(6)**:111
DIPE, *see* diisopropyl ether
dissolved oxygen ***6*(2)**:189, 207
16S rDNA sequencing ***6*(8)**:19
DNAPL, *see* dense, nonaqueous-phase liquid
DNX, *see* explosives and energetics
DRO, *see* diesel-range organics
dual porosity aquifer ***6*(1)**:59
dyes ***6*(6)**:369

E

ecological risk assessment ***6*(4)**:1
ecotoxicity, (*see also* toxicity) ***6*(1)**:1; ***6*(4)**:7
ethylenedibromide (EDB) ***6*(10)**:65
EDGA, *see* chelate addition
EDTA, *see* chelate addition
effluent ***6*(4)**:1
electrokinetics ***6*(9)**:241, 273
electron acceptors and electron acceptor processes ***6*(2)**:1, 137, 163, 231; ***6*(5)**:17, 25, 297; ***6*(7)**:19
electron donor amendment ***6*(3)**:25, 35, 51, 125; ***6*(7)**:69, 103,109, 141, 181, 249, 289, 297; ***6*(8)**:73; ***6*(9)**:297, 315
electron donor delivery ***6*(7)**:19, 27, 133, 173, 213, 221, 265, 273, 281, 305
electron donor mass balance ***6*(2)**:163
electron donor transport ***6*(4)**:125; ***6*(7)**:133; ***6*(9)**:241
embedded carrier ***6*(9)**:187
encapsulated bacteria ***6*(5)**:269
enhanced aeration ***6*(10)**:57
enhanced desorption ***6*(7)**:197
environmental stressors ***6*(4)**:1
enzyme induction ***6*(6)**:9; ***6*(10)**:211
ERIC sequences ***6*(4)**:29
ethane ***6*(2)**:113; ***6*(7)**:149
ethanol 6(1):19,167,175, 195, 203; ***6*(5)**:243; ***6*(6)**:133; ***6*(9)**:289
ethene and ethylene ***6*(2)**:105,113; ***6*(5)**:95; ***6*(7)**:1, 95, 133, 141, 205, 281, 297, 305; ***6*(8)**:11, 43, 167, 175, 209
ethylene dibromide ***6*(10)**:193
explosives and energetics ***6*(3)**:9, 17, 25, 35, 43, 51, 67; ***6*(5)**:69; ***6*(6)**:119, 127, 133; ***6*(7)**:125

F

fatty acids ***6*(5)**:41
Fe(II), *see* iron
Fenton's reagent ***6*(6)**:157
fertilizer ***6*(5)**:321; ***6*(6)**:35; ***6*(10)**:337
fixed-bed and fixed-film reactors ***6*(5)**:221, 337; ***6*(6)**:361; ***6*(9)**:303
flocculants ***6*(6)**:279
flow sensor ***6*(10)**:293
fluidized-bed reactor ***6*(1)**:91; ***6*(6)**:133, 311; ***6*(9)**:281
fluoranthene ***6*(3)**:141; ***6*(6)**:101
fluorogenic probes ***6*(4)**:51
food safety ***6*(9)**:113
formaldehyde ***6*(6)**:329
fractured shale ***6*(10)**:49
free-product recovery ***6*(6)**:211
Freon ***6*(2)**:49
fuel oil ***6*(5)**:321
fungal remediation ***6*(3)**:75, 99; ***6*(5)**:61, 279; ***6*(6)**:17, 101, 157, 263, 319, 329, 369
Funnel-and-Gate™ ***6*(8)**:95

G

gas flux ***6*(6)**:185
gasoline ***6*(1)**:35, 75, 161, 167, 195; ***6*(10)**:115
gasoline-range organics (GRO) ***6*(10)**:9
manufactured gas plants and gasworks ***6*(2)**:137; ***6*(10)**:123
GCW, *see* groundwater circulating well
gel-encapsulated biomass ***6*(8)**:35
GEM, see genetically engineered microorganisms
genetically engineered microorganisms (GEM) ***6*(4)**:45; ***6*(5)**:199; ***6*(7)**:125
genotoxicity, (*see also* toxicity) ***6*(3)**:227
Geobacter ***6*(3)**:1
geochemical characterization ***6*(4)**:91
geographic information system (GIS) ***6*(2)**:163
geologic heterogeneity ***6*(2)**:11
germination index 6(3):219; ***6*(6)**:73
GFP, *see* green fluorescent protein
GIS, *see* geographic information system
glutaric dialdehyde dehydrogenase ***6*(4)**:81
Gordonia terrae ***6*(1)**:153
green fluorescent protein (GFP) ***6*(5)**:199
GRO, see gasoline-range organics
groundwater ***6*(3)**:35; ***6*(8)**: 35, 87, 121; ***6*(10)**:231
groundwater circulating well (GCW) ***6*(7)**:229, 321; ***6*(10)**:283, 293

H

H_2 gas, *see* hydrogen
H_2S, *see* hydrogen sulfide
halogenated hydrocarbons ***6*(9)**:61
halorespiration ***6*(8)**:19
heavy metal ***6*(2)**:239; ***6*(5)**:137, 145, 157, 165, 173; ***6*(6)**:51; ***6*(9)**:53, 61, 71, 79, 86, 97, 113, 129, 147
herbicides ***6*(5)**:223; ***6*(6)**:35
hexachlorobenzene ***6*(3)**:99
hexane ***6*(3)**:181, ***6*(6)**:329
HMX, *see* explosives and energetics
hollow fiber membranes ***6*(5)**:269
hopane ***6*(6)**:193; ***6*(10)**:337
hornwort ***6*(5)**:181
HRC® (a proprietary hydrogen-release compound) ***6*(3)**:17, 25, 107; ***6*(7)**:27, 103, 157, 189, 197, 205, 221, 257, 305, ***6*(8)**:157, 209
^{2}H-tetradecane (*see also* tetradecane) ***6*(2)**:27
humates ***6*(1)**:99
hybrid treatment ***6*(10)**:311
hydraulic containment ***6*(8)**:79
hydraulically facilitated remediation ***6*(2)**:239
hydrocarbon ***6*(6)**:235; ***6*(10)**:329
hydrogen (H_2 gas) 6(2):199; ***6*(9)**:201
hydrogen injection, in situ ***6*(7)**:19
hydrogen isotope ***6*(4)**:91
hydrogen peroxide ***6*(1)**:121; ***6*(6)**:353; ***6*(10)**:33
hydrogen release compound, *see* HRC®
hydrogen sulfide (H_2S) ***6*(9)**:123
hydrogen ***6*(2)**:231, ***6*(7)**:61, 305
hydrolysis ***6*(1)**:83
hydrophobicity ***6*(3)**:141
hydroxyl radical ***6*(1)**:121
hydroxylamino TNT intermediates ***6*(5)**:85

I

immobilization ***6*(8)**:53
immobilized cells ***6*(8)**:121
immobilized soil bioreactor ***6*(10)**:171

in situ oxidation ***6*(7)**:1
industrial effluents ***6*(6)**:303, 361
inhibition ***6*(9)**:17
injection strategies, in situ ***6*(7)**:19, 133, 173, 213, 221, 265, 273, 305, 313; ***6*(9)**:223; ***6*(10)**:23, 163
insecticides ***6*(6)**:27
intrinsic biodegradation ***6*(2)**:89, 121
intrinsic remediation, *see* natural attenuation
ion migration ***6*(9)**:241
iron (Fe[II]) ***6*(5)**:1
iron barrier ***6*(8)**:139, 147, 157, 167
iron oxide ***6*(3)**:1
iron precipitation ***6*(3)**:211
iron-reducing processes ***6*(2)**:121; ***6*(3)**:1; ***6*(5)**:1, 17, 25; ***6*(6)**:149; ***6*(8)**:193, 201, 209; ***6*(9)**:43, 323
IR-spectroscopy ***6*(4)**:67
isotope analyses ***6*(2)**:27; ***6*(4)**:91; ***6*(8)**:27
isotope fractionation ***6*(4)**:99, 109, 117

J

jet fuel ***6*(10)**:95, 139

K

KB-1 strain ***6*(8)**:27
kerosene ***6*(6)**:219
kinetics ***6*(8)**:131, ***6*(1)**:1, 19, 27, 167; ***6*(2)**:11, 19, 105; ***6*(3)**:173; ***6*(4)**:131; ***6*(7)**:61
Klebsiella oxytoca ***6*(7)**:117
Kuwait ***6*(6)**:249

L

laccase ***6*(3)**:75; ***6*(6)**:319
lactate and lactic acid ***6*(7)**:103, 109, 165, 181, 213, 265, 281, 289; ***6*(8)**:139; ***6*(9)**:155, 273
lagoons ***6*(6)**:303
land treatment units (LTU) ***6*(6)**:1; ***6*(6)**:81, 141, 287, 295
landfarming ***6*(3)**:259; ***6*(4)**:59; ***6*(5)**:53, 279; ***6*(6)**:1, 43, 59, 179, 203, 211, 235
landfills ***6*(2)**:145, 247; ***6*(4)**:91; ***6*(8)**:113
leaching ***6*(9)**:187
lead ***6*(5)**:129, 145, 151, 157
lead-210 ***6*(5)**:231
light, nonaqueous-phase liquids (LNAPL) ***6*(1)**:59; ***6*(4)**:35; ***6*(10)**:57, 109, 245, 253, 275
lindane, (*see also* pesticides) ***6*(5)**:189
linuron (*see also* herbicides) ***6*(5)**:223
LNAPL, *see* light, nonaqueous-phase liquids
Lolium multiflorum ***6*(5)**:9
LTU, *see* land treatment units
lubricating oil ***6*(6)**:173
luciferase ***6*(3)**:133
lux ***6*(4)**:45

M

mackinawite ***6*(9)**:155
macrofauna ***6*(10)**:337
magnetic separation ***6*(9)**:61
magnetite ***6*(3)**:1; ***6*(8)**:193
manganese ***6*(2)**:261
manufactured gas plant (MGP) ***6*(2)**:19; ***6*(3)**:211, 227; ***6*(10)**:123
mass balance ***6*(2)**:163
mass transfer limitation ***6*(3)**:157
mass transfer ***6*(1)**:67
MC-100, see mixed culture
media development ***6*(9)**:147
Meiofauna ***6*(5)**:305; ***6*(10)**:337
membrane ***6*(5)**:269; ***6*(9)**:1, 265
metabolites ***6*(3)**:227
metal reduction ***6*(8)**:1
metal precipitation ***6*(9)**:9, 165
metals, biorecovery of ***6*(9)**:9
metals speciation ***6*(9)**:129
metal toxicity (*see also* toxicity) ***6*(9)**:17, 129
metals ***6*(5)**:129, 305; ***6*(8)**:1; ***6*(9)**:9, 17, 27, 105, 123, 129, 155, 165
methane oxidation ***6*(10)**:171, 187, 193, 201, 223, 231
methane ***6*(1)**:183; ***6*(8)**:113
methanogenesis ***6*(1)**:35, 43, 183; ***6*(3)**:205; ***6*(9)**:147
methanogens ***6*(3)**:91
methanol ***6*(1)**:183; ***6*(7)**:141, 289, 297
methanotrophs ***6*(10)**:171, 187, 201
methylene chloride ***6*(2)**:39; ***6*(10)**:231
Methylosinus trichosporium ***6*(10)**:187
methyl *tert*-butyl ether *or* methyl *tertiary*-butyl ether (MTBE) ***6*(1)**:1, 11, 19, 27, 35, 43, 51, 59, 67, 75, 83, 91, 107,

115, 121, 129, 137, 145, 153,161, 195, **6(2)**:215; **6(8)**:61; **6(10)**:1, 65
MGP, *see* manufactured gas plant
microbial heterogeneity **6(4)**:73
microbial isolation **6(3)**:267
microbial population dynamics **6(4)**:35
microbial regrowth **6(2)**:253; **6(7)**:1, 13; **6(10)**:319
microcosm studies **6(7)**:109; **6(10)**:179
microencapsulation **6(8)**:53
microfiltration **6(9)**:201
microporous membrane **6(9)**:265
microtox assay **6(3)**:227
mine tailings (*see also* acid mine drainage) **6(5)**:173; **6(9)**:27, 71
mineral oil **6(5)**:279, 289; **6(6)**:59
mineralization **6(2)**:121; **6(3)**:165; **6(6)**:165; **6(8)**:175; **6(9)**:139, 155
MIP, *see* membrane interface probe
mixed culture **6(8)**:61
mixed wastes **6(3)**:91; **6(7)**:133; **6(9)**:139
MNX, *see* 1,3-dinitro-5-nitroso-1,3,5-triazacyclohexane
modeling **6(1)**:51; **6(2)**:105, 155, 181, **6(4)**:125, 131, 139, 149; **6(6)**:339, 377; **6(8)**:185; **6(9)**:27, 105; **6(10)**:163
moisture content **6(2)**:247
molasses as electron donor **6(3)**:35; **6(7)**:53, 103, 149, 173; **6(9)**:315
monitored natural attenuation *(see also* natural attenuation) **6(1)**:183, **6(2)**:11, 163, 199, 223, 253, 261
monitoring techniques **6(2)**:27,189, 199, 207; **6(4)**:59
motor oil **6(5)**:53
MPE, *see* multiphase extraction
multiphase extraction (MPE) well design **6(10)**:245, 259
MTBE, *see* methyl *tert*-butyl ether
multiphase extraction **6(10)**:245, 253, 259, 267, 275
municipal solid waste **6(2)**:247
Mycobacterium sp. IFP 2012 **6(1)**:153
Mycobacterium adhesion **6(3)**:251
mycoremediation **6(6)**:263

N

naphthalene **6(1)**:1; **6(2)**:121; **6(3)**:173, 227; **6(5)**:1, 253; **6(6)**:51; **6(8)**:95, **6(9)**:139; **6(10)**:123
NAPL, *see* nonaqueous-phase liquid
natural attenuation **6(1)**:27, 35, 43, 51, 59, 75, 83, 183, 195; **6(2)**:1,39, 73, 81, 89, 97, 105, 137, 145, 173, 181, 215; **6(4)**:91, 99, 117; **6(5)**:33, 189, 321; **6(8)**:185, 209; **6(9)**:179; **6(10)**:115, 163
natural gas **6(10)**:193
natural organic carbon **6(2)**:261
natural organic matter **6(2)**:81, 97; **6(8)**:201
natural recovery **6(5)**:132, 231
nitrate contamination **6(9)**:173
nitrate reduction **6(3)**:51; **6(5)**:25; **6(9)**:331
nitrate utilization efficiency **6(6)**:353
nitrate **6(2)**:1; **6(3)**:17, 43; **6(6)**:353; **6(8)**:95, 147; **6(9)**:179, 187, 195, 209, 223, 257
nitrification **6(4)**:149; **6(5)**:337; **6(9)**:215
nitroaromatic compounds (*see also* explosives and energetics) **6(3)**:59, 67; **6(6)**:149
nitrobenzene, *see also* explosives and energetics **6(6)**:149
nitrocellulose, *see also* explosives and energetics **6(6)**:119
nitrogen fixation **6(6)**:219
nitrogen utilization **6(9)**:231
nitrogenase **6(6)**:219
nitroglycerin, *see also* explosives and energetics **6(5)**:69
nitrotoluenes, *see also* explosives and energetics **6(6)**:127
nitrous oxide **6(8)**:113
^{13}C-NMR, *see* nuclear magnetic resonance spectroscopy
nonaqueous-phase liquids (NAPLs) **6(1)**:67, 203; **6(3)**:141; **6(7)**:249
nonylphenolethoxylates **6(5)**:215
nuclear magnetic resonance spectroscopy (^{13}C-NMR) **6(4)**:67
nutrient augmentation **6(3)**:59; **6(5)**:329; **6(6)**:257; **6(7)**:313; **6(9)**:331; **6(10)**:23
nutrient injection **6(10)**:101
nutrient transport **6(9)**:241

O

oily waste **6(4)**:35; 6(6):257; **6(10)**:337, 345
oil-coated stones **6(10)**:329

optimization ***6*(5)**:279
ORC® (a proprietary oxygen-release compound) ***6*(1)**:99, 107; ***6*(2)**:215; ***6*(3)**:107; ***6*(7)**:229; ***6*(10)**:9, 15, 87, 95, 139
organic acids ***6*(2)**:39
organophosphorus ***6*(6)**:17, 27
advanced oxidation ***6*(6)**:157, ***6*(10)**:311
oxygen-release compound, *see* ORC®
oxygen-release material ***6*(10)**:73
oxygen respiration ***6*(9)**:231; ***6*(10)**:57
oxygenation ***6*(1)**:107, 145
ozonation ***6*(1)**:121; ***6*(10)**:33, 149, 301

P

packed-bed reactors ***6*(9)**:249; ***6*(10)**:329
PAHs, *see* polycyclic aromatic hydrocarbons
paper mill waste ***6*(4)**:1
paraffins ***6*(3)**:141
partitioning ***6*(9)**:129
PCBs, *see* polychlorinated biphenyls
PCP toxicity (*see also* toxicity) ***6*(3)**:125
PCP, *see* pentachlorophenol
PCR analysis, *see* polymerase chain reaction
pentachlorophenol (PCP) ***6*(3)**:83, 91, 99, 107, 115, 125; ***6*(5)**:329; ***6*(6)**:279, 287, 295, 329
percarbonate ***6*(10)**:73
perchlorate ***6*(9)**:249, 257, 265, 273, 281, 289, 297, 303, 309, 315
perchloroethene, perchloroethylene ***6*(7)**:53
permeable reactive barriers ***6*(3)**:1; ***6*(8)**: 73, 87, 95, 121, 139, 147, 157, 167, 175, 185; ***6*(9)**:71, 309, 323; ***6*(10)**:95
pesticides ***6*(5)**:189; ***6*(6)**:9, 17, 35
PETN reductase ***6*(5)**:69
petroleum hydrocarbon degradation ***6*(4)**:7; ***6*(5)**:9, 17, 25; ***6*(8)**:131; ***6*(10)**: 65, 101, 245, 345
phenanthrene ***6*(2)**:121; ***6*(3)**:227, 235, 243; ***6*(6)**:51, 65, 73
phenol ***6*(6)**:303, 319, 329
phenolic waste ***6*(6)**:311
phenol-oxidizing cultures ***6*(10)**:211, 217, 239
phenyldodecane ***6*(2)**:27
phosphate precipitation ***6*(9)**:165
PHOSter ***6*(10)**:65
photocatalysis ***6*(10)**:311
physical/chemical pretreatment ***6*(1)**:1, 51; ***6*(2)**:253; ***6*(3)**:149; ***6*(5)**:9, 33, 41, 53, 61, 69, 77, 85,105, 113, 121, 129,137, 145, 151, 157, 165, 189, 199, 207, 279, 337; ***6*(6)**:59, 157, 241; ***6*(7)**:1, 13; ***6*(9)**:113, 173; ***6*(10)**:239, 311, 319
phytotoxicity (*see also* toxicity) ***6*(5)**:41, 223
phytotransformation ***6*(5)**:85
pile-turner ***6*(6)**:249
PLFA, *see* phospholipid fatty acid analysis
polychlorinated biphenyls (PCBs) ***6*(2)**:39,105,173; ***6*(5)**:33, 61, 95, 113, 231, 289; ***6*(6)**:89, ***6*(7)**:13, 61, 69, 95, 109, 125, 133, 141, 149, 165, 181, 189, 197, 205, 213, 241, 249, 273, 297, 305; ***6*(8)**:11,19, 27, 43, 157, 167, 193, 209; ***6*(10)**:33, 41, 231, 283
polycyclic aromatic hydrocarbons (PAHs) ***6*(2)**:19, 121, 129, 137; ***6*(3)**:141, 149, 157, 165, 173, 181, 189, 197, 205, 211, 219, 227, 235, 243; ***6*(4)**:35, 45, 59, 67; ***6*(5)**:1, 9, 17, 41, 237, 243, 251, 253, 261, 269, 279, 289, 305, 329; ***6*(6)**:43, 51, 59, 65, 73, 81, 89, 101, 279, 295, 297; ***6*(7)**:125; ***6*(8)**:95; ***6*(9)**:139; ***6*(10)**:33, 123
polymerase chain reaction (PCR) analysis ***6*(4)**:29, 35, 51; ***6*(8)**:43
polynuclear aromatic hydrocarbons, *see* polycyclic aromatic hydrocarbons
poplar lipid fatty acid analysis (PLFA) ***6*(3)**:189
poplar trees ***6*(5)**:113, 121, 189
potassium permanganate ***6*(2)**:253; ***6*(7)**:1
precipitation ***6*(9)**:105; ***6*(10)**:301
pressurized-bed reactor ***6*(6)**:311
propane utilization ***6*(1)**:137; ***6*(10)**:145, 155, 179, 193
propionate ***6*(7)**:265, 289
Pseudomonas fluorescens ***6*(3)**:173
pyrene ***6*(3)**:165, 235; ***6*(4)**:67; ***6*(6)**: 65, 73, 101
pyridine ***6*(4)**:81

R

RABITT, *see* reductive anaerobic biological in situ treatment technology
radium **6(5)**:173
rapeseed oil **6(6)**:65
RDX, *see* research development explosive
rebound **6(10)**:1
recirculation well **6(7)**:333, 341; **6(10)**:283
redox measurement and control **6(1)**:35; **6(2)**:11, 231; **6(5)**:1; **6(9)**:53
reductive anaerobic biological in situ treatment technology (RABITT) **6(7)**:109
reductive dechlorination **6(2)**:39, 65, 97, 105, 145, 173; **6(4)**:125; **6(7)**:45, 53, 87, 103, 109, 133, 141, 149, 157,181, 197, 205, 213, 221, 249, 257, 265, 273, 289, 297; **6(8)**:11, 73, 105, 157, 209
reductive dehalogenation **6(7)**:69
reed canary grass **6(5)**:181
research development explosive (RDX) **6(3)**:1, 9, 17, 25, 35, 43, 51; **6(6)**:133; **6(8)**:175
respiration and respiration rates **6(2)**:129; **6(4)**:59; **6(6)**:185, 227
respirometry **6(6)**:127; **6(10)**:217
rhizoremediation **6(5)**:9, 61, 199
Rhodococcus opacus **6(4)**:81
risk assessment **6(2)**:215; **6(4)**:1
16S rRNA sequencing **6(8)**:43; **6(9)**:147
rock-bed biofiltration **6(4)**:149
rotating biological contactor **6(9)**:79
rototiller **6(6)**:203
RT3D **6(10)**:163

S

salinity **6(9)**:257
salt marsh **6(5)**:313
SC-100, *see* single culture
Sea of Japan **6(5)**:321
sediments **6(3)**:91; **6(5)**:231, 237, 253, 261, 269, 279, 289, 297, 305; **6(6)**:51, 59; **6(9)**:61
selenium **6(9)**:323, 331
semivolatile organic carbon (SVOC) **6(2)**:113
sheep dip **6(6)**:27
Shewanella putrefaciens **6(8)**:201
silicon oil **6(3)**:141, 181
single culture **6(8)**:61
site characterization **6(10)**:139
site closure **6(2)**:215
slow-release fertilizer **6(2)**:57
sodium glycine **6(9)**:273
soil treatment **6(3)**:181; **6(6)**:1
soil washing **6(5)**:243; **6(6)**:241
soil-vapor extraction (SVE) **6(1)**:183; **6(10)**:1, 41, 131, 223
solids residence time **6(10)**:211
sorption **6(5)**:215, 253; **6(6)**:377; **6(8)**:131; **6(9)**:79, 105
source zone **6(7)**:13, 19, 27, 181; **6(10)**:267
soybean oil **6(7)**:213
sparging **6(10)**:33, 145, 155
stabilization **6(6)**:89
substrate delivery **6(7)**:281
sulfate reduction **6(1)**:35; **6(3)**:43, 91; **6(5)**:261, 313; **6(6)**:339; **6(7)**:69, 95; **6(8)**:139, 147, 193; **6(9)**:1, 9, 17, 27, 35, 43, 61, 71, 86, 105, 123, 147
sulfide precipitation **6(9)**:123
surfactants **6(5)**:215; **6(6)**:73; **6(7)**:213, 321, 333; **6(8)**:131
sustainability **6(6)**:1
SVE, *see* soil vapor extraction
SVOC, *see* semivolatile organic carbon
synthetic pyrethroid **6(6)**:27

T

TCA, see trichlorethane
1,1,1-TCA, *see* 1,1,1-trichloroethane
1,1,2-TCA, *see* 1,1,2-trichloroethane
2,4,6-TCP, *see* 2,4,6-trichlorophenol
1,1,1,2-TeCA,*see* tetrachloroethane
1,1,2,2-TeCA, *see* tetrachloroethane
1,3,5-TNB, *see* 1,3,5-trinitrobenzene
TAME, *see* tertiary methyl-amyl ether
TBA, *see* tertiary butyl alcohol
TBF, *see* tertiary butyl formate
TCE oxidation, *see* trichloroethene, trichloroethylene
TCE, *see* trichloroethene
TCP, *see* trichlorophenol
t-DCE, *see* trans-dichloroethene, trans-dichloroethylene
technology comparisons **6(7)**:45; **6(9)**:323
terrazyme **6(10)**:345

tertiary butyl alcohol (TBA) ***6*(1)**:19, 27, 35, 51, 59, 91, 145, 153, 161
tertiary butyl formate (TBF) ***6*(1)**:145, 161
tertiary methyl-amyl ether (TAME) ***6*(1)**:59, 161
tetrachloroethane (1,1,1,2-TeCA, 1,1,2,2-TeCA) ***6*(5)**:207; ***6*(7)**:321, 341; ***6*(8)**:193
tetradecane (see also ^{2}H-tetradecane) ***6*(3)**:181
thermal desorption ***6*(3)**:189, ***6*(6)**:35
TNB, *see* trinitrobenzene
TNT, see trinitrotoluene
TNX, *see* 1,3,5-trinitroso-1,3,5-triazacyclohexane
tobacco plant ***6*(5)**:69
toluene ***6*(1)**:145; ***6*(2)**:181; ***6*(7)**:95; ***6*(8)**:35, 131
total petroleum hydrocarbons (TPH) ***6*(2)**:1; ***6*(5)**:9; ***6*(6)**:127, 173, 179, 193, 227, 241, 249; ***6*(10)**:15, 73, 115, 337
toxicity ***6*(1)**:1; ***6*(3)**:67, 189, 227; ***6*(4)**:7; ***6*(5)**:41, 61, 223, 305; ***6*(9)**:17, 129
TPH, *see* total petroleum hydrocarbons
trace gas emissions ***6*(6)**:185
trans-dichloroethene, trans-dichloroethylene ***6*(5)**:95, 207; ***6*(7)**:165
transgenic plants ***6*(5)**:69
transpiration ***6*(5)**:189
Trecate oil spill ***6*(6)**:241; ***6*(10)**:109
trichloroethane (TCA) ***6*(7)**:241, 281
1,1,1-trichloroethane (1,1,1-TCA; 1,1,2-TCA) ***6*(2)**:39, 113, 464; ***6*(5)**:207; ***6*(7)**:87,165, 281
1,1,2-trichloroethane (1,1,2-TCA) ***6*(5)**:207
trichloroethene, trichloroethylene (TCE) ***6*(2)**:39, 65, 73, 97, 105, 113, 155, 173, 253; ***6*(4)**:125; ***6*(5)**:33, 95, 105, 113, 207; ***6*(7)**:1, 13, 53, 61, 69, 77, 87, 109, 117, 133, 141, 149, 157, 181, 189, 197, 205, 213, 221, 241, 249, 265, 273, 281, 297, 305, ***6*(8)**:11, 19, 27, 35, 43, 53, 73, 105,147, 157, 193, 209; ***6*(10)**:41, 131, 145, 155, 163, 171, 179, 187, 201, 211, 217, 223, 231, 239, 283, 319
2,4,6-trichlorophenol (2,4,6-TCP) ***6*(3)**:75; ***6*(8)**:121
trichlorotrifluoroethane ***6*(2)**:49
trinitrobenzene (TNB) ***6*(3)**:9, 25
1,3,5-trinitroso-1,3,5-triazacyclohexane (TNX) ***6*(8)**:175
trinitrotoluene (TNT) ***6*(3)**:35, 67; ***6*(5)**:69, 77, 85; ***6*(6)**:133

U

underground storage tank (UST) ***6*(1)**:67, 129
uranium ***6*(5)**:173; ***6*(7)**:77; ***6*(9)**:155, 165
UST, *see* underground storage tank

V

vacuum extraction ***6*(1)**:115
vadose zone ***6*(1)**:183; ***6*(2)**:39, 65, 97, 105, 113, 155, 173; ***6*(3)**:9; ***6*(5)**:33, 105; ***6*(7)**:1,13, 61, 133, 141, 197, 205, 213, 249, 273, 281, 305; ***6*(8)**:11,19, 43, 73, 157, 209; ***6*(10)**:41, 163
vegetable oil ***6*(6)**:65; ***6*(7)**103, 213, 241, 249
vinyl chloride ***6*(2)**:73; ***6*(4)**:109; ***6*(5)**:95; ***6*(7)**:95,149, 157, 165, 173, 289, 297, ***6*(10)**:231
vitamin B_{12} ***6*(7)**:321, 333, 341
VOCs, *see* volatile organic carbons
volatile fatty acid ***6*(7)**:61
volatile organic carbons (VOCs) ***6*(2)**:113, 189; ***6*(5)**:113, 121

W

wastewater treatment ***6*(5)**:215; ***6*(6)**:149; ***6*(9)**:173
water potential ***6*(9)**:231
weathering ***6*(4)**:7
wetlands ***6*(5)**:33, 95, 105, 313, 329; ***6*(9)**:97
white rot fungi, (*see also* fungal remediation) ***6*(3)**:75, 99; ***6*(6)**:17, 157, 263
windrow ***6*(6)**:81, 119, 141
wood preservatives ***6*(3)**:83, 259; ***6*(4)**:59; ***6*(6)**:279

X

xylene ***6*(1)**:67

Y

yeast extract ***6*(7)**:181

Z

zero-valent iron ***6*(8)**:157, 167; ***6*(9)**:71
zinc ***6*(4)**:91; ***6*(9)**:79